CW01486619

Polar Exploration

This series includes accounts, by eye-witnesses and contemporaries, of early expeditions to the Arctic and the Antarctic. Huge resources were invested in such endeavours, particularly the search for the North-West Passage, which, if successful, promised enormous strategic and commercial rewards. Cartographers and scientists travelled with many of the expeditions, and their work made important contributions to earth sciences, climatology, botany and zoology. They also brought back anthropological information about the indigenous peoples of the Arctic region and the southern fringes of the American continent. The series further includes dramatic and poignant accounts of the harsh realities of working in extreme conditions and utter isolation in bygone centuries.

Voyages of Discovery in the Arctic and Antarctic Seas, and Round the World

Published in 1884 and illustrated with over 100 of his own drawings and maps, this two-volume work by the doctor and naturalist Robert McCormick (1800–90) provides an account of his voyages in the Arctic with William Parry and in the Antarctic with James Clark Ross, noting also his part in the search for Sir John Franklin. Incorporating a very detailed autobiography, McCormick's work also provides many details relating to natural history and geology. Volume 1 is mainly devoted to his Antarctic voyage (1839–43), during which he also visited St Helena, the Kerguelen Islands, Australia, New Zealand, the Falkland Islands and Tierra del Fuego. During three attempts to reach the South Pole the expedition explored the Ross Sea, described the ice barrier, and raised the British flag to claim possession. The volume ends with the abortive attempt in 1827 to reach the North Pole via Spitsbergen.

Cambridge University Press has long been a pioneer in the reissuing of out-of-print titles from its own backlist, producing digital reprints of books that are still sought after by scholars and students but could not be reprinted economically using traditional technology. The Cambridge Library Collection extends this activity to a wider range of books which are still of importance to researchers and professionals, either for the source material they contain, or as landmarks in the history of their academic discipline.

Drawing from the world-renowned collections in the Cambridge University Library and other partner libraries, and guided by the advice of experts in each subject area, Cambridge University Press is using state-of-the-art scanning machines in its own Printing House to capture the content of each book selected for inclusion. The files are processed to give a consistently clear, crisp image, and the books finished to the high quality standard for which the Press is recognised around the world. The latest print-on-demand technology ensures that the books will remain available indefinitely, and that orders for single or multiple copies can quickly be supplied.

The Cambridge Library Collection brings back to life books of enduring scholarly value (including out-of-copyright works originally issued by other publishers) across a wide range of disciplines in the humanities and social sciences and in science and technology.

Voyages of Discovery in the Arctic and Antarctic Seas, and Round the World

Being Personal Narratives of Attempts to Reach the North and South Poles

VOLUME 1

ROBERT MCCORMICK

CAMBRIDGE
UNIVERSITY PRESS

CAMBRIDGE
UNIVERSITY PRESS

University Printing House, Cambridge, CB2 8BS, United Kingdom

Cambridge University Press is part of the University of Cambridge.

It furthers the University's mission by disseminating knowledge in the pursuit of
education, learning and research at the highest international levels of excellence.

www.cambridge.org
Information on this title: www.cambridge.org/9781108071710

This edition first published 1884
This digitally printed version 2014

ISBN 978-1-108-07171-0 Paperback

DEPUTY-INSPECTOR-GENERAL ROBERT McCORMICK, R.N., F.R.C.S.

ÆT, 52.

FRONTISPIECE TO VOL. I.

VOYAGES OF DISCOVERY

IN THE

ARCTIC AND ANTARCTIC SEAS,

AND

ROUND THE WORLD:

BEING PERSONAL NARRATIVES OF ATTEMPTS TO REACH

THE NORTH AND SOUTH POLES;

AND OF AN

OPEN-BOAT EXPEDITION UP THE WELLINGTON CHANNEL IN SEARCH OF

SIR JOHN FRANKLIN

AND HER MAJESTY'S SHIPS "EREBUS" AND "TERROR,"

IN HER MAJESTY'S BOAT "FORLORN HOPE,"

UNDER THE COMMAND OF THE AUTHOR.

TO WHICH ARE ADDED AN

AUTOBIOGRAPHY, APPENDIX,

Portraits, Maps, and numerous Illustrations.

BY

DEPUTY INSPECTOR-GENERAL R. M'CORMICK, R.N., F.R.C.S.

Chief Medical Officer, Naturalist, and Geologist to the Expeditions.

IN TWO VOLUMES.
VOL. I.

London:
SAMPSON LOW, MARSTON, SEARLE, AND RIVINGTON,
CROWN BUILDINGS, 188, FLEET STREET.
1884.

LONDON :
PRINTED BY GILBERT AND RIVINGTON, LIMITED,
ST. JOHN'S SQUARE.

PREFACE.

~~~~~~

HAVING had the good fortune to be engaged in three of the most memorable expeditions of the present century —with Parry, in his attempt to reach the North Pole, in the year 1827; with Ross, in his Antarctic voyage during the years 1839-43; and having had command of a boat expedition in search of Franklin in 1852-3—affording me such ample opportunities for adding to our knowledge of these regions, both Arctic and Antarctic, I have ventured to bring before the public my own experiences, and I do so more especially at the present time, because there has been a renewal of interest in favour of Polar research and discovery since the return of the late spirited private expeditions sent out from this country, the Continent, and America.

Devoted as I have been from my earliest years to Polar discovery, in both north and south, my main object in the following pages (condensed from voluminous diaries I had been in the habit of keeping from my first entering the navy, with numerous illustrations, all from my own pencil sketches taken on the spot) is to be useful to future explorers, who may be destined to follow me into these little known but highly interesting regions of our globe, for which purpose I have given plans for reaching both Poles, North and South, as a conclusion to the Antarctic and Arctic voyages, at the end of the first volume. I have also introduced five panoramas of the coast-line of the new glaciated southern continent we had the good fortune to discover in our first voyage south, drawn on a large scale, with the Great Barrier and remarkable bergs, so as to give full effect to the appearance of this wonderful land, and which, I trust, will be found novel and striking features in the book. Nor am I without hope that the many incidents and adventures met with in the course of an eventful life, together with the popular way in which I have introduced and described the more scientific observations throughout the work in summaries, may amuse and interest the general reader. If I may judge from the deep interest I have always taken in the perusal of voyages and travels, I think I may fairly assume that such a work as the present one—which, by the gracious permission of his Royal Highness, I have dedicated to the Duke of Edinburgh, and the officers of the Royal Navy—may be useful to my younger brother-officers by calling their attention to the many opportunities and wide field of observation which the naval service offers in the pursuits of natural history, geology, geography, and the collateral sciences, such a book I should myself have considered a great boon as a work of reference on my first entrance into the navy.

Although late in life for commencing such an undertaking, the whole of these pages have gone through my own hands, and have been carefully revised by myself before the proof-sheets went to the press. In the same

way I have seen my pencil sketches and maps transferred to the stone under my own supervision, and the
cover engraved from my own design.

Having taken a very prominent part in the search for
Franklin, I have felt it due to myself, considering the
anomalous position I was then placed in, to reprint the
whole of the Admiralty correspondence, and the testi-
monials in the Appendix, after the Boat Voyage, that
my share in the search might not be left open to miscon-
struction. The departure from a strictly chronological
arrangement of those voyages arises from my having
decided to give precedence to the Antarctic voyage, as
being of fresher interest to the general reader than the
narratives of the Arctic voyages, and placing my Autobiography, being of subordinate interest, last instead
of first, the latter embracing my general services in the
navy, together with my pedestrian tours through England
and Wales, so as to maintain an unbroken chain throughout
an eventful life,—thus dividing the whole work into four
parts. The first volume embraces the voyages towards
the two Poles, and ends by my plans for reaching both
Poles. The Franklin Search commences the second
volume.

Before I bring this preface to a close I desire to return
my best thanks to the Lords of the Treasury for their
considerate kindness in permitting me to have the free
use of the blocks which had been employed in illus-
trating my Boat Voyage for the Blue Books in the Par-
liamentary returns of the Franklin Search; also to
Messrs. Pigott and Reid, of her Majesty's Stationery
Office, to whom I am deeply indebted for the trouble

they took upon themselves in searching for the blocks, and forwarding them to my publishers, through the kind and friendly intercession of my esteemed friend John Barrow, Esq. To the latter, and to Professor Owen, of the South Kensington Museum of Natural History, I owe my best thanks for their friendly encouragement during my labours in seeing my work through the press. And whilst writing this, the announcement of the promotion of Professor Owen to be a Knight Commander of the Bath, in reward for his long, laborious, and distinguished services to science, is most gratifying to me, now that, after a long friendship of nearly half a century with which he has honoured me, I have at last the pleasure of congratulating him as Sir Richard Owen, K.C.B., F.R.S.

To Messrs. Carruthers and Sharp, keepers of the botanical and ornithological departments of the Museum, I am also much indebted for their kindness in permitting their artists—Messrs. Wilson and Morgan—to make drawings of my Tartary oak and Tasmanian fossils, &c., the fossil willow, I learn, whilst these pages are passing through the press, has been named after me by the German Professor Von Ettingshausen, and for this compliment I here take the opportunity to return the Professor my best thanks. The Esquimaux dogs, the three penguins, the two lestrises, young and egg, with my remarkable duck, also drawn on the stone at the Museum by the first-named artist, will add a pleasing variety to the number of landscape illustrations in the two volumes, and are the only illustrations not sketched by myself.

To Messrs. Vincent Brooks, Day and Son, I am indebted for the excellent way in which they have produced the numerous lithographs from my sketches which have been entrusted to them; and also for the three portraits enlarged from a Daguerreotype and two miniatures.

The engravings on wood, forming the heads and vignettes of the chapters, have been executed by Mr. J. D. Cooper with his accustomed skill and care, from sketches of my own.

To my publishers I return my best thanks, not only for the kind way in which they have met my wishes in carrying out the plan of my work, but for many valuable suggestions. Also to Mr. Gordon G. Flaws, a literary gentleman, to whom, as reader, I am greatly indebted for the Index and valuable suggestions in my labours. Lastly, I have to thank Mr. J. Thompson, the manager of Messrs. Gilbert and Rivington, for the kind interest he has taken in forwarding the sheets through the press. I will add, finally, that this work, which comprises in itself three of the most memorable voyages of this century, may save the reader the trouble of having to refer to each in separate books; and should these two volumes meet with the approval of the public, I shall not regret the labour I have experienced in launching my "two-decker" on the ocean of literature.

R. McCORMICK.

Hecla Villa, Wimbledon,
January 1st, 1884.

CONTENTS.

Part I.

VOYAGE OF H.M.S. "EREBUS" AND "TERROR" TO THE ANTARCTIC SEAS.

CHAPTER V.

CHAPTER VI.

CHAPTER VII.

CHAPTER VIII.

CHAPTER IX.

CHAPTER X.

CHAPTER XVII.

CHAPTER XVIII.

CHAPTER XIX.

CHAPTER XX.

CHAPTER XXI.

CHAPTER XXII.

CHAPTER XXIII.

CHAPTER XXIV.

Part II.

ATTEMPT TO REACH THE NORTH POLE IN THE YEAR 1827.

CHAPTER I.

LIST OF ILLUSTRATIONS AND MAPS.

PART I.

VOYAGE

OF

HER MAJESTY'S SHIPS

"EREBUS" AND "TERROR"

TO THE

SOUTH POLAR SEAS,

UNDER THE COMMAND OF

CAPTAIN SIR JAMES CLARK ROSS, R.N., F.R.S.,

DURING THE YEARS 1839—1843.

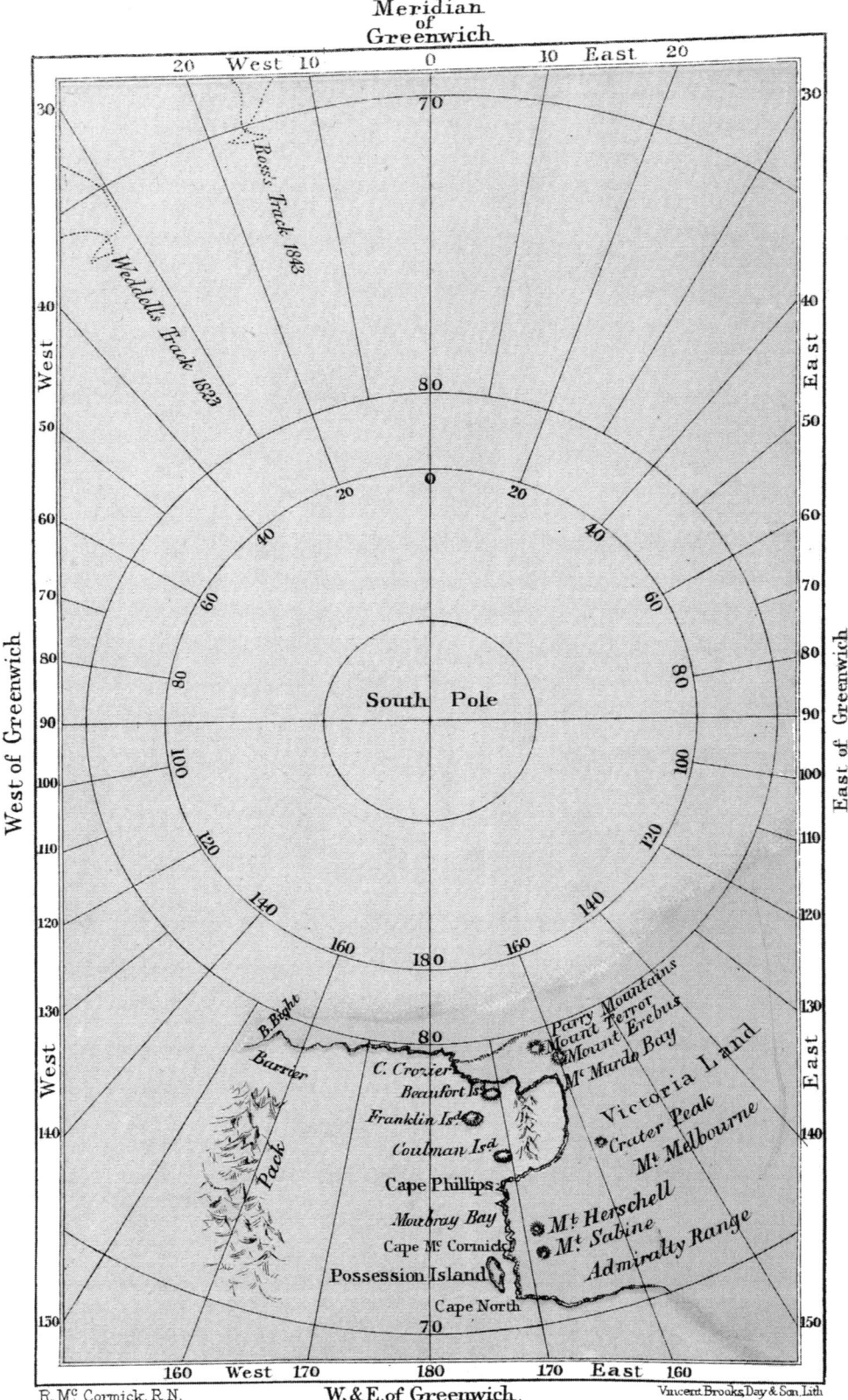

MAP OF SOUTH POLE.

R. McCormick, R.N.   W. & E. of Greenwich.   Vincent Brooks, Day & Son, Lith.

DEDICATION VOL I.

Summit of Pico Ruivo above the Clouds. *(See page* 6.)

PERSONAL NARRATIVE

OF THE

VOYAGE OF H.M.S. "EREBUS" AND "TERROR"

TO THE

ANTARCTIC SEAS,
1839-43.

———•———

CHAPTER I.

Departure from England—At Madeira—Ascent of Pico Ruivo.

Thursday, September 19*th,* 1839.—At nine a.m we proceeded down the river to Gillingham Reach, and came
to an anchor, near the *Terror,* off the church.   Here we
remained some days—to swing the ship's head round the
compass for the local attraction, and for the ship's com-

pany to receive their advance pay. On the afternoon of the 25th we weighed, anchoring for the night off the buoy of the Mouse, and on the following morning the *Hecate* steamer towed us into Margate roads. Here we were detained by westerly winds until the 30th, when we got under weigh at seven p.m., and met with boisterous weather down Channel, passing the splendid new light on the Start, flashing brightly every minute on the lee-beam. During a gale, accompanied by fog, we lost sight of our consort the *Terror*, and did not see her again until after our arrival at Madeira.

October 20th.—Between eight and nine a.m. the Desertas were sighted on the port-quarter, and the island of Madeira itself ahead, having by no means so verdant and fertile an aspect, when seen from the sea, as might have been expected.

Monday, 21st.—I went on shore, accompanied by our passenger, Lieutenant Eardley Wilmot, of the horse artillery, and after a ramble through the place, visiting the news-rooms and market-place, returned on board at 1.30 p.m.

Thursday, 24th.—The *Terror* arrived, and as soon as she had anchored I went on board of her, and from thence to the shore. After dining with Dr. Smith, a resident in the island, we rode up the steep side of the mountain above the town, as far as the church of "Our Lady of the Mountain," forming a most conspicuous object near its summit, as seen from the sea. It was a charming evening, and the view of the shipping in the roads, and the whole scene beneath us formed a magnificent panorama. On our return to the town we paid a visit to Mr. Muir, whose family showed us some remarkable encrustations from Porta de Lorenco, and gave me an invitation to join a party to-morrow for the ascent of Pico Ruivo, on the north side of the island and its highest mountain. We next repaired to the consul's, where we met Captains Ross and Crozier, joined them in a cup of coffee, and returned on board, calling alongside of the *Terror* on our way to the *Erebus.*

Friday, 25th.—I left the ship at six a.m., and at 7.15 a party of six of us started for the ascent of Pico Ruivo. The morning dawned very fine as we mounted the horses ready for us, at Mr. Muir's store, with the customary guides, the artillery-officer and the surgeon from the *Terror* having also joined our party. Six miles from Funchal was our first halting-place, at a venda called " Camacho," where we got some coarse, dark bread, and the ordinary wine of the country. Three miles farther on we passed some cascades, having a view from this of Porto Santo, Porta de Lorenco, and the Desertas. We lunched on a bank commanding a fine view of the valley beneath, and a steep rock overhanging the sea, at a place called San Antonio-de-zaza Portello, sixteen miles from Funchal. At 12.15 we reached Porta de Cruz, about a mile on, by a pretty descent. Here we rested at a small venda, with the usual fare of coarse bread and sour wine, with spirits as fiery as the wine was sour. Whilst our party were thus refreshing themselves, I availed myself of the brief interval to stroll down a winding path, through the village, to an inlet of the sea, at the base of a high rock, a quarter of a mile or so from the venda, when, from taking a wrong turn on my way back, I was very near missing my party altogether, who had started from the venda before I got back. However, falling in with one of the guides, I soon overtook the cavalcade. At 2.30 p.m. we passed Fayal, and a river winding through a deep valley, enclosed by rugged and lofty mountains, presenting scenery of a very grand description. Our course next followed a very circuitous path, steep and precipitous. Here a very remarkable rock in the shape of a pillar rises out of the sea perfectly isolated, like a lighthouse, with a heavy surf breaking at its base.

At 3.40 p.m. passed the Church of St. Anne's. I alighted from my horse and walked round the interior. Evidently no pains had been spared in ornamenting it in the usual Portuguese fashion. From this, a winding lane led us to the inn at St. Anne's, "Caza de Plaiz," which we reached at four p.m., where we took up our quarters for the night. It is kept by Señor Luiz, a Portuguese, who is also a large vine-grower, and consequently does not depend on the inn for his livelihood, it being seldom frequented excepting in the summer season. The summit of Pico Ruivo forms a grand and striking object, the most prominent feature in the landscape as seen from the window of the room in which I slept. It was at the time I saw it, 5.30 p.m., partially enveloped in clouds, the mist now and then lifting and leaving the highest peaks visible. We had a cool, cloudy day for our ride of twenty-five miles. The evening cleared up bright and starlight. The horse I rode was sure-footed, free, and spirited, although twelve years old. Our course lay through a variety of charming scenery—mountain, glen, hill, and valley, diversified by woods. The road in many places was so steep and rugged, being strewed with fragments of rock, that we frequently had to dismount and lead our jaded steeds; at other times, winding through lanes with hedgerows of blackberry, whortleberry, broom, furze, and heaths, with the magnificent hydrangea, of a bright azure-blue colour. The chestnut seemed to be the most generally diffused of the native trees of the island—and of course the vine everywhere. We passed a variety of fine waterfalls. Birds were very scarce; I only saw here and there a small finch or warbler, with a swallow or two, and in the garden of the inn was a blackbird.

We were quite ready for dinner on alighting at the inn; but no visitors being expected, our arrival evidently caused bustle and stir amongst the household, preparing rooms and beds for our reception. The fowls were still living that were destined for our repast, to which we did not sit down to table till seven p.m.

Saturday, 26th.—Arose at 2.30 a.m., and soon afterwards we mounted our horses, rode down the carriage-road to the gate, when, entering the main road, we followed a winding lane to the right, which took us to the foot of the mountain. The morning was very fine and starlight, and being so early the scene was one of extraordinary beauty, the atmosphere so pellucidly clear, that the planet Venus appeared at least double its usual diameter, shining at the same time with a brilliancy never seen in our English clime. As we looked back, ocean and sky seemed blended in one, having no visible tangible horizon save where a bank of fleecy white clouds, reflecting the rays of the sun—though that magnificent orb itself was still below the horizon—had the appearance of a wreath of snow crossed by red rays, like a bridge of burnished gold. The air was keen and chilly; but most invigorating.

The first part of the ascent is through a shrubby underwood; the acclivity a broad, barren ridge strewed with fragments of rock. On passing through a gate a singular mass of rock presents itself, standing perfectly isolated, and bearing ample evidence of having been the remnant of an ancient greenstone dyke, which, from its hardness, had hitherto resisted the atmospheric changes under which the surrounding softer rocks had disappeared in the denudation of the surface. It is called the " Homan-em-pee," and is perhaps some twenty feet high. At about six feet from the ground, a solitary bush or shrub grows out of a crevice in the rock apparently without a particle of soil for its roots to strike in or to sustain its life, beyond the rain it receives from the clouds. I brought away with me a specimen of the plant, a piece of the rock, in which it seemed to thrive, and also a rough pencil sketch of it. At six a.m. we rode along the edge of a steep precipice, overhanging a deep valley, from the bottom of which craggy masses of rock rise in the wildest confusion, some bare and sterile, others jutting out from trees and shrubs, embosomed in green foliage. This spot is the wildest and grandest piece of scenery we met with throughout our whole excursion. It is said to bear a striking resemblance to the " coral," only on a far more magnificent and imposing scale, and is called the " Torrihas." It is not far from the " Homan-em-pee." Soon after passing it, at 6.45 am., we alighted from our horses, and, leaving them in charge of the guides, we completed the remainder of the ascent on foot.

The track now became a winding one through tree-heath, with which the acclivity at its upper part is clothed; numbers of them dead and withered, some with their trunks and branches barked and blanched white, many of them trees of considerable size. Near the summit, winding amongst these withered forms of vegetation, the track suddenly divides, one branching to the left, and the other to the right, rendering it doubtful which of them was the nearest way to the top, and led to the highest peak. The young merchant, Mr. Muir, our guide, with myself, being the foremost of the party, we agreed to separate, he taking the left, and I the right, which proved to be the right one for me, for after proceeding a few paces, I found myself standing alone on the highest point of the peak, so that I was the first of our party who set foot on the summit, at 7.20 a.m. I had not long to wait before my companion's head emerged from the bushes on the opposite side ; and before long the remainder of our party assembled on the summit, which we soon found by our barometrical measurement to be 6176 feet above the level of the sea. Saw a sheep or two, with lambs white as snow, feeding on the acclivity, where there is abundant vegetation, chiefly of tree-heath with masses of rock scattered about. The highest part of the peak is green-stone, having only a few straggling bushes of the tree-heath, very much stunted in growth, of which I collected a few specimens, and also of the greenstone. The solitary note of a small bird alone broke the silence of the scene even at this elevation. A passing mountain mist soon obscured all distant objects from view, and we had only one glance of the sea through a temporary opening in the mist. At 9.20 a.m. we commenced our descent, the broom forming the second zone of vegetation below the heaths; next came a belt of ferns. At 9.45 a.m. we reached the spot where we had left our horses, when, seat-ing ourselves amidst the rocks, fragments, ferns, and brooms, we breakfasted with keen appetites on cold fowl and ham, bread-biscuit, coffee, and porter, which we had brought with us in a basket, and black grapes, bought on the way up.

At 10.15 a.m. we remounted our steeds and continued the descent, returning by another route to Funchal. Weather misty. We passed through a variety of beautiful scenery, and close to a fine waterfall of 400 to 500 feet, and across a bridge over the River Fayal. At one minute we were rounding the steep and sharp angle of a descent, and the next up an abrupt and rugged ridge. As we proceeded our road again changed, and we rode along a path of soft soil, flanked on either side by shrubs and wild flowers, amongst which flitted butterflies, grass-hoppers, and locusts. At one p.m. we alighted at the venda of "Cruzinha," to afford both guides and horses some rest and refreshment, which as usual consisted of small loaves of dark coarse bread and sour wine; I found the best thing the fresh cool water from the spring. These vendas all very much resemble each other—mere rudely-built stone huts, thatched, the bare ground forming the floor, a small counter on it, a green bottle of spirits, a tumbler, and a tin drinking-pot; behind it the cask of sour wine. The loaves of bread are arranged on a shelf in the wall; a bench between the counter and the door, serving as a seat for visitors, completes the inventory of the venda. At 1.30 p.m. we continued our journey, when some drizzling rain began to fall, shrouding the surrounding scenery in mist. At four p.m. a sudden turn in the road gave us a glance of the sea and the shipping. The evening cleared up fine; and I saw a buzzard hovering overhead. At 4.50 p.m. we alighted at the last venda, "Camento da Meis," overlooking the bay. We descended the hill by a paved road, having "Our Lady of the Mountain" on the right. We reached Funchal at 5.30 p.m., and got on board at six p.m.

The weather during our excursion proved most favourable; our horses, small as they were, carried us admirably throughout a journey of certainly not less than fifty miles in the two days, over the most rugged country, which none but sure-footed animals accustomed to such work could possibly have accomplished, yet they exist on the most scanty fare. Their owners, the guides, walked the whole of the distance, and managed to keep up with their horses in a most extraordinary manner, laying hold of their tails, and urging them onwards by a peculiar shout. They wore small blue cloth skull-caps, with a point nearly a foot in length, tapering upwards from the crown. The total expenses of our trip amounted to $64.515, or 14*l*. 0s. 6d.

The following memorandum was deposited on the summit of Pico Ruivo:—

"On Saturday, the 26th of October, 1839, a party of officers from her Majesty's ships *Erebus* and *Terror* ascended to this spot, in company with James Muir, Esq., of Funchal, for the purpose of making barometrical observations. They left St. Anne's at 3.30 a.m., and reached this station at seven. The lower ground was hidden by the clouds, previous to which the morning was cloudless. The barometer stood at a mean height of 24°, thermometer 45° Fahrenheit, hygrometer D.P. 43°, wind N.; observations taken every ten minutes for one and a half hours."

Homan-em-pée. (*See page* 8.)

The Peak of Teyde. (*See page* 14.)

CHAPTER II.

Visit to the convents —Observations round the coast—At Santa Cruz—
A Nelson recollection—Porta Praya—The baobab-tree revisited—
St. Domingo—An interior— St. Paul's Rocks—A perilous adventure
—Booby birds at home—Battle with crabs—Geology of St. Paul's
rocks.

Sunday, 27th.—I breakfasted at the Muirs, and we went
to the pretty little English church, embosomed in shrubs
and flowers. After the service, as I was returning on
board, I met the surgeon of the *Terror* coming on shore,
and joined him in a ramble to the convents. We first
visited the " Incarnation," and purchased some feather-
flowers at the revolving circular box. We next proceeded to the " Santa Clara," and were shown into an
upper room, where we had a sight of three of the nuns,
from whom we were separated by a large wire screen ;
they handed us the flowers through a revolving box in it.

The nuns were dressed in black, with large black veils bordered with white, terminating in a very narrow peak over the forehead. The celebrated Maria Clementina is still an inmate of this convent; but whether she was one of the ladies whom we saw I know not, as they did not speak a word of English. They were all tall, dark, middle-aged women. We returned on board at three p.m., when both Captains Ross and Crozier, with all the gun-room officers of the *Terror*, dined on board with us.

Monday, 28th.—Captain Ross having given me a boat for the purpose of examining the incrustation bed at Porto Lorenco, I started at 7.30 a.m., accompanied by the two artillery officers, two sons of Mr. Muir, and an officer or two from the ships. As we rounded the eastern point, Brazen Head, some wild pigeons and a hawk rose from the rocks. Here a large dyke completely intersects the cliff. We pulled along shore, passed Santa Cruz cliffs, red, and remarkably twisted into concentric circles, indicating that these rocks had undergone enormous lateral pressure, and affording striking examples of anti-clinal and other forms of stratification. Passed Machico, and at 1.45 p.m. we landed on a rock, bounding a small sandy cove, about 100 paces in length, from which the incrustation bed stretches across Point Lorenco. Two large curlews rose from the point as we landed. What is here termed the fossil-bed is nothing more than the roots and stems of the plants growing here becoming encrusted over with a calcareous deposit, from some peculiarity in the sandy soil in which they grew, the enclosed vegetable matter being in many instances charred and black.

Thursday, 31st.—Owing to the exposed anchorage for the ships, and the threatening state of the weather, Captain Ross, on returning on board, resolved upon at once getting under weigh, notwithstanding he had invited a party from the shore to dine with him. The *Terror* weighed at 12.30, and hove-to for us. At four p.m. a strong breeze brought us up with her, and we shaped a course for the Canary Islands.

Sunday, Nov. 3rd.—Hove-to off Santa Cruz in the island of Teneriffe, and a boat having been sent on shore for some beef, I took a passage in her, landing at the celebrated mole where Nelson lost an arm. We entered the principal church, and saw the colours taken on that sad mishap. They were placed very high up on the left side of the church, and were in a very faded, not to say dilapidated, condition. We met the consul on the pier, accompanied him home, and afterwards to the governor's, the Marquis de Concordia, and the commandant of marine, to whom we were introduced, and met the wife and sister of the latter, whose child was suffering from paralysis of the lower extremities, about which they were very anxious to have my advice. We returned on board with the consul in the health-boat (Sanadad). This island has a very barren aspect when seen from the sea, scorched-up, scoriaceous hills, thickly clothed with the *Euphorbia Caniansis*, the most striking feature of the vegetation.

The Archipelago of the Canaries has been called the Fortunate Isles, of which Santa Cruz in Teneriffe is the capital. (Celebrated for cochineal and the Great Dragon-tree, at Orotava Peak, sixty miles distant.) The Peak of Teyde rises from the centre of the island in the form of a cone 12,000 feet above the level of the sea. This group of islands, from their geological formation and natural productions, would appear to have been once united to the continent of Africa. According to Pliny and other ancient writers, they were supposed to have been the remnant of the submerged Atlantis, a vast continent imagined to have been engulfed by some great convulsion in the earth's crust, affecting the bed of the Atlantic in bygone ages. Columbus was for a time a resident of the Canaries, before his discovery of the New World. The wild canaries, the ancestors of our pretty yellow pets in England, are here clothed in olive-green plumage, the hen bird often quite of a brownish hue. We made sail on the following day.

Friday, 8th.—We crossed the Tropic.

Tuesday, 12th.—Saw the island of Sol on the port bow, and the first flying-fish, and pharsalia, shoals of the former.

Wednesday, 13th. Nine a.m.—The islands of St. Jago and Fogo in sight, and on the following day, at 11.30 a.m., we anchored at Porta Praya. I went on shore, and made an excursion to the westward of the town for some miles. Shot some birds for the collection, and had some of the large delicious oranges for which this place is so justly celebrated. I extended my lonely ramble so far over ridges and through ravines, that I did not get back to the town till after dark.

Friday, 15th.—I landed this morning at the rocks, and crossed over the hill to the eastward of the town, to examine the fossil shells in the calcareous strata of the cliff fronting the sea, mostly a species of oyster, the limestone overlaid by basaltic lava. From thence I proceeded to the large, deep ravine to the eastward, and which I had formerly visited when here some eight years ago. It is about four miles from the town. I put up several quail in the long grass on the plain, and shot one. On descending into the steep-sided ravine, which appears to be the breeding-place of hawks and owls, I shot two kestrels in it, and saw a large monkey. Returning, I paid a visit to my old friend the baobab-tree, in the middle of the valley, and a mile to the eastward of the town. It forms a very conspicuous object, having no other tree near it, and only a few palma-christi bushes scattered about the valley, the resort of a species of alcedo, but more allied to the jacamar in the form of the beak, and still more in its habits than the king-fisher. It preys on the locusts and other insects in the dry, parched valley, where it sits patient and silent till attracted by some passing locust, or other favourite morsel when it darts off the branch of the palma-christi in pursuit, returning to its former position almost immediately. Unlike the true king-fisher, its habits exclude it from both water and fish. On reaching the baobab-tree, I ascended it, and looked for my own initials, which I had cut, with the year 1832, on the main stem, about two-thirds up the tree, when here last in that year. Time had impressed them deeper, and they appeared larger, more marked and distinct from the contraction of the bark around. I now added the present year, 1839, beneath the former one; and after gathering some fruit, for there were no flowers, and taking a sketch of this remarkable tree, I returned at six p.m.

Saturday, 16*th.*—Started about noon, accompanied by two of the officers of the ship, on an excursion to the valley of St. Domingo. For about nine miles the road passes over an almost sterile table-land; the few stunted acacia-trees scattered sparsely over it have their branches all bent at right angles in one direction from the constant blowing of the north-east trade-wind. After we had proceeded for about two-thirds of the way, we passed through a small hollow with clusters of palma-christi, or castor-oil plant, when a slight ascent in the angle of the road brought us upon a yet higher table-land, studded with small groves of the same plant. Within about two miles of the village of St. Domingo, our path all at once abruptly descended into a lovely valley, on a much larger scale than any of the others in the island, forming a perfect labyrinth of vegetation of the richest and most peculiar kind, for some two miles in extent, the palma-christi being here superseded by a variety of beautiful shrubs and flowers. This picturesque glen is enclosed in steep and rugged lofty mountain ranges,

R. McCormick, R.N., del.

Vincent Brooks, Day & Son, Lith.

The Baobab Tree of Porta Praya, St. Jago, Cape de Verde Islands. The trunk 36½ feet in circumference.

composed of greenstone and basalt, whose frowning sterility formed a strikingly grand framework in contrast with the verdant landscape at their base. The path, winding through the centre of this valley, is strewed with loose stones like a water-course, and so narrow that the foliage of the trees, in some places, interlaced overhead.

At the extremity of this valley the road divides, the branch to the right leading to the small village of St. Domingo, consisting of a score of huts on a slight rising ground, eleven miles from Porta Praya. It was now 5.45 p.m., and, wearied with the heat of the day, notwithstanding the refreshing breeze blowing, we entered the best-looking house under the impression that it was an inn or " venda." The owner and his wife were seated outside their door. He kindly invited us in, and proved himself a most hospitable host, placing before us the contents of his larder, the remains of a fine cold turkey with some bread, bananas and other fruit, one the size of a small calabash, and the drink of the islands, a kind of white brandy. During the conversation with our host, who appeared to be a very intelligent man, we learnt that he was an officer in the Portuguese service. He could not speak English, but we managed to make ourselves understood in broken French, a language he spoke well. He was dressed in a cap and loose blue frock, belted round the waist. The young black slaves who attended us during our meal, in a row, with folded arms, were particularly clean and neat in appearance, and seemed to be most kindly treated by the mistress of the house, indeed more like their own children. After a cup of coffee, at seven p.m., we took leave of our kind and hospitable host and hostess. The night, though fine, was dark, and the road bad, so that we did not reach Porta Praya until ten p.m., and got on board about midnight. We saw no guinea-fowl during our excursion, but many ravens; and I shot half a dozen of the king-fishers, or jacamars, in the morning, passing over the hill above the baobab-tree valley.

Wednesday, 20th.—The magnetic observations at Quail Island being finished, we sailed at ten a.m. with a fine breeze.   This island of St. Jago has generally an arid and scorched-up aspect, covered with long wiry grass, abounding in locusts, yet affording sustenance to large herds of goats, scattered over the hills, and consti-tuting the chief produce of the island.   The birds are: guinea-fowl, quail, kestrels, owls, kingfishers, or jacamars, tree-sparrows, and flocks of an elegant and beautiful little African finch which abounds in the island.   There are some long-tailed monkeys and the finest oranges in the world, large, juicy, and luscious.

Sunday, December 1st, 9.30 a.m.—St. Paul's Rocks in sight to windward on the port-bow.   Beating up all day for them.   We have been for the last few days in the variables, with heavy rain and squally weather, but on the 25th ult. we had a fine trade-wind, the thermome-ter 81°; the evening bright starlight, the pole-star being only a few degrees above the horizon, and in the fore-noon of that day the planet Venus I saw very plainly, nearly on a line with the main-topsail yard-arm.

Monday, 2nd.—Captain Ross left the ship to take some magnetic observations on St. Paul's Rocks, Lieutenant Wilmot and myself accompanying him.   The weather was fine, but owing to the strong currents and the heavy swell setting in amid this labyrinth of rocks, we had no small difficulty in effecting a landing in a small creek amongst the rocks, at 8.15 a.m., on the lee-side of the rocks from which the highest peak arises.   The rebound of the surf, from the waves dashing in between the steep rocky ledges, imparted to the whole creek or basin the appearance of a boiling cauldron.   The whole group of rocks scarcely exceeds half a mile in circumference.

R. McCormick, R.N., del.

Vincent Brooks, Day & Son, Lith.

St. Paul's Rocks.

Bearing E. S. E., Lat. 0'', 56'. 0'' N., Long. 29°. 20'. 0''. W., 2 miles.

Whilst Captain Ross was employed about his magnetic observations near the landing-place, I strolled over the adjacent rocks on my geological survey of them, and on reaching the farther end of the creek, I found that I was separated from another island, on which some noddies were sitting on their nests, by a crooked strait, a few fathoms in width, through which a very heavy surf set in from the sea outside, breaking furiously on four large boulders of rock which obstructed the channel, in which a number of small voracious sharks were swimming. Having measured the highest peak in the island on which I was standing, and made it seventy feet above the sea, I was now desirous of examining the rocks and the birds' nests on the other side of the strait. But the attempt to cross this was not a very promising affair. However, I thought nothing hazard, nothing win ; so taking off my jacket, after laying down my gun, watch, and trout-basket, I at once set about it, and having gained the centre by leaping from rock to rock, found the rollers breaking here with so much force as to afford me little chance of securing a foothold on the next rock by leaping, being too far apart, so I plunged into the foaming surf and swam across without any difficulty. But I had no sooner fairly landed on the other side than I observed Wilmot, who had been following my footsteps from the landing-place, make an attempt to cross over ; when on reaching the centre rock he hesitated, on the approach of a heavy roller, then, attempting to retreat, was swept from his precarious footing into the surf, and instead of striking out for the spot where I was standing ready to assist him, and urging him so to do, he unfortunately made for the side he had just left, and was dashed with violence upon a rock, to which he clung in an exhausted state, until I had recrossed by swimming to windward of the rock, just in time to lend him a helping hand up the ledge of rocks. Captain Ross, who from a distance had been a spectator of the whole affair, becoming alarmed for his safety, sent one of his boat's crew round to us with a rope, but by the time he reached us all was right again.   Having seen my companion out of further harm's way, I crossed the noisy gulf for the third time, and ascended the " white rock " to the boobies' nests.   The old birds, however, showed no disposition to budge an inch for me from either their eggs or young, but made a most determined stand in their defence.   The nest itself is a very rude affair, a little seaweed from the rocks, the only form of vegetation found here, and a feather or two placed on the bare rock, on which the egg or eggs rested, from one to three, but mostly two, the size of a small duck's egg, having a rough, white, chalky surface.   Both male and female were equally zealous in defence of their treasures, keeping close together for mutual protection, and what is most remarkable, each pair of birds had only one young one, and I examined many nests.   This may be accounted for by the numerous voracious crabs, which swarmed on the rocky ledges, having destroyed the other egg or eggs, or from their having become rotten, for I found some in this condition. But I also have seen a crab after I had disturbed a bird from her nest carry off an egg in his claws, and that boldly before my face.   Perhaps both causes may contribute to lessen their number.   The young are about the size of large goslings, and covered with snow-white down. I brought six of them and a pair of old birds away with me for my ornithological collection, both alive and dead. Some noddies (*Sterna stolida*) formed a little colony above the boobies (*Sula fusca*), and curious enough, although on the other island their eggs were laid on the bare rock, here the noddy had constructed a regular nest of the conferva which projected from the steep face of the rock, in a rounded form, a white, calcareous fringe hanging from it.   The top of the nest forms nearly a plane surface, on which rests the single egg or young bird, for this bird only lays one egg, and, unlike the booby, flies off its nest without any attempt to defend it. The egg is about the size and shape of a plover's, but of a chalky-white colour, the large end sprinkled over with a few small brown spots. In the pools of water left on the rocks I observed some sea-slugs and a few small pilot-fish, banded black and yellow. The numerous small crabs, ever on the watch and hiding in the crevices of the rocks, were so daring, that after seizing upon an egg, after the bird had got off its nest, it would, if hard pressed in a corner, assume an attitude of determined defiance, rising up on its slender legs, and with projecting eyes and pincers and mouth open, look as savage and ferocious as it was possible for so small a creature to do. Active and quick-sighted as they were, however, I caught three or four of them.

The *Terror's* boat had now arrived with her captain and some of his officers; after landing them, the boat's crew pulled round to the island on which I had isolated myself, and I availed myself of this opportunity to put the specimens I had collected into this boat, and row to the island where I had left my gun and other things, and then returned to the observatory. I had been in my shirt-sleeves for some hours, and got dry again in the sun, after my drenching in the foaming torrents. We all returned on board at six p.m., and made sail on our course immediately. I was employed in skinning birds and stowing away specimens until two a.m. I brought on board with me a pretty little noddy, nearly fledged, for a pet, and two old boobies I shot flying overhead for specimens.

This in every way remarkable group of rocks, rising as they do in mid-ocean, between two continents so far apart, the nearest land being the coast of South America, between 500 and 600 miles distant, has been an enigma to geologists, from whom, hitherto, no very satisfactory account has appeared; the rocks of which they are composed being somewhat *sui generis,* for I know of none with which they can be fairly classified. The structure and delicate veining of some of the specimens which I collected is most singular and striking, evidently metamorphosed to a certain extent. Rising so little above the sea level, they are constantly a-wash, not even the highest peak, which is seventy feet, and the white peak of sixty-one feet in height, can escape the surf thrown over them by heavy gales. The curious white appearance of some of the summits, so striking on approaching them, is doubtless owing to a coating of guano, from their being the resort of so many sea-birds. But some of the specimens have a whitish substance interlaid sandwiched, as it were, between two hard brownish tablets, having a calcareous appearance. Others resemble some varieties of serpentine, ramifying, in very delicate fibrous veins, in all directions through the rocks. If the general structure of these is Neptunian, there can be no doubt about their having been elevated from the depths of the vast ocean by submarine volcanic agency during some great disturbance of the bed of the Atlantic, when the group of the Canary Islands, and maybe the Cape de Verdes, were isolated from the African continent. These, some half-dozen rocks, are situated nearly on the equator, in the latitude of 0° 56′ N., and in the longitude of 29° 20′ W. On a steep submarine mountain—a mile off S., 67 E.—a 500-fathom line failed to reach the bottom.

CHART OF S^T PAUL'S ROCKS IN LAT: 0° 56′ N.   LONG: 29° 20′ W.

+ *Passage I crossed.*

Nine-pin Rock, Trinidad.   (*See page* 25.)

CHAPTER III.

Trinidad—Botanical and geological features of the Nine-pin Rock—
Our Christmas dinner at the equator—Letting in the year 1840—
St. Helena—Its scenery—Visit to Longwood—Napoleon's apart-
ments—His tomb—Ladder Hill.

Wednesday, 4*th.*—Having crossed the equator at midnight, the ceremony of crossing the line was in full force
to-day ; all those on board who had not crossed before
went through the usual shaving and ducking on such
occasions.   I had already myself crossed it and the
tropics twice before.

Tuesday, 17*th.*—I accompanied Captains Ross and
Crozier to make magnetic observations on the island of
Trinidad, and after a long pull along the coast we found
some difficulty in finding a landing-place, from the heavy
surf breaking everywhere along the rugged, rock-bound
shore.   The frigate-bird (*Tachypetis aquila*) was hovering overhead as we approached the shore, and a beautiful

little snow-white tern examined us curiously with its expres-
sive, full, large dark eyes, within arm's length of us, as it
followed the boat to the shore. We at last effected a
landing in a small cove to the left of the Nine-pin Rock,
jumping out of the boat upon a ledge flanked by a dark,
frowning, perpendicular wall of greenstone on the right.
It was now ten a.m. and we found ourselves cut off
from the rest of the island, on a narrow strip of beach,
hemmed in by inaccessible hills of greenstone, their steep
sides strewed with loose fragments of rock fallen from
above, intermingled with tufts of long wiry grass and
small ferns. I lost no time in scrambling up the side of
the hill, by laying hold of the tufts of long wiry grass
for support, but all further progress was soon arrested
by walls of rock ; and whilst thus geologizing and alone,
I heard a shout from my companions on the beach beneath me for my return. When I had descended to the
beach they had already got into the boat, and were about
shoving off, with the intention of landing on the other
side of the Nine-pin, so that I had to wade through the
surf into the boat. There were numbers of white tern
and sheerwater breeding amongst the rocks, and large
land-crabs abounded. My companions had already
lunched, but there was now no time left for me to follow
their example. At two p.m. we again landed, and I at once
began the ascent of the hill, which was covered with
loose soil and fragmentary rocks, interspersed with long
tufts of cyperaceæ and ferns. I attained a height,
bounded by a large mass of rock, where a great number of
dead trees, barkless, white, and blanched, were scattered
around in wild confusion, here and there one fixed in the
soil in an erect position. I found by the hour of the day
it was now time to return to the boat, or I should have
much liked to have gained a peak still above me, and, by
crossing a deep valley to a hill on the opposite side,
have examined the only group of living trees visible, and

apparently coniferæ, occupying its summit, about a mile distant. How all the other trees became destroyed is to me a mystery, for I saw no traces of goats or other inhabitants on the island. The elegant white tern alone accompanied me in my ramble, hovering in a circle close over my head. The ships appeared immediately below my feet, close in shore. On my return to the boat, I had a most refreshing draught of deliciously cool and sparkling water from a watercourse down the rocks where we landed.

We next pulled round the Nine-pin Rock, a remarkable pillar of greenstone rising from the main ridge to a height of 850 feet; on one side is a small basaltic dyke, and on the other a large greenstone one. Portions of this greenstone exhibited a very singular, mammiform appearance, the flat surface, embossed by numerous globular, bead-like excrescences in strong relief, clearly produced by some external influences acting upon the greenstone when in fusion. I have never met with anything resembling it before. Captain Ross having finished his observations, and these not proving satisfactory from the iron in the volcanic rocks affecting the needle so much, he gave up his intention of landing to-morrow; therefore, on our return on board at seven p.m., we made all sail for St. Helena. We left a cock and two hens, brought from England for the purpose, on the beach, with the intention of stocking the island with poultry, for the use of any ships touching there.

Tuesday, 24th.—My cabin having become filled to overflowing with the Government collection of specimens of natural history, I got the second master, in charge of the hold, to relieve me of a case by stowing it away there. But our matter-of-fact first lieutenant, to whom everything connected with science is a bore and an enigma, to prove his zeal for such pursuits, ordered it up again, as having no abiding-place there.

About eleven p.m. on Christmas Eve I saw the great constellation of the southern hemisphere, the " Southern Cross," just appearing above the horizon, for the first time this voyage. But really I can see nothing very striking in the form of the cross itself ; it owes much of the interest attached to it, I believe, to our own associations with it, and still more to the vicinity of those singular and remarkable nebulæ, " Magellan's Clouds," and that strange black void, the " Coal-sack," for much of the beauty claimed for it. It cannot, I think for one moment, be compared with those striking and glorious constellations of our northern hemisphere, the " Great Bear," " Ursa Major," and " Orion." But there is undoubtedly a marvellous interest attached to the " Magellan Clouds " and the telescopic star nebulæ in this portion of the heavens astronomically, to which I may have occasion to refer again in subsequent pages.

Wednesday, 25th.— We pass our first Christmas Day at sea, with light airs and fine weather, beating up for St. Helena. After divisions we had prayers and a sermon from Captain Ross. I saw a large spermaceti whale spouting about a mile distant on the port-bow. At three p.m. we had Captain Ross and the gentlemen from the midshipmen's berth to dine in the gun-room with us ; thirteen in all sat down to table, Captain Ross in his customary place at the after end of the sofa, and I was in my own place at the foremost end, next the bulkhead. Our Christmas fare : pea-soup, followed by roast turkey and ham, and preserved meat-pie, parsnips, plum-pudding, and pumpkin-tart. I saw the " Southern Cross " again this evening.

Friday, 27th.—Had some dolphin for breakfast, caught yesterday, and again at dinner, at Captain Ross's table, with whom I dined to-day.

Tuesday, 31st.—It being a lovely night, I walked the deck throughout the whole of the first watch ; saw the old year out and the new one in. Before I turned in I went below to the midshipmen's berth, and joined them in a glass of mulled wine. To-morrow, New Year's Day, all the officers have had an invitation to dine with Captain Ross.

Wednesday, January 1st, 1840.—New Year's Day commenced with the delightful weather of these regions, a clear blue sky, bright sun, and fresh breezes. About 1.30 p.m. sounded as usual to obtain the temperature of the sea at a great depth. All hands on deck stepping out to the fiddle, whilst the line was hauling in, in lat. 28° 20′ S., long. 19° 40′. The gun-room officers and midshipmen from the berth dined in the cabin with Captain Ross to-day; thirteen in all sat down to table.

Friday, 3rd.—Saw five or six dolphins swimming round the ship, and one was caught; two or three pilot-fish playing about the rudder. The ship was hove-to for deep soundings, and it was calculated that the bottom had been reached at 2400 fathoms, but the weight carrying away the line in hauling it in, most of it was lost.

Friday, 10th.—A fine large flying-fish, weighing sixteen and a half ounces, and sixteen inches in length, fell on board during the morning watch, which I preserved in spirits.

Sunday, 26th.—At midnight I saw the planet Jupiter rise soon after the moon, and just below it, and about half an hour later "Ursa Major" appeared above the horizon for the first time since leaving the Canary Islands, and at the same time the "Southern Cross" appeared at a considerable altitude in the opposite quarter of the heavens. The night was clear and beautiful, and before the moon rose, the "Milky Way" passing through the cross, and the constellation of the ship, formed a wave-like band across the zenith, with the two white "Clouds of Magellan" remarkably distinct and bright.

Friday, 31st.—The island of St. Helena in sight

ahead, almost enveloped in clouds, five or six leagues distant. Three bonito were caught with hook and line, and we had some for dinner; white tern were flying overhead, and porpoises swimming beneath. Anchored at three p.m., having made the land at nine a.m. I recognized in the harbourmaster, who came on board, a former messmate, who had been master of H.M.S. *Tyne*, in which ship I returned to England from Rio Janeiro. On first making the island, which presents a barren, arid, and scorched-up aspect, and nothing very striking in the mountain outlines, on the summits of which a few clumps of pines are scattered, studded here and there with a white house, and signal-stations on almost every point.

Saturday, February 1st.—I went on shore, and again on the following day called on Gulliver, the harbourmaster, and with him attended morning service at the small church. Afterwards we mounted horses outside, and rode across the island to " Fairy-land," the country residence of a merchant of Jamestown, of the name of Gideon. Reached the house at three p.m., drenched in rain; the weather was overcast and threatening when we started, and before we had got half-way heavy showers fell, the mountains were hidden in mist, and the scenery entirely obscured. The first part of our journey lay over Ladder Hill, past the barracks, then winding along hills and hollows. The small deep valleys formed good pasturage, and the few cattle we saw appeared to be good-sized animals. Pretty white-washed villas studded both hill and valley, surrounded by gardens and larch-plantations, chiefly the residences of retired officers of the St. Helena corps. Hedgerows of blackberries, on which the fruit was just ripening, flanked the lanes, with geraniums and other flowers in bloom. I saw some flocks of finches.

As we approached " Fairy-land " the road terminated abruptly in a grassy knoll, and passing through a gate

R. McCormick, R.N., del.

Vincent Brooks, Day & Son, Lith·

Fairy Land and Lot's Peak
St. Helena.

by a winding path, through a plantation for a short distance, the pretty white villa, with its little gate and lawn in front, burst upon our view. Our host, little expecting visitors in such weather, had only just awakened from his siesta, but received us kindly at the door ; and after exchanging our wet clothes for dry ones, we were ushered into the drawing-room, and introduced to Mrs. Gideon and two young ladies on a visit there from Jamestown ; and on the two daughters of the house joining us, dinner was announced, and after three hours had very sociably passed, we took leave of our hospitable host and his family.

Monday, 3rd.—I dined at the mess of the 91st Regiment and artillery. The Antarctic Expedition called forth a speech from Colonel Trelawney, which was briefly acknowledged by Captain Ross. My old messmate, Gulliver, the harbourmaster, and his friend, Mr. Gideon, of " Fairy-land," were both there. About midnight, on reaching the landing-place, we could not succeed in getting a boat to take us on board, and had to sleep on chairs at " Lawler's Hotel," nearly devoured by mosquitoes.

Tuesday, 4th.—At nine a.m. I started on horseback from the mess-rooms in James Street, in company with our two artillery officers, on an excursion to Longwood. About a mile from Jamestown we passed the " Briars " on the right, once the temporary residence of Napoleon. At 9.30 a.m. we reached Captain Alexander's quarters, an engineer officer, where we breakfasted, after which he accompanied us to Longwood, passing in sight of the tomb of Napoleon, at the bottom of a valley on our left. About noon we entered the grounds of Longwood, situated 1730 feet above the level of the sea. A pasture, flanked by some straggling gum-trees, leads to Longwood " old house," and the new house, which we first visited, is about 100 yards beyond, and to the left. It is a neat and somewhat handsome structure, long and low; most of the apartments are on the ground-floor, the drawing-room, a spacious room, with the walls of the "Imperial green." In front is a balcony, with a variety of shrubs and flowers growing in great luxuriance beneath, amongst which the passion-flower was most conspicuous. Bounding the extensive prospect is Flagstaff Hill, the barn, and on an eminence across the valley stands Dead Wood. On either side of these hills the deep blue ocean appears. The day was fine, with bright sunshine, and the whole scene wore an air of peaceful retirement, partaking both of English and tropical scenery.

We next visited the old house; the main entrance is beneath a green verandah, in a sort of gable. The first room, which was Napoleon's billiard-room, is now filled with bearded wheat. This apartment opens into the bedroom, under the second window of which the great Napoleon's head rested when he drew his last breath on earth. The next two rooms were his sitting and breakfast rooms, both dark and gloomy, having little light beyond what the doors admitted—one was entirely destitute of a window. A threshing-machine now occupied one of these apartments. In the corner of the small adjoining courtyard is a willow, said to have been planted by the Imperial captive's own hand from a twig he himself brought from the valley where his tomb is now situated.

At one p.m. we remounted our horses to ascend Diana's Peak, the highest part of the island, 2697 feet to its summit, which we reached on foot, having secured our horses by their bridles to some shrubs in the valley beneath. Several beautiful tree-ferns grew along the valley ridge, over Halley's Mount, Cuckold's Point, &c., as we proceeded along a narrow pathway, flanked on either side by shrubs and dense underwood to the top, which we stood upon at 2.30 p.m., but a cap of mist sadly obscured our view. At three p.m., remounting our steeds, we returned to Jamestown by the sandy bay road, and down Ladder Hill, getting on board at five p.m.

Thursday, 6th.—Having sent a large case of specimens of natural history, my first instalment for the Admiralty, on board the *Samuel Enderby* for England, I landed on an excursion to Napoleon's tomb, which we reached at five p.m., amid bright sunshine. I had a most refreshing draught of sparkling clear water from the well at which the mighty Napoleon used to drink. I took a sketch of the tomb, over which some of my feathered friends and favourite birds, the ducks, were irreverently waddling, after which we repaired to Mrs. Torbutt's cottage, just below it, and had some refreshment, inserting our names at the same time in the visitors' book. At 6.40 p.m. we commenced our return, but it was dark before we reached Jamestown and got on board.

Napoleon's grave is covered by three large plain slabs, enclosed within iron palisades. At its head a small geranium was growing. A group of seven willows flanks the northern side, with a larger tree on the south side. The whole of the interior greensward enclosure, surrounded by a wooden fence, perhaps a hundred feet in circumference, is filled with cypress and fir-trees, alternating with each other. About fifty yards above this is the old sergeant's cottage who has charge of the keys to the enclosure, and he at once admitted us, although we had come without an order from the town-major, which is usually necessary for admission. After taking a sketch and gathering some sprigs of the willows, we drove off again, and got on board at eleven p.m. The general aspect of Longwood and its environs, the valley of the tomb, far exceeded my expectations from previous descriptions. An extensive command of prospect around, presenting a pleasing, rural, peaceful appearance, which, on a fine sunny day, with the birds singing, as was the case when I saw it, with the bold rugged hills of the Flagstaff and the Barn, and glimpses of the sea between, altogether forms a fine background to the picture.   But from every bearing the three plain slabs, beneath which repose the remains of, perhaps, the greatest man the world ever saw, girt round with their bristling palisades, these again enclosed by a green sward, encircled by cypress and fir-trees, and the group of weeping willows overhanging the tomb itself in the sequestered nook forming the extremity of the valley, produced on my mind, at the first glance, the most striking and impressive effect of any scene my eye ever before rested upon, so marked a feature does it present in this romantic glen.

Friday, 7th.—At 3.15 p.m. I went on shore with the purser, and the first thing I did on landing was to ascend Ladder Hill, to test the shortest time in which it could be done, or, at least, that I could myself accomplish it in. The ascent is at a somewhat steep angle ; the number of steps amounted to 636, each a foot apart, and of wood—one had been carried away.   There were six seats or benches at the sides, placed at intervals, as resting-places.   I reached the top in nine minutes, without resting once on the way, and I think it could be done in a shorter space of time if one were prepared beforehand with a light dress for the occasion, but I made the attempt on the spur of the moment in my ordinary dress —uniform frock-coat with epaulettes, &c., and the sun bearing a great power at the same time.   I was only four minutes in descending very leisurely.   The height of the hill is estimated at 600 feet.

At the summit I met Wilmot, our artillery officer, and an officer of the Ladder Hill station, with whom I remained about eight minutes in conversation before I descended.   The purser, who remained at the bottom of

R. McCormick, R.N., del.

Vincent Brooks, Day & Son, Lith.

Napoleon's Tomb, St. Helena,

Longwood and the sea in the distance, sketched from the road to Diana's Peak.

the ladder, timed my ascent and descent by his watch.
At 4.15 p.m. the purser and myself started for the tomb
in Mr. Gideon's chaise. The distance is little more than
three miles, but the road uphill and steep. When within
a quarter of a mile, a road branches off to the left down
the deep valley, at the narrow upper end of which the
remains of Napoleon rest. Longwood appears cresting
the ridge on the opposite side of the valley.

Saturday, *8th.*—I left the ship at eleven a.m., and
called on board the *Arachne*, now a merchant barque,
bound for England, once a corvette of eighteen guns, in
which, in the year 1830, I returned from New Providence
to England. On landing with the assistant-surgeon, Dr.
Hooker (now Sir Joseph), we started from Jamestown,
with fine weather, on our last excursion across the island,
to take leave of our kind friends at " Fairy-land." We
went by the Longwood road, passing the " Briars," a
pretty modern-looking villa, having a green balcony,
enclosed in lawn and shrubs, once the residence of the
Balcombes, when Napoleon was an inmate. Ascending
a hill, turning a sharp angle of the road overlooking the
valley of the tomb, I took a sketch of the tomb, this being
undoubtedly by far the best point of view to be had
of it. At one p.m. we continued our course along the
rocky ridge to Diana's Peak, and along a narrow lane
covered with long grass to the main road. This road
skirted a valley in which a number of wild raspberries
were growing, very tempting indeed as far as appearance
went, having a bright scarlet colour, and more like the
strawberry than the English raspberry in shape, but the
taste peculiarly insipid and mawkish. Growing on a low,
spreading bush, at a short distance they present the
appearance of small damask rose-buds that have not
unfolded, as they peer through the green foliage.

We called at " Rose Cottage," the residence of the
chief justice. On the outskirts of " Fairy-land " I shot

a canary in its native green plumage, which, with the
"aberdevate," a pretty little finch I afterwards shot a
specimen of from a flock flying overhead, constitute the
main portion of the ornithology of the island, so sparsely
are the feathered race distributed.   Here I lost my
powder-flask, and whilst retracing my steps in search of
it, found a small tortoise instead, rather a remarkable
occurrence, I was told, as the animal is not indigenous to
the island.   I took it on board with me.   Reached the
Gideons' at 4.20 p.m.   Found the young ladies at their
embroidery-work ; they accompanied us to the Flagstaff and summer-house, at the extremity of a narrow
wooded embankment, overhanging a deep valley, from
which the singular cone-shaped basaltic rock called
"Lot" arises, a bold and conspicuous object in the
scene spread out before us, which is here of the most
picturesque and beautiful description.   To the right
appears another remarkable peak, called "Lot's Wife,"
with the "Asses' Ears."   Whilst at the distance of some
two miles or so, the blue ocean peeps through an opening
in the hills, where the surf breaks on the rocks bounding
Sandy Bay.   "Lot" is a rock 1444 feet above the level
of the sea.

At 8.15 p.m. we started on our return, with six or seven
miles before us, on a dark night with drizzling rain, and
along a road we had only been once before.   The
greatest bane to the island is its extreme humidity,
the mountains being for the most part of the year
enveloped in either mist or rain.   The island is of submarine formation.

Cleft-summit of Table Mountain. (*See page* 42.)

CHAPTER IV.

A deep sounding—Africa sighted—In Simon's Bay—Excursion to Cape Town—Farmer Peck—His queer hostelry—A Dutch wine-grower and his wines—Ascent of Table Mountain—Marion's Island—Possession Island.

Sunday, 9th.—We got under weigh, bade adieu to St. Helena for a time, and stood along the land with a fine breeze, the weather fine and clear, and as I was about descending below to dinner, at three p.m., I saw the last of the land, a distant view of Sandy Bay.

Saturday, 22nd.—A number of small sepia fell on board, and I picked up one alive on the starboard-booms. It pulsated strongly in my hand, and on putting it into a tumbler of sea-water it emptied its ink-bag of dark fluid and expired. Between twenty and thirty others were picked up. I saw two sheerwaters.

Saturday, March 1st.—A fine large albatross made several sweeps round the wake of the ship, and eventually alighted in the water at some distance astern; a sheerwater or two have been following the ship for the last day or two, and this evening the wake of the ship was rendered brilliantly luminous by the pyrosoma.

Tuesday, 3rd.—The ship was hove-to to sound the depth of the ocean; two boats were lowered with reel and line, and the two captains left the ship in her. A weight of 540 lbs. was attached to a spun-yarn line of 5000 fathoms, which reached bottom at a depth of 2677 fathoms. This was by far the greatest depth we had yet obtained. The latitude was 33° 20′ 42″ S., longitude 9° 3′ E., 470 miles from the Cape of Good Hope. The line not being able to raise the weight from such a depth had to be cut, consequently the whole of the submerged portion was lost.

Monday, 9th, eight a.m.—Upon going on deck I had my first sight of the African coast appearing low-lying on the starboard-bow, standing in towards St. Helena Bay.

On the 13th, Friday, at 5.40 a.m., I saw Table Mountain on the port-bow, and soon afterwards the ships in Table Bay. The horizontal stratification of the white silicious sandstone forming the summit of the hills above their granite base, is seen to great advantage from the sea. Albatrosses and petrels were numerous; fish of four kinds were caught forwards as fast as the lines could be baited.

Tuesday, March 17th, eleven a.m.—We came to an anchor half a mile from shore in Simon's Bay. Found the *Terror* already there, and the *Melville* flagship.

Thursday, 19th.—I went on shore, and walked round the west point to the cemetery, nearly a mile from the town, proceeding along the base of Simon's Bay. Only saw a few sugar-birds, much resembling both in size and colouring of plumage the humming-birds of the West Indies ; in fact, a representative of that bird in the Old World, it having similar habits. I returned along the beach, where I met with a similar incrustation bed to the Porto Lorenco one at Madeira. It occupied some acre or two of sand, separated from the sea by a sand-dune, only studded with encrusted stems. The charred nucleus of wood here has evidently been the stem of a seaside plant now growing in great numbers on the spot. The univalve shells scattered about resemble the Lorenco ones, apparently (*Helix Ramondi*) lake shell, belonging to the Miocene period.

Friday, 20th.—Landed on a shooting excursion ; only met with a few small birds, sugar-birds, and honey-eaters. The sandy plain is scattered over with an abundant heath-growth and numerous proteaceæ.

Monday, 23rd.—Started for Cape Town. After crossing several sandy bays we passed some fences of whalebone, the "Whalebone Inn," and through a turnpike-gate. When at seven miles from Simon's Town, the well-known eccentric Farmer Peck's white gable appeared on the right, with his name in large black letters painted on it. Here we alighted at 10.40 a.m., and after ordering breakfast, had our patience put to the test by having to wait for an hour before it was got ready. Farmer Peck appears to be quite aware of his being a privileged character, and, with a certain amount of independence about him, does not seem to care very much about putting himself out of the way about anybody. He is a small, spare man, about fifty years of age, with more the appearance of a sallow-complexioned citizen than a farmer. Whilst breakfast was preparing I took a survey of the premises, which consisted of a low range of thatched buildings on the ground-floor, and most unlike all our notions of a farm, located in a barren, sandy tract, with not a tree or shrub to enliven it. Bounded on one side by False Bay, it has between it and the sea a long strip of level beach, from which it is separated by only a few yards of sand-dunes. The opposite side of the road has a background of arid hills, rising to a considerable elevation. An apology for a garden is attached to one end of the premises, an enclosure fenced in with cane, in which a few stunted fig-trees were struggling for exis-tence, with the pretty little sugar-birds flitting like rainbow-coloured gems amongst them, whales' ribs form-ing the road fence. A number of gulls were walking on the sands of the shore. In front of the door next to the main road, a signboard swings between two rough posts suspended from a cross-beam, having on it the words,—

"GENTLE SHEPHERD OF SALISBURY PLAIN."

The following lines are inscribed on either side of the lower board :—

On the left side.

"'Multum in parvo! pro bono publico.'
Entertainment for man or beast, all of a row.
Lekker Kost, as much as you please ;
Excellent beds, without any fleas."

On the right side.

" Nos patrium fugimus, now we are here,
Vivamus, let us live by selling beer.
On donne à boire, et à manger ici,
Come in and try it, whoever you be."

The small room in which we breakfasted had white-washed walls, and no other ceiling than the bare timbers of the roof overhead. The furniture consisted of four deal tables, ten dark-coloured chairs, an old cane-backed sofa, with a sanded floor. The young girl, whilst preparing the breakfast-table, handed me three curious caricatures, painted in oil on small square pieces of wood, the work of some previous visitor. One represented a grotesque figure, in cocked-hat and red coat, with " Sir Roger de Peck, of Bushell Hall, 1649," written on it. We had a chicken killed and grilled for breakfast, with a limited number of fried eggs and ham, and some very fair coffee, for which the charge was 3*s*. each.

Our road now took an inland direction, passing round the base of hills and away from the sea; and after proceeding for some distance across a sandy plain, we eventually arrived at " Little Constantia," one of the celebrated wine-farms or vineyards, amid a shady avenue of lofty trees, fine oaks. This was C. Coligne's, which we had intended visiting, but finding three carriages-and-fours standing at the entrance to the avenue, we concluded that the time of the people at the vineyard would be fully occupied by these visitors; so at 2.30 p.m. turned down the magnificent avenue of ancient oaks, whose tops interlaced overhead, forming quite an arboreal arcade. A short drive brought us to the " Great Constantia," where, entering by a wide gateway, we drove through an avenue of large trees up to a substantial, handsome-looking house, and were at once ushered into a large and well-furnished apartment, at the entrance to which stood a fine stuffed leopard, which had been shot by the owner on the side of Table Mountain about four years before. We were taken to the wine-house, and told that the proprietor would soon be with us, and we filled up the interval in rambling round a large orchard on the left side of the building, planted with pear, apple, medlar, and other fruit-trees. In a small vinery the grapes had already been gathered. On our return to the wine-house, we found Mr. Cloete, the owner of the establishment—a fine specimen of the old Dutch gentleman—awaiting us. He first showed us the wines fermenting in the vats, and

explained to us the whole process, at the same time hand-
ing us a glass of each to taste. First, the " Red
Constantia," made from the red muscatel grape ; next, the
" White Constantia," from the green kind," sold at 15*l.*
the " half-aam," of nineteen gallons ; third, the " Fron-
tignac," 18*l.* 15*s.* the " half-aam ;" fourth, the " Pontac,"
a sweet, dark red wine, and the highest in price, being
22*l.* the " half-aam ;" the fifth and last sort was the com-
monest, sold at only 7*l.* 7*s.* the " half-aam," and called
" Steen "—the " White Constantia " being considered
the finest-flavoured of all. The grapes are used when
half the bunch is converted into raisins, and fermentation
is checked from proceeding too rapidly by burning a stick
of sulphur in the cask before adding the grapes. The
wines are kept for four years in warehouse before they are
fit for the market.

We next accompanied the owner to the oldest vineyard,
a field of fifteen acres, to the right of the house. The
vines were planted in rows four feet apart, each plant
being from two to three feet in height, with from four to
five short, thick stems, each bush bearing on an average
twelve bunches of grapes, by no means of large size.
Some were red and others green, a few of the plants
having fruit not larger than the common currant. The
peculiar-flavoured, small, dark grape, which is employed
with the red muscatel in the making of the " Pontac,"
affords scarcely any juice, and is used only to give flavour
and colour ; the productiveness of this grape being so
small renders the " Pontac " so expensive a wine. This
estate is the oldest of all, dating its origin as far back as
150 years, and was named " Constantia " after the
daughter of the Dutch governor, Van der Stall, who first
founded it.

Mr. Cloete also told us that the estate had been in his
own family for a generation or two, and that the vineyard
of fifteen acres, which he had just shown us, was fifty

years old. The whole of the vinery occupies thirty-seven acres, producing from thirty to forty pipes a year, at an average price of 90*l.* a pipe. Having passed about an hour at the estate very agreeably, and not unprofitably, we resumed our journey. At the end of the lane we passed by the third establishment on the left, having the words "High Constantia" written over the gateway. It belongs to S. V. Van Renen. From this to Wynberg the road is good, and passes through this very pretty village to Cape Town, seven miles distant. The late governor was Sir Benjamin D'Urban, whose daughter and sons had long years gone by been at the same dancing-school with myself at Yarmouth, in Norfolk. General D'Urban, who was with the Duke of Wellington in the Peninsular War, (my own father at the time serving with the Baltic fleet,) was now living in Wynberg, where also resides the admiral on the station. It is a charming spot, having a cheerful English aspect; carriages and horses were passing to and fro, and omnibuses running daily. Villas are scattered about amongst the trees, and a pretty-looking church appears at a distance on the left.

The approach to Cape Town is very pleasing, commanding an extensive view over the plain, studded here and there with windmills and numerous white houses. Table Mountain rises on the left, and the town, with the bay full of shipping, spreads out in front. We passed several waggons on the road, each having from fourteen to sixteen large black oxen yoked to them, their horns curved upwards, little short of a fathom in extent.

Thursday, 26th.—Arose at four a.m., and, accompanied by the surgeon of the *Terror*, shaped as direct a course as possible for the base of the mountain, which is 3550 feet high. It was 6.15 a.m. when we commenced the ascent from the Millhouse, by a watercourse. After quitting this we had kept too much to the left, bringing

us to the edge of a ravine with a clump of trees.   Here
I parted from my companion, who continued his course
along the ravine, but I determined on a less circuitous
course, and at once struck up a steep ridge to the right,
studded with rock fragments and the burnt stumps of
trees; and, after some little labour and exertion, I gained
the usual rugged, zigzag path of the ascent, at the base
of a steep wall of rock, where the very narrow cleft
opening at the summit first comes in view.   As I approached this, the ascent became steeper and steeper, the
fragments of rock, in larger masses, piled in inconceivable
confusion.   I only saw a small bird or two in the ascent.
At 7.30 a.m. I stood on the summit, having passed
through a narrow gorge or cleft, about a fathom wide, in
the perpendicular wall of rock, rising on either side to
the height of some sixty feet.   I had been at the least
ten minutes on the top when my companion rejoined
me; he having had to cross a ravine in his more circuitous route.   We were, however, rewarded by a most
magnificent prospect.   The vast plain on the summit is
three or four miles in circumference, and spread out
beneath lies Cape Town, with its quadrangular masses
of buildings and gardens; beyond this the bay, with
some twenty or more sail of vessels.   We walked all
round the summit, and had a fine view of False Bay to
the southward, bounded on one side by Hang-lip Point,
and on the other by the hills of Simon's Bay.   We
searched for a- spring of water to wash down our slice
of cold meat and bread we had brought up with us for
breakfast, and found a very small pool beneath a large
mass of rock, just deep enough to fill a small basin from,
being only a few inches in circumference.   Two small
finches, the only ones we saw, with a frog, were drinking
at the tiny spring, a sinister hawk hovering in the
distance.   But none of the baboons, leopards, and rock-
rabbits, so much talked of, were seen by us.   The surface

of this plain is covered with tufts of large rushes, heaths, and grasses; in some spots the soil appears rich with flowering plants; in others, peat, elastic under the feet, water-worn pebbles of beautiful white quartz, form patches of shingle, looking as if they had only just been upheaved from the bed of the ocean. The sides are skirted in places by sandstone ridges, intermingled with the pebbles of quartz, and elevated above the plain. One large mass here appeared poised on another, like the Logan Rock of Cornwall.

At noon we commenced our descent from the cleft at the top, following the path down to the ravine and Mill-house. Then continuing along the watercourse, in which a number of negresses, with here and there a mulatto, were up to their knees, washing clothes. We reached the town at two p.m., and after lunch, I left Cape Town at 4.30 p.m. for Simon's Town, twenty-three miles distant. Returning in the chaise, called at the observatory, three miles from Cape Town, where I met Lieutenant Wilmot, our late messmate, whose quarters this will be for the next three years. Reached Farmer Peck's at dark, about 7.30 p.m. Had a sandwich and glass of ale here, starting again at 8.30 p.m., when we had a very intricate and difficult navigation amongst the rocks and quicksands in crossing the bays, it being high water. Into one of these quicksands the wheels of the chaise sank to the axletrees, and getting wedged between the rocks, with surf breaking over all, the harness gave way, and the horse had to be taken out before we could extricate the wheels. Whilst setting my shoulder to them I lost a gift gold ring from my finger in the surf. We had to tow our crippled craft back to the "Clarence Inn" at Simon's Town as best we could, arriving about midnight; so, being too late for a boat, we had to sleep at the inn, and returned on board at eight on the following morning in the whale-boat.

Wednesday, April 1st.—Having had an invitation from the ward-room officers of the flagship *Melville* to dine with them to-day, Captains Ross and Crozier, with the second and third lieutenants of each ship, Lieutenant Wilmot, R.A., and myself sat down to dinner with the officers of the flagship, making in all a party of twenty-one.

Friday, 3rd.—Myriads of shags or cormorants were flying about the bay to-day in phalanxes extending the whole length of the bay, all fishing, so much do these waters abound in fish.

Monday, 6th.—We got under weigh in company with the *Terror*, and beat out under cheers from the flagship, which were returned by us.

Tuesday, 21st.—As we approached the Marion Islands, the weather became unsettled, the temperature much lower, with boisterous and thick, hazy weather at intervals. The sea here teems with oceanic birds: albatrosses, petrels, Cape hens, and stormy petrels. At 1.30 p.m. I saw Marion's Island, high conical hills rising from the centre, and from these long, low points run out to seaward, the whole having a volcanic appearance. A little further on, we opened a small cove literally enamelled with penguins; they were assembled here in thousands. The slopes of the hills and long, low points presented a verdant aspect. One hill looked as if clothed with a reddish-brown moss. It was blowing too hard, with too much surf on the beach for a boat to land, and at 4.30 p.m. the ship was laid-to for the night, which was cloudy and moonlight.

Wednesday, 22nd.—The weather still continuing unfavourable, with a falling barometer, prevented our landing, as intended, on Marion's Island, so we bore up for the Crozets, going before the wind, without seeing anything of Prince Edward's Island.

Sunday, 26th.—Two of the Crozets were passed early

in the morning watch, and unfortunately I missed seeing them—a rare occurrence with me—and we continued on our course for Possession Island. The weather for the last few days has been misty, thick, and stormy, with heavy gales. The ship under a close-reefed main-topsail, and storm-staysails, rolled much in the heavy swell, quick and deep, going before the wind. The air very keen and chilly. The petrels were numerous, especially the small blue one, and stormy petrel ; an albatross or two skimming in the wake of the ship, and the first speckled Cape pigeon seen.

Friday, May 1st.—Beating up for the anchorage in Possession Island, a beautiful, clear, sunny morning, after the thick, very unfavourable weather we have had for the last few days, with only occasional glimpses of the land, half-hidden in mists, and the summits of the hills sprinkled over with snow, as we beat up between the high land of Easter Island and Possession Island. Both islands have the appearance of volcanic origin. A remarkable detached rock of hexagonal form, with apparently a hole through it, stands isolated off the north-west end. At 10.30 a.m. we hove-to off America Bay, having observed a six-oared boat putting off from the shore in answer to the signal-gun we had fired. We brought out with us a chest of tea and some bags of coffee for a seal-fishing party, fourteen in number, who have been located on the island for about eighteen months. These things had been sent from the Cape by their employers, with a letter, the contents of which seemed to disappoint the leader of the party, a fine, intelligent, sailorlike-looking fellow, who evidently had been anticipating a ship for their removal, instead of fresh supplies for a prolonged stay on the island.

Captain Ross having made up his mind not to send a boat on shore to these islands, we endeavoured to get all the information we could from these poor sailors, whose sallow aspect was by no means diminished by their unshaven beards, dirty guernsey-frocks, and woollen caps, with stockingless feet—except their manly-looking leader, who was an ideal " Robinson Crusoe " in costume, with his penguin-skin moccasins, and who answered all the questions Captain Ross put to him with promptitude and self-possession. He told us that a boat belonging to them had been lost with all her crew off Easter Island in the late gale, and on that island they had a party of eleven or twelve men. Another boat had been washed off the beach and lost. Their best fishing season, it appears, is during the months of July, August, and September. Whale-ships frequently visit the islands, the water being so abundant and easily obtained, that 100 tons may be procured in a very short time. A large seal is found here, probably similar to the Kerguelen's Land one, penguins, and some goats. A small island to the westward (Hog Island) is overrun with pigs, about seventy miles distant. This bay he considers the best anchorage close to the rocks in shore. The land at the head of the bay forms a yellowish-green coloured valley of turf, flanked by sloping hills and undulating ridges of the same colour; a ledge of black rocks divides little America Bay from it on the south, the whole having a background of lofty, rugged mountains, having a dark-brown shade, and sprinkled with snow. Many of the points and headlands look like old castles or other ruins, so picturesquely are they broken up. At a mile distant, the nearest approach we made, the columnar basalt was very apparent, and a lighter-looking rock, probably greenstone. There are also two dykes and a watercourse. We learnt from the leader of the sealing-party that he had been about eight months in Christmas Harbour, Kerguelen's Land. Having had a glass of grog each, they got their supplies into their boat, and took their departure, and at eleven a.m. we bore up, making all sail for Kerguelen's Land, run ning between the two islands. I made sketches of both. Easter Island presents a singular and remarkable appearance, its dark mountain peaks, 3000 feet high, in strong relief against the sky, towered above a belt of white clouds.

Thursday, May 7th.—At nine a.m. I first saw the high hills of Cape François to windward, with here and there a patch of snow on the summits, and a low-lying island on the lee-bow. The air was cold and pinching, the most severe we had hitherto felt, thermometer being 36°. The north-west wind blew in heavy squalls, weather hazy, with showers of sleet and snow, the blue sky only seen at intervals. It was during such weather as this that last evening we had a glimpse of the off-lying and remarkable rock called Bligh's Cap. Numbers of petrels and shags were flying about the ship. We were all day beating up for Christmas Harbour, making short tacks and little progress, so stood off the land for the night. " Old Tom," a cock brought out from England with us, who, with a hen, was to have been left here to colonize the island, died to-day, in the very sight of his intended domain; had his body committed to the deep by the captain's steward—a sailor's grave.

Friday, 8th.—Beating up all day for the land. At two p.m. saw the Arched Rock, at the entrance to Christmas Harbour. About four p.m. a barque was seen from the masthead, supposed to be our consort, the *Terror*, which we parted company with soon after leaving the Cape. Fired a gun, and soon after dark sent up a rocket as signals to her, and not being able to fetch the harbour, stood off again for the night.

Monday, 11th.—The last two days of blowing weather, with the ship rolling in a heavy sea, little progress was made; but on the morning of the 12th we found ourselves to windward, and well in with the land.

Had to work up Christmas Harbour against a strong breeze, and from the narrowness of the entrance compelled to make very short tacks, going about each time close in shore, and after all had to let the anchor go, in twenty-four fathoms, at 5.30 p.m., heavy squalls coming off the high land. Inside Arched Point we passed some fine cascades, and a remarkable white bird, about the size of a common pigeon and shape of the ptarmigan, called the sheathbill (*Chionis vaginalis*), which was numerous on the rocks, seeking its pelagic food as it leisurely walked along the ridges at the water's edge. Some penguins were swimming about the harbour, and black-backed gulls and tern flying overhead. At six p.m. Captain Ross dined with us in the gun-room.

Thursday, 14*th*—The *Terror* anchored about a mile above us yesterday at noon. A number of spermaceti whales were swimming about the bay to-day in groups of three or four together, spouting and going down with the tail uppermost. A pair of water-boots, with hose, were served out to the officers and ship's company, and I sketched the harbour all round.

Bligh's Cap. (*See page* 47.)

Basalt, or Fossil Wood Mount.

CHAPTER V.

At Kerguelen's Island—My find of fossil-wood—Remarkable geological features—First boat's expedition for exploring Cumberland Bay—Camping ashore—The Southern Cross—Exploration under difficulties.

Friday, 15*th.*—At noon the *Erebus* was warped up the harbour above the *Terror*, and at one p.m I landed in the galley with the two captains, for the first time on this, to me, most interesting island. This, and Spitzbergen, in the opposite hemisphere, constitute, I think, the most striking and picturesque lands I have ever had the good fortune to visit in a somewhat wide range of wanderings over our globe from pole to pole. Yet neither the Arctic nor Antarctic isles have tree or shrub at the present time to enliven them. The largest arborescent form in Spitzbergen is the willow (*Salix arctica*), crawling an inch or two above the surface of the ground ; whilst in Kerguelen's Island, the famous cabbage (*Pringlea antiscorbutica*) peculiar to the island, attains but a foot or two elevation above the soil.

And yet whole forests, in the form of fossilized wood, lie entombed here beneath vast lava streams; in one instance, a trunk of a tree, seven feet in circumference, has been found buried beneath the *débris* of this truly volcanic land. That singular and beautiful bird, too, so fearless and confiding, the chionis, seems peculiar to the island, to which its presence gives a charm and animation, especially to a lover of the feathered race like myself. The observatory is now being put up on the beach at the top of the bay, in the north corner. I crossed the isthmus, about a mile and a half over from the top of the bay to the opposite side, passing two lakes, soil boggy and swampy, the coast steep and precipitous. Saw Bligh's Cap and two islands in the distance. The weather squally, with very heavy gusts of wind, accompanied by drifting showers of sleet and hail, cutting to the face. I shot my first teal, with which the island abounds; its favourite food, the seeds of the cabbage, now ripening. I also shot, near the first lake, two brown-coloured lestris and five penguins, from the beach.

Saturday, 16*th*.—I landed, accompanied by Dr. Hooker, on an excursion to the summit of the black basaltic rock, ascending from the ridge at the south corner of the beach along a boggy slope, down which fell a cascade. Here I shot my first chionis, and crossing over two ridges, forming a rugged platform, we followed the west side of the base of the basaltic rock, where I had the good fortune to discover the first trace of the fossil-wood, a few small pieces, so weather-worn that they looked like the stem of a pipe from their blanched appearance. These small relics of an age gone by were loosely scattered on the surface, occupying a space of not more than a foot or two of ground. I called out to Hooker, who was within hailing distance of me at the time, to announce this unexpected discovery, and laughingly observed that we should sooner have ex-

pected to find on this spot the embers of a fire lighted
by our enterprising countryman and predecessor here,
the great navigator Captain Cook. On returning we
found larger fragments, *in situ*, beneath the black
rock, only a few feet above the talus of *débris*. Turning the flank of the mountain on the left, we ascended
by a pass between two rocks covered with ice and
snow to the summit, which we reached at 1.30 p.m.,
passing by a trap-dyke, ten feet high, and embossed
with a beautiful lichen, tinted pale green and black,
intersecting the top in a S.E. and N.W. direction;
continuing our course along the summit, which was
strewed over with rock fragments, patches of snow, and
lichens. About a quarter of a mile brought us to the
opposite side, E.S.E., but the weather becoming so
thick and foggy, with small rain—a regular Scotch
mist—drifting in our faces, with the strong breeze blowing without intermission, obscured all surrounding objects
even only a few yards off, and as the air was so keen and
cutting, we commenced our return at two p.m., reached
the observatory at four, and got on board in the dingy,
drenched to the skin, at six p.m., and ready for our
dinner. The vast and remarkable block of basalt
rises to 1000 feet above the harbour, resting on a
greenstone ridge, 600 feet in thickness, from which it
is separated by a thin bed of shale six feet, nearly
horizontal.

Monday, 18th.—At ten a.m. I landed on the south
side of the bay in the dingy, and returned on board in
her at three p.m., bringing with me two young sooty-
albatrosses from their nest.

Tuesday, 19th.—At ten a.m. I landed on the north side
of the bay, and ascended the crater-shaped hill, 1200
feet above the bay; at noon reached the summit, which
is cone-shaped, having a shallow lake, covered with ice
at the time, about thirty yards across from north to

south, contracting in the centre to six yards, and occupying the depression on the summit, around which piles of
fragmentary and prismatic basalt rise on the east and
west sides to about fifty feet, sloping towards the north
and south, where gaps are left.   Perfect basaltic columns,
some of them ten and twelve feet between the joints, are
inclined round the acclivity of the cone, and intermingled
with piles of broken fragments, exhibiting the same
prismatic structure, being generally in five or six-angled
prisms.   At a deep gorge, six feet wide, on the north
side, these columns are most beautifully arranged in a
quaquaversal dip, the lake filling the oval depression.
The ice formed five-sided prisms on the surface, like the
rocks around it, and bore my weight, though the temperature was 42°, and the sun shining.   On breaking
the ice, and then sounding with the ramrod of my gun,
I found the bottom muddy, but the water was excellent.
The same kind of elegant lichen found on the black or
fossil-wood rock on the opposite side of the harbour exists
here.   From the south side of the summit I had a fine
view of Point Pringle, Cape Cumberland, and the "Sentry-
box," with the intervening bays, and immediately beneath
me the two ships in Christmas Harbour.   My ascent
was from the east side, and I descended by the north
along a steep gorge, only six feet in width, formed by
columnar greenstone in hexagonal and five-sided pillars,
some twelve feet between the joints, beautifully fitted to
each other.   On reaching the ridge below, and proceeding in the direction of Cape François, I shot a chionis
and a brace of teal.   Saw Bligh's Cap and other islands
in the offing.   At three p.m. I had to climb over a steep,
almost inaccessible wall of columnar greenstone, before
I could gain the summit of Cape François, dividing me
from Christmas Harbour.   The horizon to windward
wore a very threatening aspect, and I had to hasten my
steps to avoid being benighted in the new course I had

adopted for my return, along ground I was wholly un-acquainted with. From the top of Cape François I had a fine view of the Arched Rock, and the ships in the harbour. Reached the landing-place at four p.m., and got on board at 5.30.

Friday, 22nd.—This morning I volunteered to Captain Ross to accompany an exploring party in one of the *Terror's* cutters up Cumberland Bay, to the weather side of the island. Landed at noon. Found some very fine specimens of the fossil-wood, and a bed of shale under-lying the black or fossil-wood rock. Here I found two young petrel in their nests ; shot two and a half brace of teal and a tern, and returned on board at five p.m.

Saturday, 23rd.—At ten a.m. I landed at the south corner of the beach, and ascended the ridge of greenstone on which the black mass of basalt rests, passing two lakes, one 100 yards in length and sixty in width, on my way to Arched Point, descending from the ridge of rocks to a level plain of shingle and alluvium, bearing evidences of having been recently covered with water. Another hour's walk brought me to Arched Point, which I reached at one p.m., crossing a saddle-like depression by which the ridge is united to the mainland, next proceeding over a columnar wall of greenstone, beyond which the ridge becomes level to the point, only strewed with a fragment or two of rock. Immediately beneath me the Arched Rock itself appeared, connected with the point by a low, narrow neck. Returned on board at five p.m. The day had been squally, with light showers of snow, covering the ground, and the drift in some of the hollows between the hills was two or three feet in depth. At intervals, through breaks in the clouds, the sun shone. I shot a gigantic petrel, and at Arched Point a young black-backed gull flying overhead, where Foul Bay runs deeply in, its star-board side extending out in a long, low spit.

Sunday, 24th.—At one p.m. a royal salute was fired

from both ships in honour of her Majesty's birthday, and divine service performed. Captain Crozier and the gun-room officers of both ships dined with Captain Ross in his cabin. . Thirteen sat down to table. We had excellent roast goose for dinner, a very agreeable change from the penguin diet of late. Captain Ross said, at the table, that Phillips, the second lieutenant of the *Terror*, and myself, with three men from each ship, were to form the exploring party to the north-west coast.

Monday, 25th.—Took one of our crew up with me to Fossil-wood Hill, to assist me in disinterring with the pickaxe the large fossil-tree I had discovered buried in the *débris* there, and I brought on board the young sooty-albatross.

Tuesday, 26th.—After skinning some birds, I landed at 1.30 p.m., and asked Captain Ross for two hands to go with me up the hill to dislodge the large tree, when, having done so, it proved too heavy, silicified as it was, for the strength of the two men to bring down the precipitous ridge, consequently we returned with a smaller one, and some loose specimens.

Saturday, 30th.—After three stormy days I went on shore about noon, shot a black-backed gull from the dingy, and a shag at the landing-place. Called on Captain Ross at the observatory, and made an excursion along the south side of the harbour to the waterfall, round the point. Shot two chionis, two gigantic petrel, two shags, and a teal, flying round the point.

Monday, June 1st.—Phillips came on board, and told me that we were to start on our exploring expedition to-morrow, when, Captains Ross and Crozier coming on board, I returned on shore with them. The latter offered to accompany me in my ascent of the Black Rock, to see the tree before I removed it, Captain Ross having given me his own gig's crew of five hands for the purpose. When we had arrived near the spot, there was an extremely narrow ledge on the perpendicular face of a rock, barely wide enough for the feet, or rather the toes, to cling to. Captain Crozier, being a somewhat heavy man, seemed rather nervous about venturing across, but, following in my footsteps, who had been so often over it in my previous visit to the fossil-tree and the *débris*, gave him confidence, and tided him over.

Tuesday, June 2nd.—I left the ship with two seamen, Fawcett and Marshall, and Barker, a marine, to join my colleague, Lieutenant Phillips, of the *Terror;* and after breakfasting in the gun-room of that ship, we all started together in the cutter at 6.30 a.m., my companion having with him also three able seamen from his own ship. The morning was fine and clear, the stars shining brightly, with a light breeze. At seven a.m. we passed the Arched Rock and a flock of the little petrel; these birds, though belonging to so different a family, bear the most striking resemblance to the little auk (*Alca alle*) of the opposite Polar regions, both in size, form, colour, mode of flight, and general habits. The main distinction is the tubular nostrils on the upper mandible, which characterize the whole petrel tribe. Several, also, of the gigantic petrel (*Procellaria gigantea*) were sailing about and skimming the surface of the sea. Having passed the openings to the two large bays between Arched Point and Cape Cumberland, both having a background of high land, we lay on our oars to take some compass bearings of the headland, within a cable's length of the shore, a low and rugged black ledge of rocks, the sea breaking on them, and full of chionis and shags. At 8.30 again out oars, and pulled to Cape Cumberland. An isolated rock, named the " Sentry-box," from its form as well as position, fronts the opening to the bay, which at 8 40 a.m. we made sail up, the change in our course making the wind favourable. After passing some drifting thin ice, we noticed what appeared to be inlets on either side of the bay. At

11.30 a.m. landed on a green bank on the port side, and whilst my colleague was getting a meridian altitude of the sun, I crossed over a ridge of rocks to examine the extent of the inlet higher up, which I soon found formed a snug bay, considerably more than a mile in depth, and about as much in breadth, the entrance being about one-third of a mile, bounded on either side by a ridge of hills, 700 to 800 feet high, forming terraces of columnar greenstone, with embedded crystals of quartz, forming large, drusy cavities. I picked up some beautiful crystals, scattered about the rocks.

After taking a hasty outline of the bay, with a set of compass bearings, from the spot where I stood on the eastern side, I returned to the boat, and at 1.10 p.m. we shoved off, and when outside I took another sketch of the entrance to it. Its distance is six and a half miles from the " Sentry-box." At two p.m. we hauled the boat alongside a ridge of rocks, separated from the main shore by a narrow channel. Here I shot a brace of teal ; and we had our dinner, starting again at 2.45 p.m., passing through much seaweed and small stream ice, very thin. Noticed some two or three small recesses in the coast-line on the south side of the bay. We were accompanied by numerous shags hovering over the boat as we pulled up the bay, so much was their curiosity excited by our intrusion on their usual solitude that we expected every minute to see them alight in the boat, as with outspread wings and drooping feet, with their long necks twisted round, they looked down upon us with prying gaze, so near that the boat's crew knocked down several with their oars for soup. At five p.m. we hauled the boat up on a sandy beach at the top of the bay, which we found to be twelve and a half miles in depth, with an average breadth of about two miles. Having secured the boat above high-water mark, and made all snug for the night, at seven p.m. we had a hot supper of teal, and the

R. McCormick, R.N., del.

Vincent Brooks, Day & Son, Lith.

Boat Encampment.
Upper end of Cumberland Bay, Kerguelen's Land.

white bird, chionis, followed by a cup of tea at nine, and then turned into our blanket-bags, amid the distant roaring of the sea outside the isthmus, indicating our close proximity to the opposite coast, the main object of our expedition.

> Far o'er our heads the cross appears on high,
> On waves of ether borne in the blue sky ;
> Magellan's clouds, and the black coal-sack near,
> That wondrous void, in which no stars appear.
>
> Magellan's great cloud like a casket seems,
> As with the brightest gems it teems ;
> And colour'd suns, of every size and hue,
> Only the telescope brings into view.

Here, as the " Southern Cross " appears to us in the zenith, the four stars of which it is composed have a position nearly parallel with the horizon, the smallest star being uppermost, and to the right the largest, a star of the first magnitude beneath it, the two others of the second magnitude forming the left side of the " Cross." This constellation, indeed, itself constitutes the " Pole Star " of the southern heavens, there being no single star correspondent with the " Pole Star " of our own hemisphere. Below the cross the two fine stars of the " Centaur," stars of the first magnitude, shine with the greatest brilliancy, looking like pointers to the " Southern Cross." The brightest of them, *Alpha Centauri*, is remarkable, not only as being the nearest of the fixed stars to us, and yet so distant that it takes above three years for a ray of its light to reach us, but it is also one of the double stars, consisting of two suns revolving round each other in an orbit so elongated as to occupy above seventy-eight years in the period of its revolution. These suns are of an orange-yellow colour.

In the course of the precession of the equinoxes—said to occupy an interval of some 26,000 years in the com-

pletion of the circuit of the heavens--the star "Vega,"
in the constellation of "Lyra," will in some 12,000 years
have become the "Pole Star" of the northern hemisphere;
when "Canopus," in the southern hemisphere, will take the
place of the "Cross" in the Antarctic heavens as the
"Pole Star."

The four stars of the "Southern Cross" are themselves said to be moving in contrary directions, and with
unequal velocities, so that this constellation, if such be
the case, will not always retain its present form. The
"Coal-sack," a pear-shaped, black void in space, situated
near the "Cross," has only one small star visible to the
naked eye; yet the great white "Cloud of Magellan"
(*Nubecula major*) under the telescope unfolds to view a
rich group of coloured stars, of extreme beauty, clustered
round the star "Kappa," a deep red, central one, amid
others of varied shades of blue and green, studding the
depths of space with brilliant gems of every hue.

Should these far-distant suns have planets revolving
round them like our own sun, what varied and beautiful
coloured days must result from these primary colours,
some days perhaps red, others blue or green, or a neutral
tint from a combination of these. How wonderful is
that abyss of space, through which it is thought our own
sun is revolving round some other vast and central unseen
orb, at a velocity of four miles a second, in the direction
of the constellation "Hercules," towards a point (Alcy-
one) in the heavens, in which the "Pleiades" becomes
the centre of the movement of revolution of the solar
system.

Wednesday, 3rd.—We turned the boat bottom upwards, stowing our things beneath it, and at 8.30 a.m.
set out upon our journey across the isthmus for the
opposite coast. Our course lay over low, rocky ridges,
by several lakes, and a remarkable stone resting on
a tripod of three smaller ones. I shot three teal at

the lakes, and at 10.30 a.m. we came in sight of the sea, on climbing over a steep rock to the left of a deep chasm, when a fine view all at once opened upon us— a bold, black, snow-clad headland, and small iceberg lying off it. Our course to the westward now lay over a steep boggy slope of soft, spongy soil, clothed with a spreading, tufted plant, like a covering of moss, on the *débris* at the base of a mountain-range of greenstone, which flanked it on one side, and the sea on the other. Cascades of water poured down the almost perpendicular wall of columnar greenstone, intersecting the bog beneath by numerous and deep watercourses. These rendered the whole of our journey very laborious, sinking up to the knees at every step, encumbered, too, as we were by our heavy knapsacks and blanket-bags, for sleeping in, strapped to our backs. At noon we rested on the bog to dine, during which I took the opportunity for ascending the columnar ridge above us at a spot where it appeared to be accessible. But just as I had reached the summit, the weather became thick and overcast with rain, depriving me of a sight of the inlet in advance of us. After descending and taking a hasty lunch, we started again at one p.m., but soon found that the work was too heavy and toilsome for our party to proceed any further with their heavy baggage.

We therefore halted at a waterfall, and I proposed taking the youngest and most active man of our party with me to explore the inlet, leaving our knapsacks with the rest of the party in charge of Lieutenant Phillips. Accompanied by Marshall, I managed to reach the entrance to the inlet, about a mile distant. Here the high rocky cliffs on its left above the low black rocky point on which we stood, and on which the surf was heavily breaking, prevented us rounding the angle so as to ascertain to what extent the inlet ran up inland. The rain now increasing, with the wind coming in heavy gusts, and with night approaching we returned to where we had left our party, after crossing some heavy torrents rushing down the steep slope to the sea. It took us an hour to get over two miles of ground.

At 3.30 p.m. the whole party resumed their knapsacks and bore up for the boat, but the night closing in upon us with wind and rain, which swelled the heavy torrents we had to cross, compelled us to pitch our tent soon after dark, at the base of a black, craggy rock, on the margin of a large lake, a cascade falling on the right. The ground being wet shingle, we paved the interior of the tent with fragments of rock, and, laying our wet clothes beneath us, got into our blanket-bags. The night was wet and gloomy, and the lake, flooded by the rains, rose within a foot of our tent, the inside of which became an atmosphere of steam from our wet clothes.

Thursday, 4th.—We rose at daylight, struck the tent, and continued our journey, ascending the steep path between the rocks on our right, and somewhat changed our course back. The weather cloudy and gloomy, but without rain. The rocks here contained numerous drusy cavities of fine crystallized quartz, of which I picked up several specimens. In crossing some streams of water down a deep descent in the rocks, our men missed us, and, having taken a more direct course, reached the boat first, whilst my colleague and myself went round a point to the eastward and came upon the beach below the boat. Here he and I also separated, as I followed up the windings of the shore, picking up shells and seaweed, and reached the boat about one p.m., having to wade through the water round a rocky point to the cove where the boat was hauled up. Here I met Barker, our marine, chasing the white birds with a boarding-pike. The shags were flying round us in great numbers, as if they were welcoming our return, for they seem to be very sociable birds. Having changed our wet clothes and dined, we launched the boat on our return-voyage down Cumberland Bay. Passed a smooth, curious-looking hill with a gently undulating outline, and having a somewhat marbled appearance in the distance, so different in aspect to the surrounding basaltic and rough-broken greenstone ridges, between which this, and many more we saw on our way down, emerged in isolated cones, and proved, on examination, to be phonolite or clinkstone.

At 3.45 p.m. we landed upon a shingly beach, on the east side of North Bay, and I walked for about half a mile to have a close examination of one of those round, conical hills, presenting so remarkable an appearance in the distance, the whole hill being a pile of laminated foliaceous fragments of phonolite. Some of the specimens I picked up were most singularly marked by what had the appearance of sea-weed impressions, the red colour of the markings on the light greyish surface of the clinkstone, or, rather, yellowish ground, were very striking, and were probably owing to an oxide of iron. On the opposite side of this bay is another of these isolated masses protruding through a greenstone range of hills, looking like a smooth, light-coloured marble saddle as it appeared between the dark trappean rocks enclosing it. At 3.10 p.m. we shoved off from the beach, crossing Cumberland Bay to our small bay on the south side, which we named South or Duck Bay, from the number of teal we met with there. But the night being very dark for finding a suitable place for hauling up the boat, we anchored her for the night off a strip of beach in the south-west corner of the bay at 6.40 p.m., and had our supper of stewed birds, made with teal, chionis, and shags ; then spread the boat's awning, and after having had a cup of tea, at 9.30 p.m., we turned into our blanket-bags for the night, which came on wet and stormy, ending in such violent gusts of wind that before the night was half spent we were compelled to lower the awning, which had been pressed inwards by a heavy fall of snow succeeding the rain. The air was cold and pinching, the

bottom of the boat wet, and we had now no other covering than what our blanket-bags afforded us stretched out
on the thwarts of the boat.

Friday, 5th.—After a hasty breakfast of cocoa and
cold tongue, we got the anchor up, and at nine a.m.
pulled up the south-west creek, landing on a fine sandy
beach on the starboard side. I then proceeded up a
valley, intersected by pools of water, bounded on either
side by a lofty ridge of hills. I shot a beautifully-
marked male teal from a flock flying over the lake,
into which it fell, but I succeeded in getting it, and in
the boat skinned it for my collection. Near this spot a
very curious bank rises a few feet from the ground, of a
rich black soil, with a bright-green covering of grass and
moss, which at a distance presented the appearance of a
small enclosure or garden on which the snow had fallen.
When on it, fancy might liken it to the deck of a ship in
shape, with a hollow bowl at one extremity. A little
further on I shot a brace of teal at one shot. They had
been feeding on the seed of the cabbage near a stream
of water, the ground being here very swampy, with some
scattered rocks. When about two miles from the boat, I
reached the margin of a fine lake, having a smooth level
beach of sand and shingle in front ; and the dark green of
the water at first gave some hope that it might have
been an arm of the sea ; but, on tasting the water, its
freshness soon dissipated the delusion. It is bounded
on either side by a lofty range of trappean hills, rising to
2000 feet in height. The length of this lake is a mile
and a half, by half a mile in breadth. At the right-hand
corner of the beach a large mass of rock and moss-
covered *débris* at the foot of the mountain enabled me,
by ascending it, to examine the rocks *in situ*, which I
found were composed of greenstone and amygdaloid.
Here I encountered a heavy storm of hail and snow,
drifting before a strong breeze, and at noon I beat a
retreat for the boat, and reached it at 12.40 p.m., shoot-

ing another teal on the way back. After a dinner of pea-soup and pork made sail down the bay in the boat, but, on reaching Cape Cumberland, the sky wore such a threatening aspect to windward, that we bore up for the top of the bay again. I shot a fine specimen of the black-backed gull (*Larus marinus*), and at four p.m. hauled up the boat on a fine sandy beach, under a rock which juts out in the form of a promontory about the centre of the head of the bay. I walked along the narrow bank of sand to the end of the beach, where a stream enters the bay. For nearly a mile inland is a low, level plain of sand, and intersected by water-courses in various directions, coming down from the hills above. Beyond this plain, to the south, is a succession of ridges, bounded on the east and west by hill-ranges, between which, in all probability, an opening to White Bay on the south exists. I returned to the boat at 5.30 p.m. The rock on the west side of it is composed of greenstone, veined with quartz. Here I shot a chionis. Had tea, followed by bird-soup, for supper, and turned in at nine p.m.

Saturday, 6th.—After breakfast, whilst the boat was getting ready, I rambled over the south ridges for some two miles, and, had my time permitted me to have continued, doubtless I should have come upon the sea. I shot a teal, and returned to the boat at 11.30 a.m. Had a fair wind blowing fresh, which carried us down and out of Cumberland Bay. At 1.15 p.m. we were off Cape Cumberland, using oars and sails alternately from the Cape to Arched Point, and notwithstanding which, from the bad weather-qualities of the cutter as a sea-boat, we were drifted so much to leeward, that had the breeze been stronger we should not have fetched Christmas Harbour. Chionis and shags lined the rocks along the coast as we passed. When about a mile from the land we fell in with one of the largest seals, wounded and bleeding from one eye. It must have received the wound in Christmas Harbour, and to put an end to his sufferings I put a ball through his head from my old double-barrel, and we passed a noose of rope over his head, but, it slipping off, we had no time to lose, and left him floundering in his last struggles. As we passed the Arched Rock I shot a brace of chionis from the boat, and landed to pick them up. As there was much swell and serf on the rocks, Phillips remained in the boat with her crew, lying on their oars at a distance, whilst I retrieved my game, and shot a teal on the rocks, till I returned in about half an hour. Having ascended the *débris* within the arch, and found some remarkable specimens of the fossil wood, with the rugged bark very distinct, and having a rugose and twisted appearance, charred in some portions, but generally very ponderous from silicification—one beautiful fragment, a perfect section of a large branch, I found embedded inside the arch, and projecting from the wall of massive basalt or lava, at about six feet from the ground—I measured the span of the arch over the uneven surface of the *débris*, making it about thirty-six paces. Returned round the narrow low neck dividing the arch from the point. Several chionis were leisurely, as usual with them, walking about it. After shoving off in the boat we hailed the ship in passing at four p.m., and finding Captain Ross on shore, pulled to the observatory to report our return, and on board afterwards.

Two bearings of Sentry-box, entrance of Cumberland Bay.
(*See page* 55.)

Remarkable Phonolite Hill in North Bay, Cumberland Bay.
(*See page* 72.)

CHAPTER VI.

Second boat-expedition—Arched Rock—Penguin Cove—Ornithological and botanical specimens—Our bird-soups—A large cabbage—Gregson Bay discovered—A hard day's work—Clinkstone Hill—A dangerous adventure in search of teal for our larder—A tempestuous night.

Sunday, 7th.—Captain Ross told me to-day that we were to start again on Tuesday, the 9th, on a second exploratory expedition to Cumberland Bay. At four p.m. I dined on board the *Terror* with my late colleague in the boat-voyage, Phillips, and returned on board at 8.30 p.m. It is intended to send a boat from each ship up White Bay to-morrow under the charge of the first lieutenant of the *Erebus*.

Tuesday, 9th.—Blowing too hard for us to start to-day, and our own first lieutenant, who, I have very good reasons for thinking, does not much relish the nature of the service his chief has nominated him for, nor the state of the weather either, has gone upon the sick-list. I had to land in the galley before breakfast, and through a heavy surf, to report him in the list to Captain Ross at the observatory. I had a delicate duty to perform, it must be confessed, and it put all my ideas of nosology to the test, and racked my brain to find a suitable term under which I could enter the indisposition of this martyr to science in my sick-book.

Saturday, 13*th.*—Employed in skinning birds, and had three boxes and two haversacks made for my boat-voyage.

Tuesday, June 16*th.*—The weather moderating somewhat, my former boat-companion, Lieutenant Phillips, called alongside for me in the *Terror* cutter at 9.15 a.m., our boat's crew being the same as before. Soon after we had shoved off from the ship, my companion discovered that he had forgotten his sextant, which we had to put back for when at 9.30 a.m. we made sail out of the harbour ; weather still gloomy and threatening, rain with a good deal of swell outside the Arched Point. On rounding Cape Cumberland at 10.45 a.m. the wind had freshened, and the sky to windward wore a black and ominous aspect, we had to lower the sail, get out the oars, and pull against a head-wind up the bay, our progress much obstructed by a quantity of seaweed lining the shore on the starboard side. After buffeting with the breeze till we had got about half a mile within the bay, we were left to the alternative of either drifting out to sea or risk getting the boat knocked to pieces in the surf on the rocks, by attempting to get her into a small rocky bend or indentation, just within Cape Cumberland, where we had noticed a small cave within a projecting mass of black rock, although a considerable swell set in upon the ledge of rocks. We fortunately succeeded, after getting everything out of the boat to lighten her, in hauling her over this bar without being stove. But had it not been high-water at the time, aided by a quantity of seaweed hanging from the rocks, we could not have succeeded. It was past midday when we had secured the boat on a narrow platform at the base of the high overhanging cliff. On our right was a watercourse down a green bank, up which I ascended with my gun in search of some teal, chionis, and tern I had seen. At two p.m. I crossed a spongy bog, intersected by numerous streams and falls from the hills above, for a distance of about a mile and a half.

I found the teal more abundant here than in any other place hitherto visited, having shot six and a half brace, two at a shot four times. The white chionis were also walking about the bog in great numbers, and I could have shot any number of them. There was also a small colony of penguins on the coast-ledges; a drizzling rain fell most of the afternoon. I returned to the boat at five p.m. The full moon sent its feeble rays through a break in the clouds, a star or two becoming visible. The shadowy, indistinct outline of the boat, as it appeared through the gloom under the shadow of the rocks, with its streak of light issuing from the triangular opening in the awning amidships, gave a momentary relief to the wild solitude of the scene, and to the feelings of one alone and drenched to the skin as I was, in approaching my place of bivouac for the night, and shelter from the boisterous elements. We enjoy most things in this life by comparison, and through sudden contrasts in our circumstances. For after changing my wet clothes, and having had a dinner of preserved meat and a supper of stewed chionis, I turned into my blanket-bag in the stern-sheets of the boat with a feeling of satisfaction, and I might venture to add, comfort not always experienced in the downy feather-bed on its ample four-poster.

Wednesday, 17*th.*—Being weather-bound in our not very cosy corner here, at eleven a.m. I started on another ramble over the bog, and saw Arched Rock Point in the distance. I shot two and a half brace more teal, and, on my return to the boat at one, had some stewed for dinner, and plum-pudding I had brought with me from the ship. At three p.m. I started on yet another excursion, over the hill and the bog to Penguin Cove, where I caught a fine penguin, the leader of a batch of twenty-two others, and shot another brace of teal.   Here I picked up some beautiful specimens of moss.   I heard a chirping sound issuing from a hole in the bank near a watercourse, evidently produced by some nocturnal petrel, for it became silent on my approach to its subterranean abode. I did not get back to the boat till after dark, through a heavy hail-storm, for the afternoon had been very stormy. As I was descending the hill to the boat, a tern flying overhead made known its vicinity by its cry on being disturbed, and to which it fell a victim; I shot it, needing specimens both for the collection and the pot.   I reached the boat at 5.45 p.m., and from nine to midnight was busily employed skinning and preserving the penguin and tern, and noting down the occurrences of the day.

Thursday, 18*th*.—The weather having moderated this morning, we struck the awning, got everything out of the boat, and, after no small labour, at 9.30 a.m. launched her over the rocky ridge at near low water; when a breeze springing up from seaward, we made sail up the bay.   At 10.15 off Black Rock Cove, at 11.30 a.m. passed South Bay, and soon afterwards those remarkable-looking phonolite hills, erupted between the fissures of the greenstone range, with their smooth, rounded, pictu-resque summits, and light shade of colour, contrasting so strikingly with the dark trappean rocks.   We sailed round the small bay in which these hills are mostly located; met with a great number of shags here, and a few small petrel.   A freshening breeze carried us to the highest part of Cumberland Bay, which, after first con-tracting itself, expands into a basin at the upper end. In attempting to beach the boat on a sandy beach at the starboard corner, we grounded in shallow water, and with the luff-tackles had to haul her up on the left above high-water mark, in a pelting hail-storm, driven by a heavy gale, which in sudden gusts blew up the bay. We at last, at one p.m., effected a landing. Whilst I was sounding for the depth of water, in getting into the boat again, the ramrod of my favourite old gun fell overboard in a fathom of water. We had some penguin-soup for dinner, which the cook for the day had spoilt by making it into a complete paste with flour, and it was five p.m. before we had completed our arrangements for the night.

I saw here a remarkable instance of the power of the native cabbage to withstand the effects of salt-water, apparently without sustaining any injury. A row of the very finest and largest of these plants, growing just above the ordinary high-water mark, about two feet in height, with stout stems bearing the scars from which former leaves had dropped off, and indicative of age, now appeared on the very spot of the sandy beach where we landed, half-submerged, as if growing out of the sea, from the unusually high tide, occasioned by the strong southerly gale blowing up the bay. This with the hail and snow-storms foreboded a boisterous night.

Friday, 19*th.*—My ramrod, I am glad to say, was found this morning in about a foot of water. Having made a hasty breakfast of cold tongue and cocoa, at 9.30 a.m. we made another start for the weather coast by a somewhat different route to our former one. Leaving our things under the boat capsized, we passed by a watercourse down a greenstone hill, where I found some stray fragments of coal and lignite scattered amongst the ligneous *débris.* At 11.15 a.m. we again came in sight of the sea through a gap in the rocks; and on rounding these we passed the margin of the lake where we encamped in the first expedition. Then continuing our course through the valley by streams of water till we reached the ridge above the bogs, we now ascended the former instead of proceeding along the latter low ground, till we arrived at the inlet at 12.30 p.m., and which had cut off my further progress last time. Being myself much ahead of the rest of the party, I had time to make a hasty sketch, and take the bearings of the bay whilst awaiting their arrival. Found the bay, which I have named after an old friend, the late Mr. Gregson of Risdon, Tasmania—-Gregson Bay—to be about two miles in length by one in breadth, bounded by a range of rugged hills, highest at the sides and lowest at the upper end, presenting a picturesque outline. The hill from which about two-thirds up I took my sketch, is, perhaps, the highest, and about 1500 feet above the bay; some low ledges of black rocks stud the entrance. At 1.30 p.m. I descended by a steep pass from the extremity of the ridge to the bog beneath, where we picked up the *cache* of a week's allowance of provisions, biscuits, preserved meat cases, and rum, we had to leave behind us in our former excursion, with a boarding-pike left as a mark to the spot. Here I shot two and a half brace of teal, and at 3.30 p.m. we commenced our return to the boat; blowing hard with sleet. Had to ascend a steep and broad watercourse, which was studded with large fragments of igneous rocks, down which rushed a rapid torrent; and on reaching the lower grounds had to wade through two deep streams. The snow was in many spots knee-deep. Finding the extra weight of the *cache* too much for our men to get down to the boat, we deposited it under a rock; and, as we approached the the top of Cumberland Bay, we passed another of the phonolite hills, and descended a steep watercourse to the beach, reached our boat at 7.45 p.m. Our poor fellows were thoroughly worn out and wearied with the hardest day's work we have hitherto had; and my worthy, staunch, persevering associate, Lieutenant Phillips, once

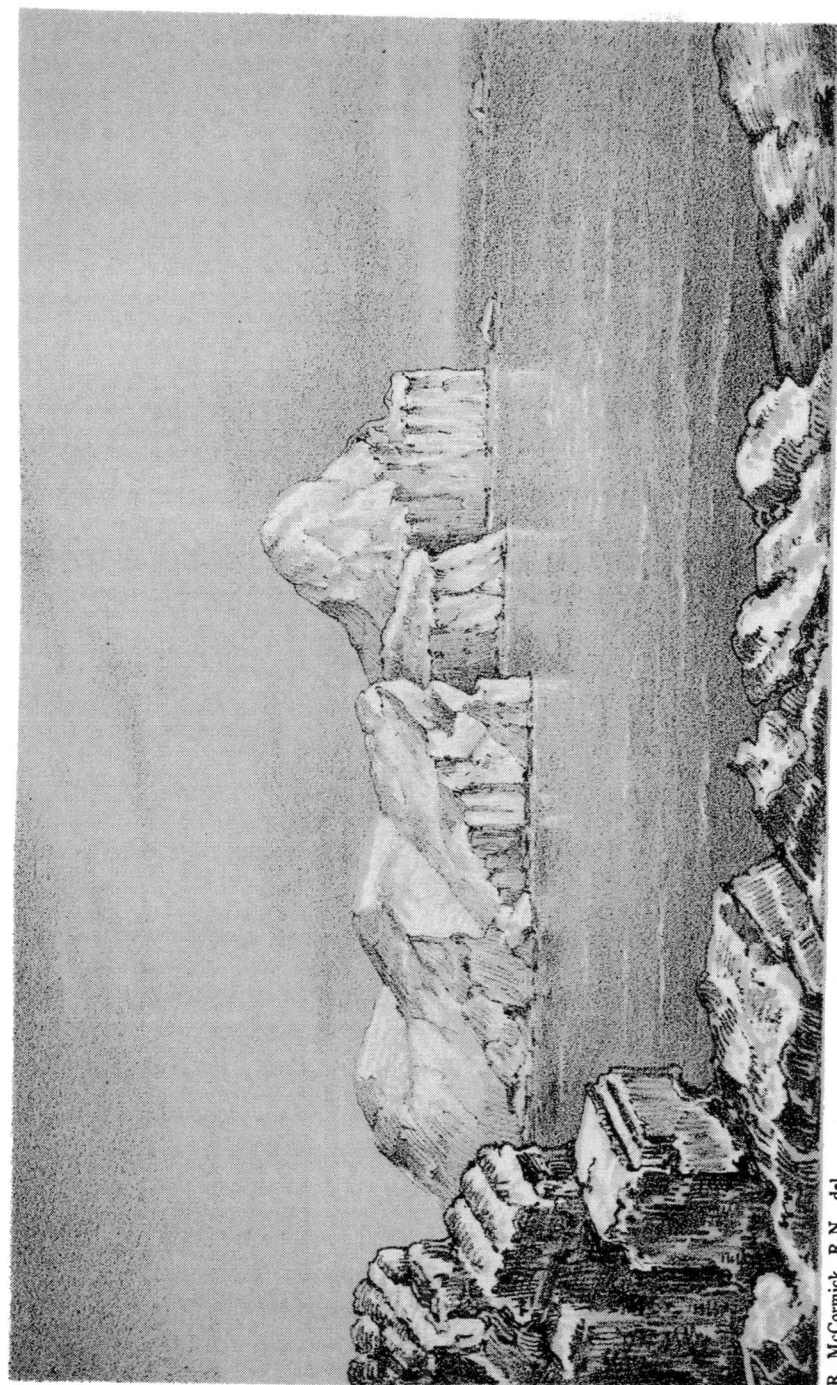

R. McCormick, R.N., del.

Vincent Brooks, Day & Son, Lith·

Gregson Bay.

Discovered during the boat expedition up Cumberland Bay, Kerguelen's Land.

got himself so embedded to the hips in a watercourse of the spongy bog, that he could not extricate himself till the cumbersome knapsack was unbuckled and removed from his shoulders; I happened to come up just in time to aid him. On more than one occasion I had myself slipped down the deep snow and ice, where intermingled with the rapidly-falling torrent, obstructed in its course by the steep and rugged rocks, and had no slight difficulty in extricating myself. Having changed our wet clothes, and lighted a fire in the boat's stove, we had some of the teal I shot on the bog roasted for supper, with plum-pudding, followed by tea, and turned into our blanket-bags at eleven p.m ; but the gusts of wind came so heavy that the boat reeled on her supports, with the awning filled and pressed down upon us by the weight of the incumbent snow, threatening every minute to capsize her.

Saturday, 20th.—We sent four of our boat's crew to bring back the *cache* of provisions left at the bog. I skinned a penguin in the forenoon; and at one p.m. took my gun for a stroll along the beach, and shot three and a half brace of teal, returning to the boat at three p.m. Had roast and stewed teal for dinner. Started again at four p.m. along the beach to the creek; shot three more teal, two at one shot, and returned at 5.45 p.m. I tasted some of the stewed mussels to-day, and although I have no taste for shell-fish, found them remarkably well-flavoured and of large size; they abound in the rocks about the bay. Had tea at eight p.m., and afterwards skinned and preserved two teal, and turned in at 10.30 p.m.

Sunday, 21st.—Rose at eight a.m., and after breakfast launched the boat on our return down Cumberland Bay, and, before I left, pulled up the finest of the large cabbages from the phalanx, a little forest in miniature, which at high water-mark flanked the top of the bay. From the appearance of its rugose, knotted stem, and being a perennial, it would seem to have braved the storms and lashings of the sea at every high tide, throughout a somewhat prolonged term of existence. I placed it in the stern-sheets of the boat for preservation, as the largest specimen I have met with in the island, and still have it in my collection ; indeed, it was the largest plant of any kind met with. At 10.45 a.m. we commenced our homeward voyage down the bay, sounding with the lead-line as we proceeded. At noon I landed at a slight indentation of the greenstone range, assuming the columnar form opposite the hills of phonolite on the south side. After crossing over a swampy level, strewed with foliated fragments of clinkstone, I ascended the hill on the right, and went over the whole length of the summit of the ridge to the swampy plain below, about a mile from the boat, which I reached at two p.m.

This hill of clinkstone emerges from the greenstone range of hills, 700 or 800 feet in height, and from which it is separated by a ravine, and itself attains a height of 300 feet, having a north and south direction. Four and six-sided, or hexagonal columns of the phonolite, in places, emerge from the confused heap of broken schistose fragments with which the mountain is for the most part covered. Some of these slaty foliaceous specimens I collected are curiously marked with red concentric lines on a yellowish, sandy-coloured ground, but none with the singular seaweed-like impressions found on the hill in North Bay.

We shoved off in the boat, through rain and fog, after our customary meal, now of teal roasted, and teal stewed by way of change ; for any variety or change in the diet, however trifling, has at all times a beneficial effect on the general health. We fetched into North Bay, and hauled the boat up on a sandy beach, at the south-west corner, beneath a phonolite hill intersecting the trap. The long line of sandy beach is backed by a level plain of swampy ground, covered with mosses, dwarf, tufted plants, and strewed over with stones—a watercourse flowing through it. Had tea, and some of the fine-flavoured stewed mussels, and turned in at 8.30 p.m., the wind blowing in heavy gusts, accompanied by rain, the boat's awning flapping violently, threatening to unroof us every moment.

Monday, 22nd.—We passed a most uncomfortable night, everything about us cold, damp, wet, and chilly. After a cocoa and grilled-pork breakfast, I left the boat at 8.50 a.m. to ascend and examine the structure of the remarkable hill above us, following the watercourse where it lies in contact with the greenstone, or line of junction. The ascent to the summit, 600 feet, is steep, the sudden gusts of wind rendering the foothold very insecure. About two-thirds up, the phonolite appears in prismatic columns through the loose, fragmentary *débris*. It came on to rain before I reached the boat, at 11.20 a.m., when we shoved off in a fresh gale for the opposite side of Cumberland Bay, and the wind proving fair, at 12.30 p.m. we hauled the boat up upon the beach, at our old quarters, to the left of the small promontory at the head of South Bay. Being now reduced to short allowance of provisions, in short, to a piece of salt beef and a flour-dumpling for our boat's-crew's dinner, after the boat was secured for the night, I took my gun for a ramble along the beach in search of some teal, whose plaintive low whistle I had heard, for an addition to our larder, little thinking when I left the boat the adventure I was entering upon, and which was so near leading to a night out, without the slightest shelter, in one of the most tempestuous nights I ever experienced.

For, on reaching the watercourse formed by the falls of fresh water, brought down in torrents from the ridge of hills bounding the top of the bay to where it debouches into the sea, I found that the rains had so flooded it, both in depth and breadth, that, in attempting to ford it, so strong was the rush of water, I was nearly carried off my feet bodily into the sea, and when swept off my foothold only escaped by striking out for the opposite bank, and grappling with it. My gun, having become submerged, neither barrels would go off at the only teal I met with, at the further end of the beach, where a few chionis were searching amongst the seaweed for their marine food.

Being thus well drenched, and my heavy water-boots filled with water, and gun useless, I would have returned to the boat at once by the way I came, as the shortest route, not more than half a mile; but the stream, which had every minute been increasing, and getting more impassable, now, on my return to it, rushed along in such an impetuous torrent as to render any further attempt to cross it hopeless in the extreme. The only alternative left me was to make a long circuit of the ridge of hills bounding the top of the bay, or rather plain, which was now intersected in every direction by rapid streams of water; yet at other times I had walked over it without wetting the soles of my shoes. But, before I had completed one half of the distance, my progress became arrested by cascade after cascade rushing down the steep rocky declivities in foaming torrents.

Darkness had now set in to increase the difficulties of my situation, and the increasing wind threatened a most tempestuous night. My only chance now was to make the circuit of the ridge higher up, where, possibly, the torrents might become narrower, so as to enable me to leap over them. First disencumbering myself of my haversack, containing some of the finest specimens of quartz crystals in large drusy cavities, weighing in all some fifty pounds, and which I had picked up soon after crossing the first torrent, I reluctantly made a *cache* of them on a rock, at the edge of a foaming torrent, taking a cross bearing of the spot, as well as the darkness would permit of, by an observation of the most striking point and depression of the mountain-outlines, on either side, as they stood out in strong relief against the sky.   Thus lightened, I hoped to have cleared all obstacles to my further progress, when reaching the lower portion of a black, rugged mountain, my ears were all at once saluted by the disheartening sound of a large, foaming torrent, which soon burst upon me, dashing down a deep gorge in the rocks, in a white, foaming cascade, forming eddies and whirlpools amongst the loose rocky fragments, as they obstructed its course.

The darkness of the night, only relieved by the fitful glare from the white, foaming spray, the torrents sent upwards, the terrific gusts of wind, accompanied by a deluge of rain, combined, together with black, overhanging, frowning precipices, to form a scene of the wildest description.   I began to relinquish all hope of reaching the boat, and to reflect on the best course to be pursued; whether to attempt to dash through this torrent by swimming, as it was impossible to ford it in any direction, or to seek some cavern in the rocks for shelter through such a night in such a country, when on sounding the torrent with my gun, to ascertain its force and depth, it was all but twisted out of my hand.   The crossing, therefore, was hopeless.   It was now past six o'clock, and I followed the torrent upwards in the vain hope of finding it narrower, approaching its source, groping in the dark amidst crags and hollows, rendered so slippery with the rain, that on more than one occasion I was only saved from being hurled down some yawning precipice, through the friendly aid of my old double-barrelled gun, with the butt-end of which I sounded my way, which becoming steeper and steeper, and despairing of getting nearer its source, I resolved upon trying to follow it in its descent to the plain: a tortuous route, but eventually I succeeded in fording it, where it spread out on some sands, lessened in depth.

As I felt that I could not be now at any great distance from the boat, I hailed as loud as I could, in the hope that a pistol might be fired as a guide to its position, and, after a few minutes had transpired, a flash lighted up all around, although the boat itself was invisible; and after crossing a small, boggy hillock, for I had been wading up to my knees in water, I saw the light of a port-fire, which illumined the whole of the swampy space separating me from the boat, and on my rounding the promontory behind which she was lying, she suddenly came in sight, the light between the folds of the awning indicating her position.

At about seven p.m. I had the inexpressible gratification of finding myself in the stern-sheets, under the cover of that awning, after wandering for upwards of two hours in the dark, and drenched to the skin for double that time. After having changed my wet clothes I had some tea, with some stewed chionis our thoughtful, kind-hearted boat's-crew had caught during my absence, and kept for my supper, for they had been most anxious about me, and had been watching in the bow of the boat for hours, listening for any signal from me, hence their so speedily hearing and answering my hail. They had expected to hear the report of my gun, which, unluckily for me, had early become useless; and in such a place and night of weather they knew not where to look for me.

At 9.30 p.m. I turned into my blanket-bag; but our rest was of short duration, for scarcely time was allowed me to congratulate myself on my narrow escape from passing the night out—and such a night—in these wilds, when the gale increased to such a hurricane, and the heavy rain beat upon us with such violence, that the boat's awning was soon flapping about our ears in all directions. We had to strike it, and roll ourselves up in its folds for the remainder of this dismal and freezing night. So ended this to me memorable excursion in search of teal for the pot.

Tuesday, 23rd.—This morning various things blown out of the boat during the late tempestuous night were picked up along the beach, some of them, improbable as it may seem, half a mile from the boat, and such was the force of some of the gusts of this tornado, that one heavy box of my rock-specimens was hurled out of the stern-sheets to a considerable distance. The air was chill and intensely cold. These sudden floods are caused by heavy rains melting the snow and ice on the summits of the mountain-ranges.

Coal Seam. Duck Bay, Cumberland Bay. (*See page* 81.)

CHAPTER VII.

New bay discovered—Club-moss Bay—Seals for supper—Find a coal-
seam—A wet bed in a wintry night on the rocks—Piovisions
exhausted—Return to the *Erebus*—Packing the fossil tree—
Burrowing for "night petrels"—Farewell to the island.

WHEN I was in this bay last I came to the conclusion,
from the appearance presented by the ridges above it, and
their direction, bounded by high land on either side, that
a southerly course would take me to White Bay, the next
bay running parallel with Cumberland Bay. To-morrow,
being the day fixed for our return to the ship, I determined to settle this question, and at 11.30 a.m., accompanied by Fawcett, our boatswain's mate (who, poor fellow,
subsequently perished with the old ships in the ill-fated
" Franklin Expedition "). Our way lay over the scenes
of my last night's adventure, first across the flat, sandy

plain between the boat and the ridges, which, swollen by the floods, then was one sheet of water, but was now in so short a space of time drained nearly dry ; and the watercourses down the rocks above it no longer foamed in torrents, but could be easily crossed from stone to stone.   In short, no greater contrast could be conceived than between last night and this morning.

At 12.40 p.m., when three and a half miles from the boat, on the top of one of the ridges, I was gratified with a glance of the sea at a distance, and, pushing on, in about ten minutes a fine bay opened to the left, bounded by a bold, rocky headland.   At 1.30 p.m. I stood on a steep ridge overhanging a low, swampy valley beneath, looking green, and skirted by a pretty smooth sandy beach, laved by the waters of the bay, distant from the boat about five miles.   I descended a steep declivity between two waterfalls to the beach, where we found four seals basking ; I shot one with a ball, and another with a charge of small shot, for their hearts and livers, which we brought away with us for our supper.   The remaining two only rolled themselves a little higher up on the grass on which they had been sleeping, and made no further attempt to escape, and we left them there uninjured, to condole with each other on the fate of their companions. Numerous skeletons of seals were strewed about the beach on which the chionis were walking amongst the seaweed ; I killed two of them at one shot.   Saw several black-backed gulls, and picked up a few Pholas shells and some specimens of echini (sea eggs) and coralines ; a few tern were hovering overhead.   In the centre of the beach is a very remarkable square hummock near high-water mark. At 3.15 p.m. we started on our return.   Weather fine, but cloudy.   Struck through the valley, and over ridges by a trapdyke.   E.N.E. ½ E.   Just before we reached the boat, at 5.15 p.m., a little after dark, when crossing the sandy plain, which I have too good reason to recollect, it became thick with a hail-storm. We had the seals hearts and livers for supper, and turned into our bags at 7.30 p.m. Our new discovery appears to be a bay inside Point Pringle, near the entrance to White Bay, which I have named Club-moss Bay from having discovered a *new* species of that *lycopod* in its vicinity only.

Wednesday, 24th.—As it was blowing too hard to put to sea, I filled up the spare time by making an excursion up the ravine above the creek, at the south-west corner of South or Duck Bay. I left the boat at 9.40 a.m., and was accompanied by my marine, Barker. Lieutenant Phillips at the same time started with his own three men for the bay I discovered yesterday. At 10.40 a.m. we arrived at the margin of the lake, my furthest in this direction in the first expedition. Only saw a solitary teal there. We next shaped our course along the left side of the lake, over the rocky *débris*, forming a talus at the base of a high range of hills, and in one or two places a shingly beach. As we were rounding a sharp angle of the mountain we encountered a tremendous gust of wind, even for this stormy land, whirling with all the fury of a tornado, and were only saved from being blown into the lake by my clinging to the projecting crags, or, as my companion did, by throwing his Herculean form on the ground in a horizontal position. At 11.25 we reached the upper end of the lake, which is a mile and a half in length. We were now five miles from the boat, following the zigzag course of the swampy valley, with high ridges of trappean hills on either side. Four miles above the lake we ascended the ridge near a cascade which fell over a precipice in front of a cavern, beautifully frosted over by the congealed spray. From this spot hills arose in every direction to obstruct all view further ahead; and after passing another frosted fall, and seeing no prospect of the valley speedily terminating on the coast, which it had been my object to reach, and I hoped might be the case,

R. McCormick. R.N., del.

Vincent Brooks, Day & Son, Lith.

Club-moss Bay, Kerguelen's Land.
Discovered during the boat expedition up Cumberland Bay.

we rested by the side of a waterfall, to take a hasty lunch, and then descended by the fall to the valley. At one p.m. we ascended an isolated rocky hummock in the centre of the valley, which afforded a view for about a mile ahead, the last bend of the valley being to the S.S.W. We commenced our return journey at 1.30 p.m., and, passing the lake, I picked up on its shore a solitary piece of fossil-wood, the first I have met with in Cumberland Bay, and a piece of coal, with some red fragments of a hard-baked argillaceous clay. We reached the boat at five p.m., after a journey of about eighteen miles there and back. We saw scarcely any birds. Phillips and his party had already returned from their excursion, and at Club-moss Bay had slaughtered no less than fourteen seals, to leave them to rot, a feat without any justifiable object, and consequently much to be regretted.

Thursday, 25th.—This morning, after breakfast, accompanied by one of my boat's-crew, I went in search of the rock crystals I left behind me on Monday night, when beset by the floods. After a search I found both lots, and a few more, notwithstanding the ground was covered with snow, and the weather thick with hail-storms, having to face a sharp drift. At 3.45 I visited the ravine where I was arrested by the torrents on Monday night. Here I discovered a small seam of coal or lignite. It crops out in two places at the bottom of a watercourse, where it is intersected by a trapdyke three inches in width. The adjacent mountain is composed of green-stone and amygdaloid, and about half a mile above the Boat Promontory. Reached the boat at 5.15 p.m., had a penguin and pea-soup supper, and turned in at eight p.m.

Friday, 26th.—Still stormy, with hail. Our supply of provisions getting short, cocoa all expended, and only tea and biscuit enough for breakfast. I made another visit to the coal seam. Found it again cropping out on the opposite side of the watercourse, twelve feet across and about twenty yards up the ravine, fifty feet above sea level. Returned to the boat at noon. In the afternoon I made an excursion to the Phonolite Hill, north of the coal seam, and found another outcrop on the opposite side of the hill, where, indeed, I first discovered some also on the side of the watercourse, and it was a foot in depth and ten feet in length on the surface, the most beautiful specimen I have yet met with. Black and glossy, very bituminous, with the brittle shining fracture of anthracite. Direction of the seam south-east and north-west, and, like the first seam, which was two feet by twenty feet, immediately overlaid by amygdaloid and greenstone, a three feet wide gorge winding round a mass of prismatic phonolite, and led to the summit where the phonolite was separated from the greenstone by a basaltic dyke, three feet wide, running south-east and north-west. A lake, 200 yards long by 150 broad, fills the crater-like depression on the summit, narrowing towards the north-east end, and widening at the southern extremity; depth, three feet; the surface in places covered with ice, marked with hexagonal lines. This lake is encircled by an irregular wall of greenstone from five to twenty feet in height, and empties itself by a watercourse and cascade down the hill on the south-east, descended by another of those gorges in the columnar clinkstone. Reached the boat at 4.15 p.m., through a heavy snow-storm. Had cormorant soup for supper.

Saturday, 27th.—Launched the boat amidst heavy snow, freezing sharp. Pulled along the west side of the bay. The hollows in the moss-covered *débris* forming a talus at the base of the hills, presented a very beautiful appearance from the brilliant display of icicles hanging in large festoons in front of them. For about half an

R. McCormick, R.N., del.

Vincent Brooks, Day & Son, Lith.

The Cumberland Bay exploring boat bearing up in a gale of wind off Christmas Harbour, Kerguelen's Land.

PAGE 82—VOL. I.

hour we pulled alongside in a small creek in the rocks, and had our breakfast of soup and biscuit. At noon we made sail, and ran before a fine breeze down Cumberland Bay. The weather, after we had rounded Cape Cumberland, looked threatening, and we attempted to beat up against a strong head-wind and short broken sea for Christmas Harbour. Between Foul Bay, however, and Arched Point, we shipped so much water, both over the bows and to leeward, as to nearly swamp the boat. Being up to the thwarts in water, we were compelled to lower the sail and bale her out; and still further to lighten her and increase her buoyancy, my worthy colleague ventured, though reluctantly, to propose that my boxes of specimens should be thrown overboard as a sacrifice to the storm. But this was expecting too much from me after all the toil and risk I had undergone in collecting them, and my well-meaning, excellent boating companion saw himself that this would be too great a sacrifice on my part, and to be resorted to only as a last resource. So they retained their place as ballast at the bottom of the boat, for I felt that the time had not yet arrived for such an extreme measure, and so the result proved.

After pulling against a short head-sea amid heavy squalls, without apparently gaining in the least upon the Arched Point, we were really drifting to leeward. At three p.m. we bore up when within half a mile of the point, and scudded under the mizzen, set forward for Cape Cumberland, the sky at this time looking most black and threatening. We had a tough pull round the cape against the sudden squalls, and were nearly blown out to sea. At last, at four p.m., we secured a berth in our old corner, under the lee of the land. It being now low water, we could not, as before, haul the boat over the ledge of rocks, so after getting everything out of her, we moored her outside to take her chance for the night;

and on the swampy, snow-clad ledge of rocks we rigged up a tent with the mast and sails, supported on a tri-angle at one end, and by the face of the cliff on the other. It was quite dark before we had completed all our arrangements for the night, after which we had some cold preserved meat and biscuit for supper, and turned into our blanket-bags, which were as wet as our clothes.

Having got the stove within the tent and arranged the lately all but doomed specimen-boxes as pillows, we were nearly driven out by the smoke from the stove, and when the fire was allowed to go out the snow it had melted froze again from the severity of the intense cold. In the morning, when attempting to rise, I found my outer garments frozen to the ground beneath me. The night-petrels on the rocks above our heads did their best to lull us into slumber by their melancholy moaning, kept up through the greater part of the night.

Sunday, 28th.—Rose at 7.45 a.m. and struck the tent, getting everything into the boat, and took our final departure from Cumberland Corner at 8.45, having a moderate breeze and smooth water. Nearly all our pro-visions were exhausted; the tea, cocoa, flour, and some other articles had been out for several days, and the last of the grog, the sailor's comfort, was served out to the boat's-crew this morning; forty-eight hours of preserved meat alone remained; so that had we been longer weather-bound, we had intended making an overland journey to the ships. However, we now pulled round Arched Point, and got on board the *Erebus* at 10.30 a.m.; found Captain Ross and the ship's company at divine service, and, as soon as it was concluded, Captain Ross shook me heartily by the hand, and seemed to be very glad we had returned.

I had scarcely got through the very essential ordeal of a shave and entire change of things, after nearly a fortnight's absence without either, when the White Bay party's boats were reported in sight, and at 1.30 p.m. arrived. They had left the ship two days after us.

When our coals fell short in the boat we cooked our meals with the island coal I discovered in South Bay, which burnt remarkably well, so that we should not have been at any straits for fuel, and my gun would have furnished us amply with game, to which the native cabbage added an excellent antiscorbutic vegetable very palatable. The groceries we should have missed the most.

Wednesday, July 1st.—Blowing weather, since my return, preventing communication with the shore, I employed myself in arranging and stowing away my specimens of natural history.

Thursday, 2nd.—I landed in the gig and crossed the isthmus to the lakes, shot a brace each of teal and chionis.

Friday, 3rd.—At 8.30 a.m. the whale-boat landed me at the corner of Arched Point Bay; when about the centre of the bay a seam of very fair coal, of course, like all found cropping out in the igneous rock-formation of this island, of lignite origin. It occurs here four feet and upwards in thickness, forty feet in length near the surface, and about thirty feet above the level of the sea. Above the coal, just within reach of my geological hammer, a fossil-tree is imbedded in a vertical position, in the face of the perpendicular cliff of greenstone, which rises to 600 feet above it. After collecting a few plants, and having shot three chionis and a shag, I returned on board in the gig at five p.m., amid heavy squalls and large quantities of ice falling from the cliffs.

Saturday, 4th, 8.30 a.m.—Left the ship on an excursion round Cape François; following the line of beach over the rocks was a toilsome journey, rendered still more so by my having to face a sharp, cutting snow-

drift during the heavy squalls that were blowing. About
the middle of the bay is a cave, hollowed out of a bed of
shale, to the extent of twenty feet in depth and twelve in
height, the entrance to it about thirty feet in width. A
thin vein of coal only a few inches in thickness underlies
the shale. It may also be seen on either side of the cave,
where the rocks have been undermined. A number of
gulls were assembled in the bay, feasting on the remains
of a seal and some dead penguins. Of the latter bird I
did not meet with a living one throughout my excursion.
I shot two large petrel (*Procellaria gigantea*), two black-
backed gulls, and two shags. At the base of Cape
François is a perfect chaos of rock-fragments and black
ledges of rock jutting out, on which a heavy surf
was breaking. About noon I began my return at a deep
cleft or chasm, and reached the observatory at 3.30 p.m.,
where I found Captains Ross and Crozier at dinner, and
joined them.

Sunday, 5*th.*—Captain Ross came on board and performed divine service, and at four p.m. both captains and
the gun-room officers of the *Erebus* dined with the gun-
room officers of the *Terror*. We had an excellent dinner
for this land of desolation, and excellent cabbage, with
fish, soup, boiled mutton, and roast pork, pastry, &c.

Monday, 13*th.*—The weather having been so unfavourable, blowing and raining for the last few days, I had
only landed twice for a short time, but employed my
time in skinning birds, laying out plants, and packing the
large fossil tree in a box the carpenter had made for it,
accompanying it with a full description in writing. At
eleven a.m. I landed, but the weather being so threatening, with a heavy surf breaking on the beach, throughout
the morning, I did not venture out of sight of the
Terror's boat, employed getting her observatory on
board, and it was fortunate for me that I did not, or I
should have had to remain on shore all night, and alone,

as her people with the tents were all taken on board in the last boat, which I got a passage in after running down to the beach just as she was about shoving off, shooting a tern which was attempting to make headway against the squalls as it flew over my head on my way down to the boat. I had been digging with a shovel for the night-petrel through the holes made by them in the ridge above the beach, but did not succeed in finding any. The snow on the ground had drifted in wreaths, into which I sank to the top of my boots at every step. Soon after I got on board a heavy gale came on.

Wednesday, July 15th.—At nine a.m. I landed here for the last time, for the purpose of searching the burrows of the night-petrel for specimens, as we intend putting to sea so soon as the weather will permit. To-day much of the snow was melted by the rains of last night. On first landing, I shot a seal by discharging both barrels, with only small shot, at his head, near the stream which enters the bay at the south end of the beach; and Captain Ross sent one of his gig's crew to secure the skin. I afterwards occupied myself for above three hours, till 1.30 p.m., in digging on the ridges just above the beach, opening several holes, where marks of the impression of feet, droppings, and feathers, indicated the presence of birds; but found them all empty, with the exception of a small, soft-shelled, rotten egg of a white colour in two of the holes. However, on descending a little down the ridges, I at last succeeded in unroofing four very fine specimens from two holes in a greenbank, the entrances to which bore no marks of there being birds within. I dug out the first pair at noon, and shortly afterwards another pair from a hole just beneath the first. Its narrow entrance was partially concealed by the stalk of a cabbage growing in front of it. I first ascertained the presence of the birds by sounding with the ramrod of my gun, which caused them to move and utter a low note. The burrow, for the bird is evidently the architect of its own dwelling, is about three feet in length, generally winding to the left, and ending in a circular dome, six inches high and eighteen inches in diameter. In the centre of this is a slightly raised mound of soil, strewed over with a green leaf or two of the tufted plant which so generally forms a carpet to the soil here, and a blade or two of grass. On this little ottoman sits the bird itself, and in both holes they sit apart from each other, on either edge of the mount opposite. What instinct, nay, wonderful design, do these little creatures exhibit in their system of drainage, by the construction of a canal two inches deep and the same in breadth, on an incline, for letting off the water. No skilled artificer could have surpassed them, for whatever the floods may be around them, here they reposed in dryness and comfort. Whilst I was digging above their heads not a sound escaped them, silent as the grave till they were touched, and on being handled, uttered a sharp, shrill cry, emptying their stomachs at the same time, the contents of which were the remains of crustaceæ, and the beak of a small cuttle-fish was thrown up from one of them. The bird has a broad, flat head, and, like most nocturnal birds, very large, full, dark eyes, a deep brown approaching to black; the plumage is of a sooty black. On returning to the beach I found Captain Ross busily employed in superintending the pulling down of the observatory.

Being desirous of adding a specimen or two more of that highly interesting white bird, the chionis, to my collection, I rambled for about half a mile along the rocky *débris* at the water's edge, where they seek their pelagic food amongst the seaweed on the south side of the bay. But so scarce have they become from the long presence here of the ships, that I only met with one, which I shot, and brought on board, with a few specimens of seaweed, the conglomerate rock, with limpets, and the native cabbage; getting on board at 4.30 p.m., when it came on to blow and rain soon afterwards. In the evening I laid out my seaweed.

Thursday, 16*th.*—Blowing too hard to-day to unmoor ship, a boat was sent on shore with an inscription in copper case, to be deposited on the site of the observatory, as a memorial of our visit to the island.

Boat Refuge, Cumberland Corner. (*See page* 84).

Arched Rock, Christmas Harbour. (*See page* 92.)

CHAPTER VIII.

Geological *résumé* of Kerguelen's Island.

VIEWED geologically, this and the island of Spitzbergen, in the opposite hemisphere, are the two most deeply interesting and remarkable islands on the face of the globe, and the least known.

Kerguelen's, isolated amid the vast southern ocean, in lat. 50°, and long. 69°, was first discovered by M. de Kerguelen, a lieutenant in the French navy, in the year 1772; and subsequently our own renowned navigator, Captain James Cook, was the first to anchor his ships in Christmas Harbour, and give to the land the name of Desolation Island. Christmas Harbour is in somewhat the shape of a foot, and named after the French frigate, *Baie de l'Oiseau.*

The island is entirely of volcanic formation, upheaved from the bed of the ocean, most probably during the tertiary epoch. Ancient forests have been entombed within the lava currents which once flowed from sub-marine craters long extinct. The silicified trunks of

trees, lignites, beds of anthracite, and coal, so abundantly
embedded in the basaltic rocks, amply testify ; where,
now, not a vestige of aqueous formations occurs, and
neither tree nor shrub are to be found, the native cabbage
(*Pringlea antiscorbutica*), about two feet in height, being
the largest plant existing on the island. Yet these fossil-
trees, chiefly of the coniferæ, remain as monuments of
their past existence in a land long ago submerged, to be
again upheaved by some subsequent disturbance of the
bed of the ocean. The trunks of the trees being subjected
to great pressure beneath a body of water, and in the
process of slow-cooling, would become infiltrated with
the silex in solution ; also, under the slow-cooling of the
submarine lava streams, the beautiful, crystalline, drusy
cavities in the rocks, and prismatic structure would take
place whilst shrinking in the melted state ; and the
characteristic formation of the schistose trachyte, forming
those phonolite domes, upheaved between the greenstone
ranges in Cumberland Bay, have arisen under the ex-
pansive force of the elastic vapours. The red marking
in the foliaceous phonolite is owing to oxide of iron.

The general aspect of the island is picturesque in the
extreme. The trappean terraces, rising one above the
other, studded here and there with a conical, crateriform
hill, rising above all ; the coast-line, intersected by bays
and creeks throughout, leaving only very narrow isth-
muses between one coast and the other. Small lakes
are scattered over the surface of the valleys, and countless waterfalls or cascades rush in torrents down the
steep and precipitous black basaltic cliffs, their white
foam in striking contrast with the black surface of the
lava. Cumberland Bay contains some excellent commodious harbours, affording safe anchorage to ships or
boats ; its shores much diversified by the many smooth-
looking, cone-shaped hills, from three to four hundred
feet high, of phonolite or clinkstone, both prismatic and

foliated in structure, so characteristic of the trachytic
formation, the light-yellowish colour of these rounded
cones appearing in striking contrast with the dark shade
of the greenstone ranges between which they emerge.
Some of the fragments I collected were singularly marked
with concentric circles of red oxide of iron. One beautiful specimen, in particular, so resembled a bright red
seaweed occurring on the shores of the bay, both in
colour and the delicate tracery of its form, that
Captain Ross, to whom I showed it, could scarcely
believe that it was not the fossil-weed itself, preserved
in foliaceous sandstone, which, indeed, it very much
resembled.

Christmas Harbour, the northernmost and finest bay
between Cape François and the Arched Point, is about
a mile in breadth at its entrance, bounded on either side
by trappean terraces one above the other. Table Mount,
having a crateriform summit, filled by a lake, is the
highest hill, attaining an altitude of above 1200 feet on
the north side of the bay; whilst, on the south side of
the harbour, a vast block of black basalt, some 400 feet
in thickness, rests upon a greenstone ridge of 600 feet,
the whole attaining an altitude of 1000 feet above the
harbour, a horizontal bed of shale, six feet thick, dividing
them.

It was here that I discovered the first specimen of the
fossil-wood, when ascending the mount on the day after
anchoring in the harbour, and subsequently the trunk of
the large silicified tree, seven feet in circumference,
embedded in the *débris* beneath, with some remarkable
specimens in the face of the Arched Rock, at the
entrance to the bay. And in a small cove near this a
seam of brown coal or lignite, four feet in thickness,
and forty feet in length, crops out above the talus of
débris, thirty feet above the bay, underlying the lava or
basalt, which rises 500 feet above it.

This discovery of coal in so many localities of the island might be turned to good account as coaling-depôts to the numerous steamers that are employed in our East India and Australian trade, where they might replenish supplies when about mid-voyage out or home, without having to go far out of their course in doing so. The seams of coal, too, for the most part occur near the surface. All that would be required would be to tunnel horizontal galleries through them. The coal, I know from experience, burns well, having cooked our food with it when away with the boats in Cumberland Bay, at a time when the coals brought with us from the ship had run out. All it wants is to be brought under the notice of the Government.

The rise and fall of the tide in Christmas Harbour is very slight indeed, only between one and two feet. The climate is stormy and tempestuous in the extreme. We made the land, Bligh's Cap, in the month of May, amid fogs and squalls, and, on our departure from it, towards the end of July, we encountered a succession of tremendous gales in the passage to Tasmania. It is not a little remarkable that the general direction of the mountain range should have the same bearings, north-east and south-west, as the Tasmanian and great Australian mountain-dividing ranges.

The highest land I met with when away in the boats up Cumberland Bay attained an altitude of 2500 feet, and the disintegration of the hills must be going on very rapidly, judging from the enormous accumulation of *débris* at their bases, forming steep slopes of talus from 300 to 500 feet, down to the black ledges of basalt upon which the sea breaks along the whole weather-shore, which, with the numerous watercourses rushing down from the trappean mountain ranges above, to the carpet of vegetation beneath, render it an almost impassable bog, in which our party sank knee-deep at every step, and then finding an inlet bounding it and all further progress to the westward.

The vegetation is chiefly limited to mosses, lichens, and grasses, and a tufted heath-like plant, which forms quite an elastic covering to the soil everywhere. The native cabbage, an excellent esculent and antiscorbutic, formed a wholesome addition to our bill of fare. It is a perennial, with lateral flowering stems. The delicious teal, blue-winged, is, I believe, peculiar to the island, and also the white bird, or chionis. This species of sheathbill, too, is limited to this land; the others were mostly sea-birds, sooty albatrosses, gigantic petrel, night-petrel, tern, shags, and penguins. The skua and black-backed gulls frequented the bays, seals of two or three kinds occasionally came into the harbour.

When we come to reflect that, at the present time, not a vestige of any arborescent form of vegetation has existence here, we may well ask from whence came all the fossil-wood. A far greater extent of land must at some period of time have existed to account for such a vast accumulation of vegetable remains, altered, indeed, into lignites, fossil-wood, and coals, but which must once have constituted whole forests, and have been submerged long enough to convert the vegetable structure into coal and lignite, and silicified the wood, which would have the effect of enabling it to resist the heat of those subaqueous lava streams which poured out and overwhelmed them in deep water, when the zeolites in Christmas Harbour were formed.

At what period of our earth's history all this took place must necessarily be somewhat a matter of conjecture, in the absence of all sedimentary formation as evidence to guide us. But the brown coal and lignite, however, may afford us some clue, inasmuch as they greatly resemble the brown coal and lignite found under very similar circumstances in the newer tertiary deposits of continental Europe, a period of geological time when lava streams, submarine craters, and their characteristic lignites and fossils, so much abounded.

The ancient forests of the land, at whatever period of time they may have existed, must have consisted mainly of conifers, from all the fossil-wood found here having the glandular structure, or resin-duct, peculiar to all the coniferæ or pine tribe, a class of plants which, from their resinous composition, are so well adapted to resist the extremes of cold of an Arctic or Antarctic climate; this resin being formed by the air slowly entering into combination with the carbon, a non-conductor of heat, and from the glutinous resin not freezing but concentrating all its heat within, most effectually seals up its vitality.

Hence the power of the coniferous types—so conspicuous in all the carboniferous forests of bygone ages —to resist the great changes and vicissitudes of temperature to which they have been subjected, and thus enabled to exist as a large family of trees throughout so many epochs in the earth's history.

Compare the family of the lycopodiaceæ of the old red sandstone, or Devonian epochs, such as *Lepidodendron Sternbergii*, which attained the height of forty feet and upwards, with the small but elegant and beautiful species of lycopodium, or club-moss, at present existing, a new species of which I was fortunate enough to discover living up Cumberland Bay, when away on my boat expedition; a mere creeping plant, scarcely showing itself above the scanty surrounding herbaceous vegetation, and, remarkable enough, was the only specimen found during the stay of the expedition at the island. I picked it up with a small new species of fern in the same tuft, and never met with either again, although I directed my

boat's crew to keep a good look-out for any other
straggling specimens there might be.   We discovered an
extensive bank, named after the ship, " Erebus Bank," and
a reef, named " Terror Reef," off the island.

Cumberland Bay and Sentry-box, with Foul Bay between the Arched Rock
and Cumberland Point, as seen from Summit of Crater Peak.

Crater Peak, Christmas Harbour.

CHAPTER IX.

Our boatswain drowned—Sighting the South-west Cape—Severe gales—
Our sails blown into ribbons—Reach Tasmania—At the theatre,
Hobart Town--A complimentary dinner—An invitation from Sir
John Franklin, governor of the island—The first house built in the
colony—Richmond—A kangaroo hunt—A horse-race—Going to
Government House Ball under difficulties.

Friday, July 17*th.*—Unmoored after breakfast; but the
weather again coming on thick and threatening, we
remained at single anchor. I finished arranging and
packing my birds of the island, some half-hundred, and
took a sketch of the fossil-wood rock of basalt.

Saturday, 18*th.*—Still detained by boisterous weather.

Monday, 20*th.*—We got under weigh early this morning, sailing out of the harbour at eight a.m., and took our
final leave of Kerguelen's Land; *Terror* just ahead of us.
I took a sketch of the Crater Hill as we passed out.
Sky very lowering and threatening, with a rapidly-flying
scud, the land soon becoming concealed in thick gloom.

Both ships under double-reefed top-sails, storm stay-sail, and jib, going about six knots. Shipped a sea or two and made all snug for the night by close-reefing the top-sails. As we passed out of the harbour I noticed that four ridges superimposed the coal-seam forming the summit of the cliff; these ridges, both at Cape François and Arched Point, dipping to the north at an angle of 15°.

Wednesday, 29th.—The continued blowing weather we have experienced since leaving the harbour increased to a heavy gale from the westward this evening, before which we scudded under close-reefed fore and main-top-sails.

Having seen nothing of the *Terror* since last evening at seven o'clock, when she was some seven miles off, we had to lay-to for her from eight a.m., and did not bear up until 3.30 p.m.

Thursday, 30th.—At 2.40 p.m. we met with a sad loss in our boatswain, Mr. Roberts, who, whilst employed with two men near the lee-gangway on the port side, was struck by the staysail sheet, and whirled overboard by a sudden lurch the ship took at the time. The life-buoy was immediately let go, and several oars thrown overboard, both quarter-boats—the first and second cutters—were lowered as soon as the way in which they had been secured against the bad weather of late would permit of; but unhappily too late, poor fellow, to save him. I happened to be walking the quarter-deck when the melancholy accident occurred, and saw the whole affair. He swam very strongly and high out of the water for some time, but the ship having been going through the water at the rate of six knots at the time, he was rapidly dropped astern before she could be hove-to. The last I saw of him was as he rose on the top of a wave, where a gigantic petrel or two were whirling over his head, and might have struck him with their powerful wings or no less powerful beak, for he disappeared all at once between two seas. Our first boat returned, having picked up his hat and one of the men's caps. The other boat, sent to pick up the life-buoy, was near losing two hands, a sea having struck her and washed two of the crew overboard, but they succeeded in regaining the boat again. It was 4.20 p.m. before the boats were, at a great risk, secured again.

Wednesday, August 12th.—At nine a.m. saw the land to windward on the port bow. South-West Cape, being seven or eight leagues distant, appeared very indistinctly through the haze, and a bank of clouds in the horizon in the form of a line of hummocks, Hobart Town being dead to windward. Yesterday my little tortoise from St. Helena I found in the morning dead on the floor of my cabin. From Kerguelen's Land to Van Diemen's Land, a distance of between 3000 and 4000 miles, we had a succession of heavy gales and boisterous weather, shipping heavy, green seas, in which the ship rolled tremendously, getting her decks flooded, accompanied by smart hail-storms, some of the hailstones measuring half an inch in diameter.

The Southern Cross now appears in a reversed position, having the " pointers " above. The last remaining goose was killed the other day. This afternoon, coming on to blow hard, the ship was hove-to, with her head to the southward and westward ; and before the evening closed in the gale increased to a perfect hurricane, which, with the mountainous sea running, raising the foam and spray from the surface, drove like mist or smoke, obscuring everything ahead of us ; the barometer, rising and falling rapidly, only a little above 28° at the highest. Whilst I was on deck, from seven to 8.30 p.m., we shipped some very heavy seas ; one vast, swelling, green mountain of a sea came rolling up astern, threatening to engulf us, sweeping over the starboard quarter-boat, in upon the quarter-deck, which it deluged, drenching me to the skin, as I clung to the mizen-mast, catching hold of some gear

to avoid being washed overboard, for as she surged
everything on deck found its way into the lee-scuppers or
overboard.   The sails were taken in with great difficulty,
till we were reduced to a close-reefed main-topsail.   When
about eight p.m., in the height of all the fury of the blast,
the main-topsail sheet was blown away, and the sail shivered
into ribbons, the battered, torn fragments flapping in the
wind with a sharp, snapping sound, like the report of
so many pistols.   A new main-staysail—the only sail left
on the ship—soon shared the same fate, leaving us
under bare poles; in such a tempest not a stitch of
canvas had a chance even of standing against it.   The
night was moonlight, and the scud was flying rapidly over
the surface of the pale orb.   Captain Ross maintained his
position on the weather-quarter by having three turns of
the mizen-topsail halyards round him for support.
Towards the close of the first watch the gale began
somewhat to abate, but still blew hard throughout the
night.   On the following morning a very heavy sea was
running, but there was less wind.   We had the hatches
battened down all day, with lighted candles in the gun-
room.

Monday, August 17th, 1840.—Yesterday we made the
land to windward, working up Storm Bay, but could
not fetch into Adventure Bay.   In the evening saw the
Iron-Pot Lighthouse, and after firing guns for a pilot,
anchored about midnight some four miles below the lighthouse; and this morning at 9.30 we weighed anchor,
and with squally weather worked up the Derwent.   About
noon we hoisted the Jack and fired guns for a pilot.
At 12.30 the pilot came on board; we passed to the left
of the Iron-Pot Lighthouse, a square tower on a small
barren, sandy rock.   Betsy Island to the right, covered
with wood.   The banks, both of Storm Bay and the
River Derwent, are densely clothed with wood, but the
colour of the foliage being of one uniform dull green,

with the absence of the varied tints of light and shade, gives to the whole landscape a monotonous aspect, notwithstanding the richness, and even luxuriance of the vegetation. The gum-tree and other forms of *eucalyptus*, so prevalent here, shed their bark, instead of their leaves, the latter appearing with the under-surface upwards, or with the leaves occurring transversely, one above the other. The lofty white, barkless trunk of the gum-tree, shining like the silvery stem of a birch through the dense foliage somewhat relieves the otherwise sombre appearance of Australian forests in the distance. The entrance to Storm Bay is between Tasman's Head and Cape Raoul, these headlands being about thirty miles apart. The breadth of the bay where the Derwent begins is about twelve miles across, and the river is six miles wide at its entrance from the lighthouse, from which Hobart Town is twelve miles distant. The course of the Derwent is about N.W., averaging from two to five miles across as it approaches the town.

On the port-side we passed in succession several beautiful glades in the woods, shelving down to the water's edge, with pretty farmsteads surrounded by highly cultivated patches. One villa-like residence especially, and evidently of recent erection, made its appearance some four or five miles from Hobart Town, between which lies an excellent road. The young corn looked everywhere very green and promising. The wind failing us at 5.15 p.m., we came to an anchor in twelve and a half fathoms, off Fort Mulgrave. The approach to Hobart Town is very picturesque, the houses appearing thickly scattered amongst trees and green slopes, on the sides of undulating hills studded here and there with a windmill. We found that our consort the *Terror* had arrived at ten p.m. on Saturday; her captain and several of her officers came on board, bringing us letters. Several of us went on shore to the theatre, it being the last night of the performance for the season. The play was " Rory O'More," and the afterpiece " The Irish Tutor." It concluded at 12.30 p.m., and we returned on board in the guard-boat.

Tuesday, 18*th.*—Weather very fine, the ships were warped alongside of the Government paddock, and moored in from five to seven fathoms. Having answered a letter from Sir William Hooker, in reference to his son's (Dr. Joseph Hooker, the assistant-surgeon) state of health and fitness for the arduous service he had embarked in, I went at noon on board the Australian packet-ship getting under way for Sydney. Afterwards several officers of the 51st Regiment quartered here came on board of us, and amongst them, our old acquaintance, Captain Foreman, whom we had met at Chatham. At 3 p.m. I landed at the dockyard, and also called at the Colonial Hospital; then, after taking a stroll round the town, returned on board in a pilot-boat at six p.m. Saw in the *United Service Gazette* the report of the Commission on the new Medical Regulations for the Navy.

Wednesday, 19*th.*—Dr. James Wingate Johnstone, of the *Asia*, convict-ship, came on board, and several officers of the 51st Regiment. It was a beautiful day, and at six p.m. we left the ship, to dine at the mess of the 51st, to which the captains and officers of both ships had received invitation; with three or four civilians from the shore, we made up a party of thirty-five. I fell into a seat on the left of the president, Captain Foreman, and for the first time in my life tasted kangaroo soup and the bronze-winged pigeon of the island. The major gave a toast, accompanied by a complimentary speech in honour of the expedition, which was replied to by Captain Ross.

Friday, 20*th.*—Mr. Anstey, a barrister of Hobart Town, came on board, and invited me to dine with him on the following day, at 6.30 p.m., where I met a surgeon, a Mr. Bedford and his wife. Left at 10.30 p.m. and returned on board, having about midday lunched at the barracks, where I saw the eagle, tiger-cat, and kangaroo of the island.

Sunday, 23rd.—At eleven a.m. I attended divine service at the town church, and afterwards called at the Derwent Tavern, on Dr. Johnstone, who, with another brother officer, Peter Fisher, and Captain Foreman, after a stroll round Sandy Bay, dined on board with me.

Monday, 24th.—Dr. Clarke, the Inspector-General of the Forces, with his brother-in-law and Mr. Estridge, the traveller, lunched on board with me. I received an invitation card from the Governor, Sir John Franklin, to dine at Government House to-day, at 6.30 p.m. Twenty sat down to table. I was introduced to Sir John and Lady Franklin by Captain Ross. At eight p.m. we adjourned to the library, to hear a paper read by the Rev. Mr. Lillie, it being the night of the meeting of the Tasmanian Natural History Society, which takes place on alternate Mondays.

Tuesday, 25th.—After breakfast, Captain Foreman called alongside for me in his boat, for a trip up the river to Risdon, about four miles distant; Hooker, the assistant-surgeon of the *Erebus,* accompanied us; the weather fine. On our way up the river, Mr. Gregson, the gentleman we were to visit, was riding on the left bank for Hobart Town. We landed, and I was introduced to him, who subsequently became one of my most valued friends. He had been one of the guests at the dinner given to us by the mess of the 51st. Captain Foreman having introduced me to Gregson, he continued his journey on to Hobart Town, and we ours to the ferry at Risdon, where we landed and walked on to the house, about a mile from the ferry, most picturesquely situated on a rising knoll, embosomed in trees, and perfectly isolated, approached by a winding road skirting the creek, where his boat is kept, for most part of the way having a thickly-wooded hill on the right, upon a foundation of argillaceous rock. After partaking of some refreshment with the family of our newly-made friends, we took a ramble round the grounds, and in company with our host's eldest son, John, I shot several small birds for the ornithological collection, with Foreman's gun, not having my own with me at the time. We returned to the house just before dinner-time.

Wednesday, 26th.—We did not retire to rest till one o'clock this morning. I arose at eight a.m. and strolled to the bottom of the garden, sketched the old ruins of the first house ever built on the island, by its first governor, Colonel Collins. For it was on this spot the first colonists established themselves, attracted to it by the fine stream of fresh water running near. It now forms the summer-house of the family, and the hives for their bees. After breakfast my host's eldest son accompanied me in the ascent of Mount Direction, a remarkable hill, forming a very conspicuous object, as seen from the ship, bounding the curve of the river in the distance. We made our way up its steep sides, through woods of gum-trees, and the black and silver wattles; many of the trees appeared much blackened and charred near the base, and on the summit withered trunks and branches were strewed about; a rude signal-post, or what had originally been intended for one, crowns the highest peak. From this spot I had a fine view of the windings of the Derwent below, with its many little points and promontories jutting into the stream on either side. The New Norfolk steamer was passing at the time; the day was fine, and so warm that I perspired from every pore. Saw several paroquets, and shot two blue titmice and a lark returning.

In the afternoon we started for the township of Richmond, nine miles distant, in Gregson's car, along a beautiful and picturesque road called the Simplon, forming magnificent curves and deeply-wooded glens; these, again, backed by hills densely clothed with gum-trees to the summits. At about half-way on the left is Grass-tree Hill, on which are two or three huts, forming the quarters of an officer and his party stationed there. Dined at Mr. Breton's, the police magistrate of Richmond, a half-pay lieutenant in the navy. He showed me his collection of fossils and other productions of the island. Here I met the first lieutenant and surgeon of the *Terror*. We slept at the Lennox Arms Hotel.

Thursday, 27th.—Rose at eight a.m., and walked round the township, up a hill by the small coal-river, where the coal has been worked; returned to the inn for breakfast; when Hooker and myself drove in Gregson's carriage to Grass-tree Hill to see a kangaroo hunt, with Gregson's pack of hounds. Found him there on horseback, accompanied by Colonel Elliott, of the 51st Regiment. The weather was fine, and two kangaroos were started. I saw one of them leap across the road at some distance, followed by the barking and yelping dogs; they, however, gave the dogs a long run, taking the hill side into the dense woods; both were soon out of sight and hearing. We learnt subsequently that one had been killed some miles off by the dogs. About noon we returned to Richmond, and at one started for the race-course, to see a match between " Randolph " and " Waterloo," two celebrated horses here. The latter ran away with his rider, and actually went four times round the course before he could haul him in, and then darted into the woods, but fortunately without injury either to horse or man. Lunched at Lieutenant Breton's with Colonel Elliott and several other officers, and some gentlemen of the neighbourhood. When on the race-ground, the weather was showery. Amongst the number of country people present, one young girl in a riding-habit, having the appearance of a yeoman's daughter, was thrown from her horse, but regained her saddle instanter, as if nothing had happened, so coolly was it done. About five p.m. we left the inn with Mr. Gregson, for Risdon, and after dining there, I very reluctantly left a most agreeable, pleasant party to pull four miles in Gregson's boat, against a head-wind and beating rain, in a dark, tempestuous night, to attend a ball at Government House, given by Sir John Franklin in honour of the arrival of the Antarctic Expedition, from which, of course, there could be no excuse for my absence. At 8.30 p.m. young Gregson accompanied me to the top of the creek, where his boat was ready to receive me, with two hands. Soon after I had embarked, the creek being narrow, about half a mile in length, we took the ground in the mud on the port side, but soon got her afloat again. It was pitch dark, and on getting clear of the creek, the wind shifting, blew up the river in our teeth, so that we did not get alongside the *Erebus* before 10.30 p.m., first making for the *Terror* by mistake, in the cimmerian darkness of the night. Having changed my dress for full uniform, I got on board, and landed again at 11.15 p.m.

I entered the ball-room at a much later hour than I had anticipated when I left Risdon, it being exactly midnight. Sir John Franklin and his aide-de-camp received me very kindly at the ball-room door, and Captain Ross, coming up to me, offered to get me a partner in the dance, a waltz then going on, which I declined, as I do not dance. Many of the company had already left, and carriages were in waiting at the door. The night proving so unfavourable, there were many absentees. I left at two a.m., and slept at the Derwent Hotel. In the ball-room I received two invitations; one from the solicitor-general, and the other from the commissary-general.

Sunday, 30th.—I attended morning service at the

town church, and made a call at the barracks ; met the solicitor-general there, from whom I received an invitation to dine this evening, to meet Captain Foreman and another military captain.

Monday, 31*st.*—Dined at the military mess with the assistant-surgeon of the regiment ; after which we all attended a concert held in the theatre.

Tuesday, September 1*st.*—Called on Anstey, the solicitor-general, and Dr. Clark ; from the latter I received an invitation to dine with him at the mess. Visited the Union Club. Received a note from Lieutenant Breton with a parcel of birds' skins from Richmond, and a message from Captain Ross, to get ready the specimens of natural history for embarkation on board a ship for England *viâ* Sydney.

Wednesday, 2*nd.*—I was employed arranging the specimens in Captain Ross's cabin, and had several visitors ; the Rev. Mr. Lillie, Lieutenant Breton, and Captain Foreman ; all of whom dined on board with me, and I asked Dr. Hooker to meet them. Mr. Anstey and a naval lieutenant also formed part of our dinner-party.

Thursday, 3*rd.*—Employed all day packing specimens and writing descriptions of them, till 6.30 p.m., when I dined at the military mess.

Monday, 7*th.*—At seven p.m. I dined at Government House, fourteen sitting down to table ; Sir John Franklin in the centre of the table, between Captain Ross and myself. On adjourning to the library a paper on the geological structure of a hill on the island was read by Mr. Lillie ; another on New Zealand by Mr. Estridge.

Monday, 21*st.*—Dined at Sir John Franklin's. Twenty guests present. At eleven p.m. three papers were read at the Natural History Society's Meeting.

Saturday, October 3*rd.*—At three p.m. Lady Franklin, accompanied by her niece, Miss Cracroft, and her daughter and governess, visited the ship, escorted by the two captains, calling alongside the *Terror* on their way.

Monday, 5th.—Dined at Sir John Franklin's. Met Lieutenant Breton at table, and afterwards I read a paper on the geology of Kerguelen's Land at the Natural History Society's Meeting.

Tuesday, 6th.—Judge Montagu came on board to kindly offer me a seat in his carriage, with another for my friend and boating companion, Lieutenant Phillips, on a drive across the island to Launceston, as he was about attending the assizes.

Fossil-tree, Macquarie Plains.   (*See page* 119.)

CHAPTER X.

A drive across the island to Launceston—Over Jordan and through
   Jericho—The Vale of Jerusalem—Campbell Town—Epping Forest
   —-Ornithological specimens— Excursion down the Tamar to George
   Town—Return on horseback—Convicts on a sheep farm—A fossil-
   tree.

Sunday, October 11*th.*—At seven a.m. I went on board
the *Terror* for my old friend, Lieutenant Phillips, and,
after breakfasting with him, we landed at the Paddock,
and at the old wharf Judge Montagu's carriage, with a
pair of greys, was coming down the hill from his house.
The morning was fine, with a fresh breeze.

   At eleven miles from Hobart Town passed the " Black
Snake Inn," next over a causeway, and across Bridge-
water Ferry ; a limestone quarry at the causeway dips
at an angle of 15° to the south-east.   " Castle Inn "
stands on the left.   Passed by Brighton and the River

Jordan, a narrow stream, at 11.15 a.m. We made our
first resting-place at " Maule's Inn," where we lunched
on some cold turkey. At the plains of Bagdad, seventeen miles from Hobart Town, we called on Lieutenant
Foster, R.N., the police magistrate, living in a newly-
built residence, Green Ponds. At twenty-seven miles
we alighted, and walked over Constitution Hill, on the
left of which is a thickly-wooded glen, and on the
right a bed of sandstone dipping at an angle of 15° to
the north-west. The road forms a fine sweep round
the glen below, strikingly resembling the Simplon at
Richmond, passing through trappean rocks. On re-
entering the carriage, we crossed the plain of Green
Ponds. Here is the oldest inn in the colony, called the
" Royal Oak," where I had a glass of Hobart Town ale
handed me by a very pretty landlady. At thirty miles
passed Cross Marsh, having a number of cattle feeding
in a good pasturage. Mount Vernon, the residence of
Mr. Kemp, appears at some distance to the left of the
road, embosomed in woods and pasture-lands. We next
passed through a rich and fertile tract, well named
Lovely Banks, from the beautiful, soft landscape of green
pasturage, studded over with lightly-wooded knolls and
sloping banks clothed with shrubs and trees. At present
" Lovely Banks Inn " is not anything more than a dilapidated, deserted, old house, on the left side of the road.
Thirty-six miles from Hobart Town at 4.15 p.m., we
took up our quarters for the night at the " London Inn."
Spring Hill, four miles farther on; I walked through the
garden at the back of the house, up the hill, the summit
of which was composed of trappean rocks. We had
kangaroo cutlets for tea, and I afterwards looked for
opossums by moonlight up the hill, turning in at
10.30 p.m.

Monday, 12*th*.—Resumed our journey at five a.m.
Mr. McDowall, the attorney-general, whom we met at the

inn last night, is going the same road as ourselves, to attend the assizes at Launceston. Half a mile from "Spring Hill Inn," and about half a mile to the left of the road, is a deep gulley, at the extremity of which stands a small hut, in which a horrible murder was committed by the natives a few years ago. Two miles farther on we passed Jericho, about a mile to the left of which stands an estate forming a group of buildings at the base of the hills, which had belonged to my friend, Mr. Gregson. Near the road is the church; and to the right the Vale of Jerusalem appeared flat, and studded with clumps of wood, bearing a striking resemblance in its general aspect to some parks of Somersetshire. On Fourteen Tree Plain I saw a great number of parrots and black and white magpies. The charred stumps and fallen trees, with withered, standing ones, presented a desolate scene, flanking a part of the road to the right. Forty-three miles from Hobart Town is Lemon Springs, and Anstey Barton, the residence of Barrister Anstey's father, lying amongst woods to the left. I saw the summit of Table Mountain, bearing north-west. At Oatlands, fifty miles, we breakfasted at the "Oatlands Inn," and there are four others in the township, which is situated on a plain. Whilst breakfast was getting ready, I strolled over the marsh, which is some miles in circumference, a complete mud-level, having the appearance of a drained lake in a sandstone formation. At 8.15 a.m. we started. The face of the country has now undergone a change. Instead of the hilly, rugged, trappean rock formation, we come upon undulating, sandstone slopes, on which large flocks of sheep were grazing. From Antills, fifty-three miles, the road to Tunbridge lies through beautiful, park-like plains, glowing with the golden-yellow hue of the wattle blossom, a beautiful shrub, very abundantly distributed here. Gorze, or furze-bloom, even rivals the wattle in the richness and depth of its yellow colour, scenting the atmosphere all round with its sweet fragrancy.

On the left side of the road, a few miles from Tunbridge, Mr. Harrison's elegant-looking villa, with its verandahs and doorways entwined in flowers and evergreens, the whole embosomed in its gardens well stocked with shrubs, and backed by densely-wooded hills, have a most striking effect. The yellow wattle flourishes here in the richest luxuriance, numerous flocks of sheep feeding on the green pastures. Saltpan Lakes lay to the right of us. At 10.15 a.m. saw the mountain called Quamby's Bluff, bearing north-west from the inn, beyond which lie some plains. In passing through the small township of Ross, we crossed a bridge over the river. Just beyond this is a Mr. Horn's estate, whose son was barbarously murdered by the natives some time ago. Between Ross and Campbell Town is a gravelly plain and open valley, with a fine view of the mountain, Ben Lomond, in the distance. Here, for the first time, I saw two of the large eagle-hawks hovering overhead. We alighted at the " Fox-Hunter's Return," a comfortable inn in Campbell Town, seventy-six miles from Hobart Town. Crossed a bridge over the Macquarie River, and strolled through the township—a mere village. Four ranges of hills seen from this have a remarkable appearance, table-topped, and sloping to the north-east; Quamby's Bluff conspicuous. After lunching at the inn, we started again at 3.10 p.m.

A sudden turn in the road brought us to Epping Forest, the entrance to which is striking and beautiful. An excellent turnpike-road, flanked on either side by lofty, forest trees, taking a straight course for a considerable distance, presents a fine, extended vista in the prospective. But the very sameness throughout so many miles becomes tedious and monotonous before we get clear of it. I shot three magpies, three parrots, and a miner.

All these birds were abundant, crossing the road in flocks. Near the extremity of the forest we took up our quarters for the night at the " Eagle," a small inn on the left side of the road, and the only one in the forest. We alighted at seven p.m., and having had tea and eggs at eight p.m. —it being a fine, moonlight night—I went up the hill at the back of the house into the dense forest, opossum shooting ; but, after searching the tall trees for some two hours for a considerable distance, I only met with one, which I shot as it was sitting on the stump of the stem of a lofty tree, to which it made an effort to cling by its tail on finding itself wounded, but in a few seconds, relaxing its hold, it fell to the ground ; and had it not been shot through the head dead, I should most certainly have lost it.

Tuesday, 13*th.*—Passed through the township of Perth ; we crossed the bridge over the South Esk, and, approaching Launceston, Cocked-Hat Hill, so named from its supposed resemblance to that shaped hat ; it is a very conspicuous object. Launceston appeared skirting the sides of a valley, half-concealed in a volume of white mist and vapour suspended over it, and filling the valley. We entered the town at 8.30 a.m., alighting at the judge's quarters, Government Cottage. I walked up Signal-Staff Hill, and lunched at about noon.

We shoved off from the harbour in a boat manned by four hands. Fine weather on leaving the wharf. The river is formed by the junction of the North and South Esks, and makes a narrow curve in the first reach, afterwards expanding out. At five miles from Launceston is Pig Island. At Fresh-Water Point it contracts again to from half, or even a quarter, of a mile across ; reached this point, eight miles, at 3.15 p.m. The water thus far is quite fresh ; beyond this it widens out to two miles, having the Nelson Shoals on the right and Ship Channel on the left (eleven miles). At 4.15 p.m. passed Cemetiere Point and the " Mermaid Inn." In the next reach is Swan Bay, and at twenty-one miles Swan Point with Egg Island. Mount-Royal signal-station on the hill to the right very strikingly resembles Mount Direction on the Derwent, near Risdon. It is the half-way mark to George Town. At Spring Bay (twenty-four miles) the river again expands considerably. The recent clearance in the woods here, with a log hot or two, marks the commencement of a new settler's life, with some cattle feeding. Beyond this, at thirty miles, is Whirl-pool Reach, which we reached at 6.15 p.m. The worst part of the navigation of the river, being very narrow, leaves bare room for ships to pass. Buoys point out the danger, a ledge of rocks in the centre, having a foaming eddy and ripple over them. After passing several other points and Redwood Island, we entered the last reach to George Town (Moriarty's), thirty-two miles, at 6.45 p.m. The river now opens out to one and a half miles in width, with Middle Island. The evening just closing in and getting dark before we reached our destination, little could be seen of our last stage in the voyage.

The course of the Tamar is far less varied and interesting than the Derwent. The wooded hills, ave-raging 500 feet in height, present the greatest sameness and monotony throughout. The only birds I saw on the passage were four black swans, a few gulls, and a shag. At 8.30 p.m. we arrived at George Town, and passed the evening at Lieutenant Friend's, to whom we had an introduction; and at 11.15 p.m. took up our quarters for the night at Mrs. Wilson's hotel.

Wednesday, 14th.—Strolled into the woods and shot two of the native cuckoos, and, after breakfasting with Lieutenant Friend, R.N., mounted horses and rode out to the lighthouse, four miles, along a rugged, very primitive sort of road, by the banks of the river. As we were approaching a gate, the horse I rode, a wild, untrained, and high-spirited animal, made a desperate effort to run away with me, and, after nearly unhorsing me, I had no small difficulty in curbing him in his headlong course before reaching the gate against which he appeared to be very indifferent about dashing himself and me. We, however, reached the lighthouse on Lowhead at 9.20 a.m. safe and sound. On ascending to the top of the lighthouse, which is situated on a low, sandy neck of land, barely three-quarters of a mile in length, we had a fine view of Bass's Strait, Barren Island, and the coast to the right and left, from the gallery outside of the lanthorn, consisting of fifteen lamps forming a circle in three divisions of five in each. The Tamar is here three miles across.

Soon after our return to George Town, Mr. Friend and his wife accompanied us in their own boat up to Launceston, transferred, however, at Spring Bay to the harbour-master's boat, or sloop, in which we sailed to Launceston, with a fresh breeze, arriving at six p.m. We met Count Strelecki on landing, dined at the "Steamboat Inn" with the Friends, and slept at the Government Cottage.

Thursday, 15*th.*—Breakfasted at Government Cottage, called on the Friends, and at Dr. Pugh's, where we found Count Strelecki at breakfast. I took a stroll through the town; a neat, quiet place, containing some good shops, an exchange and news-room, both of which I visited. It is a smaller and less bustling town than Hobart Town, but situated in the midst of a fertile country. Vessels lay close alongside the wharfs. Judge Montagu having finished the duties of his circuit, we lunched at one, and left Launceston on our return at two p.m. The morning showery, but it afterwards cleared up fine. Had a glass of ale at the "Perth Hotel." The bridge here is built entirely of greenstone. We alighted at the "Eagle Inn" in Epping Forest at 5.30 p.m., where I had a young dead kangaroo given me.

We arrived at the " Fox-hunter's Hotel," Campbell Town, at 11.45 p.m., where we took up our quarters for the night.

Friday, 16*th.*—Started at 6.30 a.m., passing the late Dr. Pearson's estate, which the judge told us was for sale at 50,000*l.*, and that 45,000*l.* had been offered for it. We reached Tunbridge at 8.30 a.m., the stage-coach passing us on the road. Breakfasted at the inn here, when the judge most kindly proposed to send his carriage home and mount horses ourselves, and ride back by the lake region, thus affording us a change of route and an opportunity of seeing more of the country by a return over the Western Tiers.

I mounted a favourite grey pony of the judge, named " Polly;" Phillips a large black horse, called " Cossack;" and Montagu himself bestrode " Comet," a very fine blood horse. The weather was cloudy, with a fresh breeze. Crossing a branch of Blackman's River, we shaped a westerly course, and through a wood, where a flock of black cockatoos were flying heavily along, slowly flapping their large wings, not unlike the lopping flight of our own lapwing, now and then sending forth a harsh scream, as they changed their position from one tree to another. I alighted from my horse, and was tempted to give chase to them, in the hope of securing a specimen or two for the ornithological collection, but I found them far too wary to allow of my getting within shot of them in the brief space of time I could spare to follow them up. On remounting my steed, I had to ride at a smart pace to enable me to overtake my companions, a stern chase being proverbially a long one. We now struck off through thick woods of lofty trees, and up a range of hills of con-siderable height, having a base of greenstone, upon which rested a stratum of sandstone, again capped by green-stone, which formed the summit, and this we reached at three p.m. Our fine weather seems to have deserted us for thick mist and drizzling rain, denoting that we were entering upon a mountain region. Saw a kangaroo in the woods. At 3.15 p.m. we skirted the margin of Lake Sorrell, a fine sheet of water some five or six miles in length, and about the same in breadth at its widest part. Towards its southern extremity its form becomes irregular, from the number of wooded promontories jutting into it. Off the south-west shore is a small islet, on the opposite side to that on which we made the lake. Saw two of those majestic birds, the black swans, swimming on it. The margin of the lake is wooded all round ; and as we rode round to the right, sometimes passing through woods, at others over fences, or along the little, sandy beaches or coves of the lake, eight more black swans (*Cygnus atratus*), with a few ducks and divers, made their appearance. Between its southern extremity and Lake Crescent is a flat, marshy tract, not half a mile across at its narrowest part, covered with swampy grass, through which runs a very narrow stream which connects the two lakes. Lake Crescent is about four miles in length and two in breadth, lying across the southern extremity of Lake Sorrell.

We now had to ride through a tangled mass of trees and bushes and over log fences to Mr. Kemp's stock-hut, at the south-west corner of Lake Sorrell, which we reached at six p.m., amidst heavy continuous rain, drenched to the skin ; the weather being thick, muggy, and heavy. The hut is a small, low building of logs, with shingle roof and a small square window or two, formed of four small, square panes of glass each. Opposite was a stable, in which we put our horses for the night, amid the loud barking and yelping of some ferocious dogs.

Saturday, 17*th.*—Having taken a hasty breakfast, we continued our journey at 8.45 a.m. Our track lay through a swampy ground of grass. We now shaped a south-westerly course, through a wild and thickly-wooded

country, for the township of Bothwell.   Near a stock-hut
we crossed a shallow river, and finally lost all traces of a
track; for a few minutes we were undecided what course
to follow, when we came upon a cart-track, the impression
of the wheels barely perceptible on the grass.   We had
now to leap over log-fences, formed by trunks and stems
of trees placed horizontally on each other, frequently
having to remove the uppermost one before the horses
would take them.

At 12.30 p.m. we descended a very steep hill of trappean
formation, down which we had to lead our horses, and
with great caution, to prevent their slipping down.   We
crossed the river Clyde in our course.   The ground was
generally swampy, and clothed with long grass; in many
places the trees stood so thick that it required some
tacking and manœuvring not only to steer clear of the
overhanging branches, but at the same time to look out
that the knees did not come in contact with the trunks as
we trotted through the woods.   We entered the " Bothwell Inn " at two p.m., most thoroughly drenched, it having
rained incessantly throughout our last journey.   We had
ridden not less than thirty miles.

Sunday, 18*th.*—I took a stroll through the little township, which consists of some half a hundred houses, half
a dozen of them respectable-looking ones.   There is
another good inn, the " Crown," in the place, opposite
to the " Bothwell Castle."   The church is a white edifice.
The township lies in an almost circular valley, bounded
by a range of hills of moderate height.

At 10.15 a.m. we again mounted our horses.   The
surface of the country now much improved in appearance,
swelling into picturesque and wooded hills; we saw a
number of green paroquets.   At 2.30 p.m. we alighted
from our horses on the summit of a fine, green, swelling
hill, commanding a charming prospect around; the
township of Hamilton spread out in the centre of the

green valley beneath us, with the road winding down it, the church being the most prominent object. I rode down the descent on " Comet," having exchanged horses with the judge. It was quite a pleasure to ride this high-spirited blood horse, as he dashed with me ahead of the party. At 3.15 p.m. we lunched at the " Hit or Miss Inn," and left it at five p.m. From the hill above had the finest view of the grounds of Laurenny stretching along a beautifully wooded valley, terminated by a large mansion at an angle of the hills. At dusk, 7.20 p.m., had tea ; and took up our quarters for the night at the " Woolpack Inn."

Monday, 19*th.*—Rose at 6.20 a.m., and after breakfast continued our route, and paid a visit to the celebrated fossil tree at the extremity of a narrow ridge of scoriaceous rock some seventy feet above the river, a stream about twelve feet across, winding through a wooded ravine 100 yards wide. The tree is of silvery white, with a yellowish, brown, resinous colour at its base. It is six feet in height, seven feet three inches in circumference, fifteen inches in diameter at the top, having an exterior covering of a loose, flocculent, siliceous substance, and embedded in vesicular lava. The rock has been excavated around it, and on either side are a few stunted bushes.

The second tree had been completely enclosed in the scoriaceous rock, forming a steep cliff, a short distance only to the right, twenty feet above a curve in the river, which here is not more than ten feet across, winding through a grove of trees in a valley sixty yards wide. Only a portion of this fossil tree remains at the top of the chimney-shaped orifice in the rock. It has a very opalescent appearance ; the lower portion has been removed, leaving the impression on the surface-sides of the lava for about seven feet downwards, and a foot in diameter. Below this again, on the shelving portion of the rock, a fragment of the lower part was embedded in the soil, the whole trunk having been in a vertical position. The summit of the cliff itself was about forty feet above the river, having a green sloping hill opposite. Regaining the road, we continued our route for nearly a mile further to Mr. Barker's, the proprietor of the fossil trees. His house is situated in a valley where the winding of the river forms a peninsula in its reach. At 11.15 a.m. Mr. Barker accompanied us over the green hill above the house to a reach of the Derwent, about two miles distant, to see the spot where two fossil trees had formerly been embedded in lava on a low bank ; both had been in a vertical position. A long, low, wooded island occupies the middle of the river here, lying parallel with its banks. We returned to the house at one p.m., where we dined. At 2.15 p.m. we departed for New Norfolk along the bank of the river nearly the whole way. At 3.20 p.m. passed a sandstone cliff twenty feet in height, on the left side of the road, dipping to the westward at an angle of 40°. Here very singular masses of greenstone, nearly cylindrical in form, and from eighteen to twenty inches in diameter, and six feet in length, lay embedded in the sandstone, pointing downwards on the road, like a tier of cannon in a battery. In some spots dark shades, like black dust scattered over the cliff, would appear to indicate traces of coal.

I rode " Comet " into New Norfolk, where we took up our quarters at five p.m. at the " Bush Inn," having crossed the river over the floating wooden bridge. New Norfolk, although pleasantly situated on the banks of the Derwent, looks like a deserted village, so dull and quiet is it. The steamer and the stage-coach arrived here about the same time from Hobart Town.

Tuesday, 20th.—Rose at five a.m, and mounting " Comet," we started for Hobart Town; the road diverges from the river for about a mile, then again following it closely for most of the way, passing several sections of sandstone and limestone on the cliffs to the right. Passed the " Black Snake Inn," and several villas, through New Town down the lane to the river, and across the paddock to the observatory, where we alighted, and took our leave of the worthy judge, through whose kindness we have had an opportunity of seeing so much of the country in the short space of nine days. We much enjoyed our excursion, notwithstanding the very unfavourable weather returning. Met Sir John Hammett, a brother officer of mine, at the " Derwent Tavern." Captain Ross having a party of friends on board to dinner, we could not get a boat to land us till 6.30 p.m., and were consequently too late for dinner at the army mess, to which we had been invited.

Wednesday, 21*st.*—Showery weather. I breakfasted with the Friends at the " Freemason's Hotel," and at ten p.m. went to the governor's ball, and left at one a.m., night fine. Called alongside *Terror,* and returned on board at two a.m.

Friday, 23*rd.*—Fine day. Went on shore and ordered a new gun-stock to replace the one I had found broken at one of the inns on our journey, but how I knew not ; it vexed me, being my old pet stock, which I had before broken at Spitzbergen, in knocking down my first reindeer, and had had repaired on board. Having received through Captain Ross an invitation from Sir John and Lady Franklin to accompany them in their yacht on an excursion to Port Arthur, I went to Captain Moriarty's, the harbour-master, and after taking tea with his family, at seven p.m. accompanied him on board the yacht from his own house. We got the yacht under weigh, but the breeze failing us, made no progress.

Tesselated Pavement, Eagle-Hawk Neck, Tasmania. (*See page* 124.)

CHAPTER XI.

Excursion to Port Arthur—Inspection of convicts—Dog sentinels—The tesselated pavement—The coal-mines—Back at Hobart Town—A funeral procession of boats—Complimentary ball—Laying the foundation-stone of the new Government House—Farewell to Tasmania.

Saturday, 24th.—At 8.30 a.m. we breakfasted; and at 9.10 a.m. I accompanied Captain Moriarty in a boat to board the *Janet Willis*, coming up the river from England, which she left on the 8th of July; but had nothing for us in the bag of letters and newspapers we examined on her deck. After lunch, the sea-breeze taking us to Norfolk Bay, we all left the yacht and crossed by the railway, a very narrow road cut through the thick woods, along which we were pushed in small carts by a number of convicts. I rode part of the way and walked the remainder. At six p.m. we passed the halfway house, and it was dusk when we embarked in the boat on the opposite side, where the lights of Port

Arthur appeared; landed at eight p.m.; went to Captain Booth's, the commandant, where we dined. The Governor inspected the whole of the convict establishment here, and we accompanied him and the commandant all round. The prisoners were arranged in tiers along the sides of the room, each narrow crib being separated by a low board. A light is constantly kept burning, and an overseer on watch throughout the night. What lamentable specimens of poor humanity, crime indelibly stamped in the vile physiognomy of each! Several of us visited Lieutenant Kelly's (of the 51st) quarters, who is in charge of a detachment of the regiment here.

Sunday, 25th.—In the afternoon several of us walked round Point Puer, and I examined the argillaceous cliff here, containing several kinds of fossil shells; and in returning near the signal-station, I had an opportunity afforded me of seeing the beautiful, pure white hawk of the island in his wild haunts flying overhead. Here were several of the shrubs called the native plum, with a variety of wild flowers. At seven p.m. I dined at the commandant's with Sir John Franklin and suite.

Monday, 26th.—Started at 6.45 a.m. from the wharf, where a guard of soldiers were drawn up to receive the Governor. We embarked in three boats, and at 7.30 a.m. landed, where the railway-cart took us to the half-way house. I walked occasionally and saw a great number of wild flowers along the sides of the road, and a flock of black cockatoos. The woods were very dense, some of the trees very lofty, having trunks as erect as a ship's mast. At nine a.m we again embarked in the boat, went alongside of the schooner-yacht, and shoved off again in about ten minutes for Eagle-Hawk Neck, a low, narrow, sandy isthmus, connecting Tasman's with Forester's Peninsula, having on one side the sea and Pirate's Bay, and on the other an inlet from

Norfolk Bay. Up the latter inlet we pulled the boats;
it is wooded on either side.

Reached Eagle-Hawk Neck at ten a.m., and landed at
a long wooden pier, amid the loud barking of eleven
dogs, nine of which were chained to stakes in the ground,
and two chained on a platform, erected over the water
on piles, forming a line of ever-watchful sentinels across
the low sandy hollow, for about 100 yards or so, the
breadth of the neck between two ridges. They were
the most ferocious-looking brutes I ever saw. Dashing
at us, as far as the length of their chains would permit
them, with deafening howls, and barking as we passed
them to the opposite beach—a hard, fine, white sand,
to examine the so-called—by the colonists—" Tesselated
pavement." This freak of nature forms indeed a most
remarkable specimen of her works. Turning to the left,
along the beach of Pirate's Bay, at 10.30 a.m., we reached
the argillaceous cliffs, abounding in fossil shells, and at
their base the *débris*, instead of forming a talus, had
been, by the action of the waves, where the sea breaks
over the fallen rocks, formed into a platform. This plain
surface of the argillaceous rock is divided by lines, arranged with the most geometrical precision, forming the
siliceous clay, of which it is composed, into symmetrical
slabs, varying in length and breadth, frequently having
their margins bordered in strong relief. The dimensions
of those I measured were from three to nine feet in
length, and from four to eight inches, or even six feet in
breadth. Others formed squares of eighteen inches.
These divisional planes had a general bearing of E. by
N., with a perfectly geometrical parallelism in relation
to each other. The curious structure here displayed
may probably be due to the agency of some electro-
magnetic forces, acting upon the atoms or molecules of
matter whilst passing into a solid condition in the process
of cooling; thus, giving a definite direction to the

ordinary partings which argillaceous deposits so frequently present, when contracting under sudden changes of temperature, during their consolidation. At the first glance its marvellous symmetrical appearance, so like a work of art, is very striking. Spiriferæ and other similar fossil shells are most profusely embedded in this deposit, forming quite a mosaic pavement.

At 2.45 p.m. we landed on the west side of the bay, to visit the coal-mines. The first is open between basaltic rocks, assuming the columnar form. The shaft is sunk to the depth of fifty-two yards, down which I descended in a basket. The galleries are very low and narrow, swampy and muddy, and worked underground for about 300 feet. The seams of coal are from four to six feet in thickness, and overlaid by a stratum of sandstone. Sixty men are employed in this mine, and the average daily quantity of coals raised from the pit amounts to forty chaldrons.

We walked round to the second mine, lately opened near the beach; the coal here is so near the surface no shaft has been sunk, but a tolerably wide tunnel has been excavated in a straight line for forty-seven yards in the sandstone and argillaceous rocks, beneath which lies the seam of coal, of a better quality than in the first mine. It has not yet been worked to its entire depth.

Tuesday, 27th.—Off the Iron Pot Lighthouse this morning; reached Hobart Town at 9.30 a.m. As I left the yacht, at ten 10.15 a.m., passed nine of our boats in a line, forming a funeral procession, with the body of our poor old captain of the hold, who had unhappily lost his life by falling into a water-tank in the hold, being suffocated in the foul air. I received an invitation to dine at Government House on Monday. The *Mary Hay* sailed for England to-day, with our Natural History specimens on board.

Thursday, 29th.—Mr. Gregson of Risdon, and Lieutenant Breton, R.N., of Richmond, dined on board with me ; and after landing them at nine p.m., I went to the ball given by the inhabitants in the new custom-house rooms, to the captains and officers of the Antarctic Expedition. I entered the ball-room at ten p.m., soon after which the Governor, Sir John, and Lady Franklin, and suite arrived, and with them the captains. The ball-room was lined with flags on the left side, two colours, with the arms of Captains Ross and Crozier.

We breakfasted at nine a.m. at Lieutenant Breton's, and saw his fine collection of fossils and other native specimens of the island.

In the afternoon we sailed in the whale-boat for the *Erebus,* getting on board at 5.30 p.m. Gregson and Breton dined on board with me, and Dr. Clarke, the Inspector-General of Army Hospitals, a party of thirteen in all sitting down to table in our small gun-room to-day.

Monday, November 2nd.—Landed with Gregson, who slept on board in my cabin last night. We went to a Mr. Crombie's, a friend of his in Hobart Town, from which I accompanied Mr. Gregson and his family in their carriage to the Paddock, where I had a boat in readiness to take them off to see the ship. At 3.30 p.m. I took them on shore again in the galley, and, after showing them the Observatory, I returned on board to dress for dinner at Government House, which took place at 6.30 p.m. My friend Gregson was present, and also at the Natural History Society's Meeting, when a paper on skulls was read by Mr. Bedford, and the leader on Magnetism in the last *Quarterly* was read by Sir John Franklin. I got on board again at one o'clock in the morning.

Thursday, 5th.—Saw the first stone of the new Government House laid by Sir John Franklin in the Paddock. Both the captains were present, with the military, and a large concourse of people. The two ships were dressed in flags, and salutes fired from them ; after which there was a dance and supper at the Pavilion.

Saturday, 7th.—The *New Norfolk* steamer was taken by the Governor to-day, for the purpose of conveying a large party up the river to see the foundation-stone of the new college laid. The two captains and the officers of the expedition had tickets given them for a passage in her, Sir John and Lady Franklin, with their daughter, proceeding by land. The steamer started from the old wharf at nine a.m., with her decks crowded, and having a fair proportion of ladies. Sir John Franklin went through the customary form of laying the stone, not far from the Government Cottage, making an appropriate speech on the occasion. After the ceremony was concluded a large sheet of paper was laid upon the table, for the signature of those disposed to annex their names, so I added my own autograph to the rest. Returned to Hobart Town at 7.30 p.m.

Sunday, 8th.—At 2.30 p.m. I started for Risdon, across the Paddock. Reached the ferry at 3.30 p.m., and Gregson's at four p.m. The beautiful situation of this mansion, on a rising knoll, commanding so fine a view of the Derwent, the grounds of New Town, on the opposite bank, with the fine background of Mount Wellington, the smooth water of the creek, like a lake, beneath the house, and flanked by woods, together with the fineness of the evening, induced us all to have our tea on the lawn.

Tuesday, 10th.—The ship was yesterday unmoored ready for sea, but thick weather and calms prevented our sailing. Sir John Franklin came on board and dined with us in the gun-room. We had rather a large party, Captains Ross and Crozier, Mr. Gregson, and Captain Foreman, being of the number.

Wednesday, 11th.—The state of the weather detaining us at anchor, at 4.15 p.m. I landed and walked through the Paddock to the ferry, which I reached at 5.25 p.m., and Risdon at six p.m., to spend my last evening with my excellent friend Gregson and his charming family, under whose roof I had passed so many happy hours, and looked upon as a home, in the brief intervals of leisure my own multifarious duties in such an arduous service as we are engaged in would afford me. I found my friends at tea, after which we passed a delightful evening in music and singing, and after supper I took my departure at 11.30 p.m., with a present of some home-made raspberry vinegar and ginger-beer, young Gregson accompanying me down to the head of the creek, where his boat was awaiting me to take me on board. I got alongside the *Erebus* soon after midnight, a beautiful moonlight night, and the wind changed in our favour.

Thursday, 12th.—We at last got under weigh at five a.m., and hove-to off the town for Sir John Franklin, who came on board about seven a.m., and we stood down the river, accompanied by the *Eliza* yacht. At 1.30 p.m., when the pilot left us, the Governor took his leave of us, and as he descended the ship's side into his boat our rigging was manned, and three loud cheers given him, which, on his getting on board his yacht, was returned from her. To-day we commenced our sea-hours of dining at three p.m. Weather fine, but cloudy till evening set in, when rain fell.

We have now fairly bid adieu, for some months at least, to the shores of this lovely island, and with deep regret. Its fine climate, beautiful and varied scenery, together with the boundless hospitality of its inhabitants, had made our sojourn here a delightful break in our voyage.

An almost impervious Ravine in Auckland Islands. (*See page* 132.)

CHAPTER XII.

First voyage towards the South Pole—Auckland Islands—Enderby Island—Excursion up an almost impervious ravine—Botanical features—Ewing Island—Sandy Bay—Albatross eggs—Ornithology of the islands—Departure.

Friday, November 13*th*, 1840.—We have now entered upon the first interesting portion of our expedition, to pass our summer, or what goes by the name of summer in this hemisphere, amid huge packs of icebergs and a glaciated land. Our future for the next few months is so exceptionally novel, so full of interest, so promising in the prospective of great discoveries in a region of our globe fresh and new as it was at creation's first dawn, and to the more sanguine and enthusiastic there is the possibility at least of unfurling our flag at the ice-girt Pole itself.

Sunday, 15*th.*—Night fine. Saw the aurora australis shooting upwards to the zenith during the first watch.

Friday, 20*th.*—Saw the land, about seven a.m., beating up along Enderby Island, one of the Auckland group distant between 800 and 900 miles from Hobart Town, and at one p.m. came to an anchor in a large harbour called "Sarah's Bosom," having a deep creek running up to the left. At three p.m. I landed opposite to the ship in company with two of my shipmates. On first landing I shot two small ash-backed gulls at one shot. Saw only two small green-finches in the dense thickets which clothed the hill-sides, through which I made my way with great difficulty, rather by crawling than walking, for some two miles. My two companions, soon tiring of this sort of work, not having the same spur to exertion which my own natural history pursuits afforded me, returned to the beach.

The trees scarcely exceeded twenty feet in height, and were so interlaced and matted together at the summits as almost to exclude the light as well as the sun's rays. Beneath was a rich, dark, deep, elastic, vegetable mould, clothed with a rank and dense underwood, creeping plants, grasses, mosses, and lichens. Some of the larger trees were growing in a nearly horizontal position, the trunks covered in the greatest profusion with lichens and mosses, beautifully embossed by them. I descended a ridge on emerging from this tangled forest inland, returning along a valley on the right clothed with thickets of underwood shrubs and long grass; reached the beach at 4.30 p.m., and as it now began to rain, I returned on board with my two shipmates, whom I found awaiting my return on the beach.

Monday, 23rd.—Landed at 7.45 a.m. to examine the extent of the large creek on the left side of the harbour, called Laurie Harbour. I first made my way through the densely tangled thickets over the point, reaching the creek at nine a.m. The trees here became much taller, and the trunks of greater size, perfectly embossed with the richest covering of cryptogamic plants, both mosses and lichens abounding in the greatest profusion I ever saw. Ferns too grow in the utmost luxuriance. Here and there large trunks of trees completely encased in cryptogamic plants lay prostrate over the narrow, muddy channels of water which empty themselves into the bay, forming natural bridges. On ascending the hill-side from the inlet, the trees became replaced by bushes, in some places forming almost impenetrable thickets, denser than the most tangled white-thorn hedge. From the spot I had reached I saw our boat up the inlet, and at once made my way through the belt of wood to its shore.

Throughout this excursion I met with scarcely any signs of animal life; it was almost the silence of an Arctic solitude; all so still save the rather pleasing, plaintive, low note of a small olive-green finch and blackish-looking bird, about the size of a starling, sitting on a bush at too great a distance to discover its species, but probably belonging to the promerop's family. On the beach I shot a black-backed gull just as I was about to enter the boat. A brown-coloured duck and a merganser frequent the harbour. After pulling down the creek to the bay, we made sail on board, reaching the ship at noon.

Thursday, 26th.—Blowing fresh, with squalls of sleet and snow at intervals. At nine a.m. I left the ship in the cutter, landing first at Pig Island, where we had landed our porkers. The rock is basalt, highly magnetic, and clothed with underwood. From this I crossed over to Point Deas Bluff, and landed there at 10.10 a.m. Ascended to the summit from the south-west side. It is formed of prismatic columns of basalt, highly magnetic, and only accessible through the help afforded by the long tufts of coarse grass growing out of the fissures in the rock, and these I found rather treacherous to trust to. A clump of trees and bushes, intermingled with ferns and other herbaceous plants, crowned the summit. One very pretty purple blossom, a gentian, peeped out from the rocky fissure.

I had a fine view of an inlet in the coast on the opposite side of the island, flanked by a black headland. I returned by an easier descent on the other side of the head of the bay. Shot a shag, a black-backed gull, and a fine falcon on the beach. I met with a few limpets and mussels, and returned on board at three p.m.

Friday, 27th.—Our two cutters having left the ship at eight a.m., to haul the seine for fish up the inlet, I availed myself of the opportunity it afforded me to land at the upper end of it, without having to make my way through the dense thickets over the point. My intention was to follow the narrow stream or river which apparently runs from the top of the creek along the centre of the ravine to the opposite coast, for which purpose I left the boat at 9.30 a.m., and shot one of the small olive-green birds on landing.

I now followed the course of the running stream, which at its broadest part is only fifteen feet across. For the first half-mile it runs through a thicket of trees, ferns, grass, and rank vegetation. My course after this kept diverging from the banks of the rivulet till I came to some very thickly tangled bushes, indeed as impervious as a thicket of thorns to get through, and so matted and interlaced at the top that I sometimes had to roll myself bodily over their flat summits. I crossed numerous small feeders trickling down from the hill-sides to enter this small river, winding its devious way through the bottom of the valley, sometimes heard gushing along in narrow channels or gullies, quite screened from sight by the overhanging, thickly tangled bushes, ferns, grasses, and mosses, and only warned from slipping into the stream beneath by the murmuring sound of the running water. This sound, indeed, was often my only guide in finding its tortuous curves, concealed as it was beneath the superabundant vegetation, but heard at some distance. When I had made my way with no small difficulty as far up the valley as between two and three miles, I crossed an open space of bog, a black peat, free from bushes, for a short distance, and having the blackened aspect of coarse grass burnt to the roots. This again was succeeded by long grass and thickets of bushes. When I had proceeded for about three miles, I saw before me spread out for some distance onwards a lovely orange-coloured patch produced by a large bed of the wild tritoma (*Chrysobactron Rossii*) in full bloom, the rich golden-yellow flower spikes contrasting with the deep leek-green of their long liliaceous leaves and the framework of ferns, grasses, and short thorny scrub in which they were set, combined to produce a charming effect, much heightened, too, by the vicinity of an umbelliferous plant having flowers of a purplish hue. The bottom was of so swampy a nature that I sometimes sank knee-deep in the morass, rendering all progression most laborious and toilsome.

After crossing a low wooded ridge I came upon a bend of the river on my right, foaming and murmuring along a deep channel, some six feet across, but entirely concealed from view by the overhanging densely tangled thickets. At eleven a.m. I plunged into an impenetrable bush, which appeared to extend for at least a mile ahead, but seemed bounded by a hill; crossing the valley, this was partly enveloped in mist. It being now noonday, with the uncertainty of being able to reach the opposite side of the island by this course, and the utter impossibility, even if I did, of reaching the ship before nightfall; this prospect, together with the heavy rain which had been incessantly falling since I started, accompanied as it was by a cold, chilling state of the atmosphere, which all along had been overcast, thick, and gloomy, determined me on bearing up, and trying some more eligible route on some other day to carry out my plan.

I was now literally drenched to the skin, with the rain and the water constantly falling on me from the trees and bushes, as I forced my way through them ; my heavy water-boots, too, were sodden, and filled with water from frequently sinking above their tops in wading through the swampy bog. In returning I followed the course of the river as nearly as I could, reaching the top of the creek of Laurie Harbour at 2.30 p.m., and after wading waist-deep across the mouth of the river, entered one of our boats, and got on board at 3.45 p.m.

Sunday, 29th.—Accompanied Abernethy, our worthy gunner, on shore for a ramble, when we fell in with two large hogs in the thickest bushes, and I saw a falcon, ringed plover, two larks, some other small birds, and a number of gulls on the point. The pigs, originally left on the island by Captain Bristow, are now very numerous, although exceedingly difficult to find from the dense cover of vegetation. They live chiefly upon the roots of the *Pleurophyllum criniferum* and the *Arabia polaris*, a very remarkable plant bearing large umbels of waxy flowers, both plants abounding all over the island. We landed on the island at the point beyond Deas Head, and returned on board at 4.30 p.m.

Monday, 30th.—Left the ship at seven a.m. to examine the breeding-places of the albatross (*Diomedia exulens*), landing at Ocean Point. We hauled the boat up a narrow creek in the basaltic rocks, which have the columnar form. I found a gull's egg, and shot a green bird and a shag at the point. At nine a.m. sailed for Ewing Island. The rocks were covered with shags. I climbed over these precipices, through the thick bush-grass and rank vegetation, to a low ledge of rocks, on which the sea breaks, and shot an albatross flying overhead, which fell winged among the bushes on the hill above ; and on going up the ridge to pick it up, I found three of the young of the great petrel, and shot a mature one and some shags. At eleven a.m. I started for Enderby Island, and there caught two old albatrosses, sitting together among the long grass flanked by bushes; they were doubtless about selecting a nesting-place. These birds are now in my own collection.

I afterwards landed at Sandy Bay, a pretty white sandy beach of crescentic form for a quarter of a mile, bounded by trappean rocks and crowned with wood. Above the centre of the beach is a hollow, filled with long grass, growing in a rich, boggy soil in such rank luxuriance as to be up to the hips, and flanked by a sand-hill clad to the summit by the same kind of grass, having the whole skirted by a thicket of trees and bushes. I found snipe in the long grass, which were very difficult to flush, and then rising close under the feet, darting down again immediately only a yard or so away, rendering it a somewhat difficult matter to secure specimens without having them shattered to pieces. I shot a brace, however, and saw a paroquet or two. Started at 3.30 p.m. and landed at the "Bluff" on the left, upon a low ledge, covered with chitons, mussels, and limpets. Shot a shag. All around is an inaccessible wall of basalt.

Tuesday, December 1st.—Started in a boat under sail for Sandy Bay. Saw a penguin, two paroquets, and a snipe there. Shot three small larks and black cap (*sylvia*), and a brown gull. Found a new plant.

Thursday, 3rd, 7.20 a.m.—I landed at the observatory, and dug out a blue petrel and egg from a hole in the bank, under some short bushes. I returned on board in the boat with the seine; the two cutters afterwards started at 10.15 a.m. for Sandy Bay to haul the seine, and I took a passage in the first cutter with Abernethy, and on landing I shaped a W.N.W. course through the thickets, following a watercourse over the hill till I reached the

grassy summit above the brushwood. Here I saw a
pair of albatrosses sitting near a nest without an egg,
and crossed to the opposite side of the island, coasting
it round to the right along the long grass, with here
and there a few flowers. In an hour and a half I had
picked up a dozen albatross eggs ; only in one solitary
instance did I find two eggs in the same nest, and these
two eggs were of an exceptional shape, one being more
rounded and the other much more elongated than in the
normal form, and were evidently twins, and laid by the
same bird—no unusual occurrence amongst mammalia
having one offspring as the ordinary number. The
great naturalist, Cuvier, had been under the erroneous
impression that the albatross laid more than one egg ; the
opportunity these islands have afforded me of examining
so many nests during the season of incubation enables
me definitively to assert that, like the petrels, to which
they have a strong affinity through the form of the beak,
one egg is the normal number. The fine white head and
neck of the bird, appearing above the grass as she sits on
her nest, would lead to its discovery at a great distance,
even did she not betray the having an egg beneath her
by so pertinaciously refusing to quit it till pushed off,
bravely defending herself and her treasure with her beak,
snapping the mandibles together with a sharp, ringing
sound, making no attempt at flight unless driven off by
force, when she waddles off in the most awkward and
grotesque manner, floundering with outspread wings
amongst the long grass, often rolling over and over if
any ruggedness of the ground obstruct her progress. I
returned to the beach through a tangled thicket of
trees, bushes, and long grass down the rocks, at no small
risk to my freight of eggs, each averaging above a pound
in weight, carrying such a fragile cargo for about a quarter
of a mile, till I came upon the hollow of long grass, the
haunt of the snipe, and where I shot three, and a paroquet

flying overhead, which, falling in the long grass, I had some difficulty in finding. I reached the boat at 3.15 p.m., where I had to wait till 7.45 p.m. before we got all our party together, blowing and raining all the time, during which I shot a small gull and two green birds on the beach, and then had to pull against a strong head-wind, coming in sudden gusts, with thick weather and a short head-sea breaking over the boat, the moon just making her appearance over the land. We got on board at 9.45 p.m., having made a good day's work of eggs.

Monday, 7th, 5.30 a.m.—I landed in the whale-boat, and crossed the island from the observatory to the opposite bay, through the usual tangled thickets and long grass; reached the bay, which is divided in two by a ledge of rocks. At 6.15 I passed a number of pig-tracks, and found abundant traces of them everywhere in the woods, in the turned-up soil, and fresh leaves of plants just torn off and strewed about. The impressions of their feet in the mud of a watercourse above the way, formed a regular track made by numbers of these animals. Yet I did not hear or see a single one. The steep rock on the right side of the bay was hollowed out into a cave, and full of gulls. I crossed the stream and ascended the wooded hill on the opposite side, returning to the cliffs along the coast, which, being covered only with burnt grass, presented a black and charred aspect. I now shaped my course round the extreme point of the island, by the Arched Rock, through a rank ground vegetation of ferns and umbellifera to Deas Head. There I shot a specimen of the Tui, or parson-bird; and a few hundred yards inland of this and the bay I passed a small lake or pond, some twenty-five yards in length, in the centre of an almost impenetrable thicket, a narrow, deep water-course running from it into the bay. I got on board through wind and rain at between five and six p.m.

Saturday, 12th.—At 5.40 a.m. we took our departure

from the Auckland Islands.   I had been employed for
the last few days in skinning albatrosses and other birds
for the Government collection.   In going out of the harbour the fine perpendicular columns of basalt appeared
very conspicuous and to great advantage on either side
of Sandy Bay.

CHAPTER XIII.

Campbell Island— Both ships aground—Albatross and lestris shooting—
An albatross wooing—A dull Christmas Day—Our first iceberg—
Ocean birds—Bagging specimens while under full sail—Sunrise at
midnight—In the ice-pack—Our Twelfth-cake—Man overboard.

Sunday, December 13*th.*—At eight a.m. Campbell Island
in sight. Had to beat up in thirty-six tacks to the top of the
harbour, which is very deep, and there being barely room
for going about in some places. We grounded before
we reached the top in two or three places ; thus losing
two hours before we got afloat again, by warping off the
shore. We anchored at five p.m. ; albatrosses and gulls
numerous.

The *Terror* ran aground off the point, and did not get
afloat again till the middle watch this morning, Monday.

14*th.*—I landed in the second cutter on the right
point of the creek, and walked to the head of it along
the beach, which was thickly strewed over with quartz
pebbles. Here a narrow stream, which may be leaped
across, pours its waters into the creek. The valley
through which this stream runs in its descent is clothed
with withered-looking, blackened bushes. Ascending a
hill on the left, through a thick fog, at 10.45 a.m., I
reached the base of a steep rock, and under the lee of
this the conspicuous white head and neck of the albatross appeared through the fog only a short distance off,
and I found the first egg on this island here. The nest
is a very simple affair, formed of withered grass and leaves

matted together, and intermingled with a mound of soil eighteen inches in height, two feet and a quarter in diameter at the top, and six feet in circumference at the base. The albatross is frequently found sleeping on its egg, with its head under the right wing. After taking the egg from the nest, I scattered some Kerguelen Land cabbage-seed in it, as the soil appeared to me favourable for its germination. I found six eggs hereabouts, and descended to the other side of the hill, to the base of a ridge of rocks. The fog just now beginning to lift, the harbour appeared below me, and the inland ridge of hills which appear to separate it from the opposite coast, for which I directed my steps at one p.m., and saw the sea with only a deep valley between; I found five more albatross eggs here. A brown skua-gull, or lestris, alighted close to my feet, and appeared to be very desirous of sharing the spoil with me. This bird, I have no doubt, both from its audacity and manner, is the albatross's worst enemy, never losing an opportunity of robbing the nest when the albatross leaves it, for however brief an interval. I have myself noticed it always on the alert prowling around.

Here I passed a small lake or pool of water, some thirty feet in length, with a rivulet running under the thickly interwoven short scrub of ferns and long grass to the river, abounding in those beautiful orange-yellow spikes of the tritoma which I met with at the Auckland Islands, together with the large purple umbels. Indeed, the whole flora bears a striking affinity, and many of the species are the same. Descending the ridge through the grass-covered valley, I reached the shore at 2.45 p.m., a steep, perpendicular cliff overhanging the sea.

The whole outline of coast presented a bold, wild, rugged chain of precipices, with points jutting into the sea; six rocks rising above the surface of the water beneath, of which the two nearest would appear to have fallen, from corresponding excavations in the cliffs on which I stood. To the left a small bay runs in, flanked by a remarkable barn-shaped hill, having a gable-like end ; another hill on the right, having circular terraces to the summit, much resembled the trappean terraces of Kerguelen's Land. At three p.m. I returned over a saddle between two hills, where I found four more albatross eggs, and saw four of the birds grouped together ; and one flying overhead, being out of the reach of shot, I brought down with a ball one of the barrels of my gun was loaded with, by breaking its wing, needing it for the collection. At 4.30 p.m. I reached the boat, and here shot three of the albatross's foe, the lestris, and got on board at six p.m.

Wednesday, 16*th.*—Landed at the foot of the hill on the starboard side of the bay, opposite to the ship. I reached the summit of the hill in about half an hour, a grassy slope studded over with a few stunted bushes. Passing along a perpendicular wall of rock, I reached the peak which crowns the summit of the range ; this is a bare, inaccessible pinnacle of basalt overhanging the deep valley on the opposite side, a bay running in from the sea on the right. I continued along this ridge, or backbone of the range, in the direction of the sea. The albatrosses were breeding in great numbers in the long grass covering the land. Their beautiful white necks seen above the grass as they sit on their nests, studding the hills in all directions. Now and then the male bird may be seen standing by the nest alongside of its mate, but more frequently the female is alone, and not unfrequently sleeping. In all the nests that I examined here, and they were not a few, each contained but one egg.

On the summit of the ridge I had an opportunity afforded me of witnessing an albatross wooing. A group of thirteen of these birds had assembled together, evidently for the purpose of selecting their future partners, and so intently were they engaged in this ceremony, that my own approach was quite unheeded by them; I silently stood for some time within a few paces watching them. It was quite a pantomime of billing and twisting of necks into various curvatures, without a note or sound escaping them. Several others were flying overhead as they arrived from the sea, rapidly cleaving the air with their huge pinions, anxious to join the coterie below them. It was a lovely day, a brilliant sun in the clearest blue azure sky, with the wide expanse of the surrounding ocean of the deepest blue. I passed by two or three small lakes, or rather lakelets, about, perhaps, 100 yards across, and had a fine view of the harbour with the ships beneath me, from a cliff near this. Descending from the ridge for about a mile, through long grass, bushes, and short scrub, the ground in places carpeted, with an elegant small fern, which several albatrosses had here selected as a cover for their nests, often seen sitting in groups of two or three, and within a few yards of each other.

On the ridge several lestrises fiercely attacked me, circling round my head, and darting with open beak at my face, so close as nearly to strike me with their out-spread wings. This very clamorous proceeding left me in no doubt whatever as to my close proximity to their nests. Indeed, after a very short search, I found a young one, covered with down of a yellow colour, and squatted amongst the long grass; and not far from it, in a slight depression of the ground, a solitary egg, a valuable find, as I believe it to be a new species; and to have the whole complete needed the addition of the parent birds, which the cries of the young one, on being captured, rendered desperate. Indeed, to get rid of their annoyance, in self-defence, I may say, I was induced to shoot, not only the old pair, as I supposed, but a third bird, that became

Edwin Wilson, del.

Vincent Brooks, Day & Son, Lith.

Antarctic Lestris or Skua (Stercorarius Antarcticus),
with its young and egg.

The bird standing is a new species
or variety of the same.

PAGE 142--VOL. I.

too pertinacious. I now returned along a deep grass-grown hollow to the bay, near to the spot where I ascended the first hill. Here I met with some of the ship's company, who had like myself been egg-hunting, but had not been equally fortunate in getting their eggs safely through the dense thickets, having broken most of them, as the condition of their outer garments but too plainly betrayed, and from the effects of which they were endeavouring to extricate themselves when I fell in with them.

Thursday, 17th.—At 9.15 a.m. sailed with a fine fair wind to the southward, accompanied by albatrosses, blue petrel, black-backed gulls, and a solitary lestris.

Friday, 25th.—A disagreeable, overcast, rainy Christmas Day, the thermometer 57° Fahr., latitude at noon 62°, longitude 170° 20'. Captain Ross and the midshipmen dined with us in the gun-room at three p.m., ship lying-to. No divisions or divine service. I have been engaged ever since we left the land in skinning birds, drying plants, and arranging and stowing away natural history collections.

Saturday, 26th.—Fine day, ship laying-to, and drifting to the eastward. On the following day, Sunday, we had neither divisions nor divine service, lying-to in a gale of wind, with snow at intervals. Just as we were sitting down to dinner, our table was swamped by two seas coming in quick succession down the gun-room skylight, deluging the table-cloth and everything on it, and I happened to be the unlucky president on the occasion.

Monday, 28th.—Fine day, with a bracing fresh breeze. Going six knots through smooth water. At 7.15 p.m. land was reported ahead, which, however, proved to be a large iceberg on the lee-bow, the first we have as yet fallen in with. It was five or six leagues off when we steered for it, and at 9.40 p.m. we passed within a quarter of a mile, and to leeward of it. It was more than 100 feet in height, of a white colour, in places faintly tinted with blue, which became of a deeper shade when the sea washed its base. It might be compared to a large frigate just launched, without her masts and spars, calmly and majestically floating with bow to leeward. A large petrel was flying near it. Two more bergs soon made their appearance, one of them resembling the roof of a house covered with snow. *Terror* about a quarter of a mile astern. At midnight, when I left the deck, I counted six more, three to windward and two to leeward, and one ahead, with a whale blowing. Thermometer 31°, latitude 63° 22′, longitude 174°, a fresh breeze, and smooth water. Although the darkest portion of the twenty-four hours, the thermometer could be readily read off by daylight.

Tuesday, 29th.—Fine morning. I counted sixteen pieces of ice around the horizon. Becalmed during the middle of the day. Afterwards passed a line of bergs to windward, six of them of large size. Since we entered amongst the ice the blue petrel seem to have deserted us. The sooty and small black-backed albatross, with several whales about us, the latter blowing and sending up jets of vapour and spray, and then descending with the tail uppermost; a berg to windward, resembling a cottage in shape, having a chimney at its gable, and another might be likened to a chariot and pair of horses, quite white. The thermometer has now fallen below the freezing-point, being 29° to-day, the air keen and piercing, latitude 64° 7′.

Wednesday, 30th.—Being a fine day with light winds and calms, we sounded in latitude 64° 34′, longitude 173°, in 1560 fathoms, having 5000 fathoms on the reel, formed of 300 fathoms of strands of whale line, 700 of 9 yarn, and 360 of 6 yarn, spun yarn in the portion run out. Observed a petrel we have not hitherto met with, of a lighter colour, and somewhat larger than the Cape petrel.

Thursday, 31st.—Fine day, with a fresh breeze. Going four and five knots through the water to the southward. South-easterly winds, thermometer 30°, latitude 66° o′ 25″, longitude 171° 53′. Passed an iceberg about three miles to windward, resembling in form a high-prowed galley, with several others of large size. The south-west horizon to leeward presented a long yellowish-white streak of light, like an ice-blink, surmounted by a dark bank of clouds. The same two kinds of albatrosses we have seen for the last few days still accompanying us, and also the light ash-backed petrel, with a solitary stormy petrel flying in the wake. The surface of the sea was studded over with small bits of ice. This morning, tor the first time, I saw the beautiful and elegant white petrel (*Procellaria nivea*), its black beak and feet forming a striking contrast to its pure, unsullied, snow-white plumage, rivalling in its whiteness the snow-clad berg itself; several were hovering round the ship, rising higher as they swept to windward, in their rapid and graceful evolutions. After skimming along the surface of the sea for a time they will frequently fly round the ship in pairs.

Being anxious to secure an early specimen of this rare and beautiful bird for the collection, I seated myself in the galley, on the port-side of the quarter-deck, with my old double-barrel gun in my hand, and during the forenoon watched for an opportunity when the bird was hovering over the mast-head, well to windward, so that when shot dead it might fall on board. After a little practice, by taking into calculation the force of the wind and velocity in the flight of the bird, I became very successful in thus bagging my specimens; the eye and hand soon acting in concert, in estimating the angle at which the bird should be fired at, to secure its falling on board. At the fourth shot I had the satisfaction of examining this lovely bird in my hand, it having fallen dead on the taffrail to leeward, striking against the mizzen-trysail in its descent. This was now my only chance of obtaining specimens for the ornithological collection, as the ship's course could not be delayed for lowering a boat to pick them up. Saw a large berg ahead, and a smaller one on the lee-bow, resembling the hull of a ten-gun brig. I saw the old year out and the new one in, under somewhat novel circumstances, the sun rising exactly at midnight, in the S.E. by E. quarter, gliding slowly along the horizon so that the lower limb was scarcely clear of it when I left the deck at 1.40 p.m., to turn in. A streak of red diffused itself along the horizon, surmounted by a bank of clouds, very beautifully striated with the same colour.

Friday, January 1st, 1841.—We commenced the new year by crossing the Antarctic circle, but light winds and a quantity of loose stream-ice prevented us from making much progress within its confines; the thermometer at 31°, with a fall of snow, and a westerly wind. Several icebergs in sight, and whales spouting; the white and the ash-backed petrel flying about us with a solitary stormy-petrel. The crow's-nest was got up to the fore-topmast-head to-day, and I went up into the fore-top to have a look at the ice. At four p.m. the officers from the gun-room and the midshipmen's berth dined with Captain Ross in his cabin; the fare, fresh roast beef; and bullock's heart. During the first watch I went up to the crow's-nest. A box-cloth jacket and trousers, a pair of water-boots, two pairs of hose, two comforters, a red frock, and a Welsh wig were this morning served out to the officers and ship's company.

Tuesday, 5th.—At 9.15 a.m. we entered the pack with a fine fresh breeze in our favour, having yesterday passed several icebergs, one resembling a village church, and another like a cottage, with a third having all the appearance of an old hulk, with a lighter alongside. To-day we are going at the rate of four and five knots through the pack, with fine, clear sunshine. Passed two more bergs, the one nearest to us was sixty to seventy feet in height. The pieces of ice through which we passed were loosely packed, seldom exceeding thirty or forty feet in diameter, with a smooth, flat surface, having here and there small irregular prominences or hummocks, the whole of a snowy whiteness; the sea-edge of the ice not more than six to eight inches deep. Now and then we had to haul out of the way of a heavier and higher hummocky patch; at times we passed through a mile or two of open water. It required strict attention to the helm in working through the narrow lanes and openings, to avoid coming in collision with the larger masses, some of which had cracks in them, tinted with the finest azure blue. Looking towards the horizon, the pack in the distance presented a uniform white surface from the dark lanes of water between being concealed. The birds about us were chiefly the white petrel, that harbinger of ice, never met with beyond the vicinity of the pack, a gigantic and a stormy petrel, and a pair of penguins; the latter, sitting on a piece of ice on the port-bow, plunged into the sea as the ship passed. In the afternoon I went up to the fore-topsail yard and saw open water in the horizon, with a dark water sky, so far favourable for us.

Wednesday, 6th.—Being Twelfth-night, all the officers took tea in the cabin with Captain Ross, and partook of a Twelfth-cake, which had been given him in a tin case, and was to have been opened on the 6th of January, 1840, but had been reserved for entering the ice. It was accompanied by the customary painted figures on paper and sugar, with enigmas to solve, which afforded us all some amusement and laughter; to aid which we had a glass of cherry brandy each. Captain Crozier was a guest; and Captain Ross afterwards, with some of the

L 2

officers, returned with him on board the *Terror*, to spend
the remainder of the evening. At nine p.m., as the boat
was being hauled up alongside, to take them, one of
the *Terror's* crew fell overboard from the gangway; the
life-buoy was immediately let go, and the port-quarter
boat lowered, into which, being on deck at the time, I
jumped as she was in the act of being lowered in the
falls, and afterwards from her into the *Terror's* boat,
which had just picked up the poor fellow only a short
distance astern of the ship. After changing his wet
clothes, and getting him between warm blankets, and
into a hammock, he soon recovered from the effects of
his more than usually cold bath. We fortunately were
hove-to at the time, in an opening of water amongst
the ice. During the first watch I went up into the crow's-
nest, to have a look at the leads of open water, and discovered the appearance of one to the southward. Abernethy, our able and experienced ice-master and gunner,
was on the look-out in the nest at the time I went up.
The sun is now constantly above the horizon; thermometer 30°, and wind westerly.

Thursday, 7th.—Hove-to, and occasionally filling and
tacking amongst the loose ice. Saw a seal swimming
on the port-bow, the first I have yet seen here. This
afternoon I shot the first penguin, on a piece of ice on
the port-quarter, and Captain Ross permitted a boat to
be lowered to pick it up. Several others, generally in
pairs, I saw to-day. At seven p.m. the captain gave
me the second cutter with six hands, to pull along the
edge of the ice, whilst the ship was tacking off and
on; when I shot four more, two at one shot, and
also a white petrel. I had thus an opportunity afforded
me of landing on a piece of Antarctic ice for the first
time, to pick up a penguin. An opening in the ice to
the southward appearing in our favour, the recall pennant was hoisted, and I returned on board at eight

p.m. This evening I saw a gigantic petrel, and two stormy petrels of a larger size than the common kind; they flew very high, hovering like swallows or martins over the mast-head.

Saturday, 9th.—On going on deck after breakfast this morning I found the ship laying her course to the southward, in a clear, open sea with a strong breeze, going four knots. We have now run about 134 miles through the pack. "Billy," our young goat, the sailor's pet, exhibited a very ludicrous performance on the quarter-deck last night, staggering about and committing the greatest absurdities, whilst under the influence of some port wine which had been given him in the gun-room, and he has consequently been all day stowed away in his cask on the starboard-side of the quarter-deck, paying the usual penance for his debauchery.

Blowing a fresh gale, with the ship rolling heavily and thick weather. I heard the hoarse, harsh "Qua" of the penguin above the howling of the gale, as they passed alongside during the night. Many of the dark petrel, with the white wing-coverts, flying about us. Very unexpectedly during the first watch I found our poor young goat "Billy" dying.

Sunday, 10th.—At eight a.m. we attained the seventieth degree of latitude. Gale abated; sea gone down, and quite free from ice, not a particle anywhere to be seen. I saw three or four penguins in a group, several white and other petrel, a large grey one, and a solitary stormy petrel. We are now in the longitude of 174° 43.

Yesterday there was a continuous fall of very small, fine snow for twenty-hours. This large space of open water occurring in so high a latitude looks very promising for us—may it continue to the Pole, or a continent discovered!

CHAPTER XIV.

Discovery of a southern continent—My first view—Take sketches—
We land on Possession Island—Penguins by the million—We hoist
the British flag—Return to the ships—A magnificent scene—High
peaks—Mounts Sabine and Herschell—Within 500 miles of the
Magnetic Pole—Mount Melbourne—Franklin Island, &c.

Monday, January 11*th,* 1841.—At the early hour of
2.30 a.m., of the middle watch, land was reported ahead
from the "crow's nest," and little more than an hour
afterwards, at 3.45, the officer of the watch called me, as
all the officers in charge of the different watches were so
well aware of my habit of coming on deck at any hour of
the twenty-hour, day or night, to take sketches of any land
of interest in sight, or objects of natural history worthy
of record ; and upon such an announcement as this, in the
very high latitude we were in, it may be well imagined not
a moment was lost in the present instance on my part in
leaving my comfortable bed, even in such a climate as
this, for the deck.

This newly-discovered land at first appeared very
indistinctly through a light haze, and a few light clouds
skirting the horizon. It was best seen on the port-bow,
where I could just trace the faint outline of a somewhat
conical summit of a lofty mountain, having a steep escarpment, longitudinally streaked white with snow. After
the lapse of about an hour it became so intermingled
with the hazy, cloudy horizon, as to give rise to doubts
in the minds of some as to its being in reality land at all.

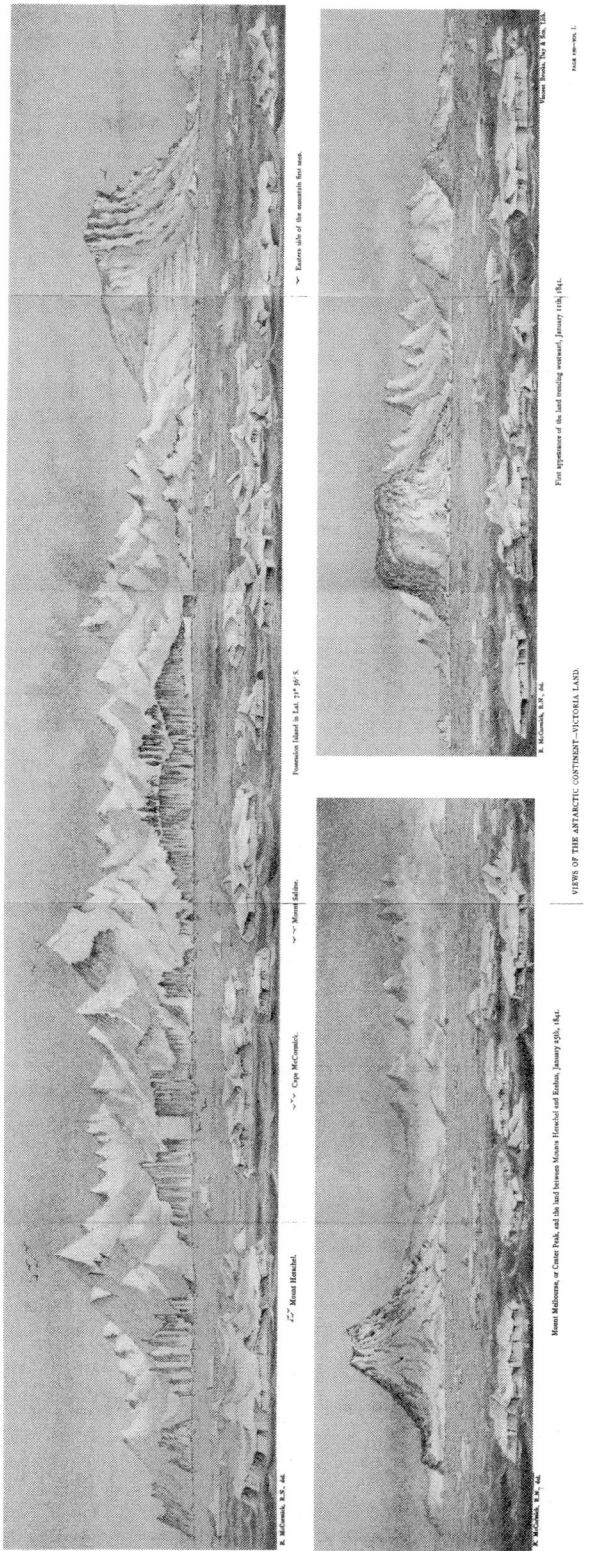

VIEWS OF THE ANTARCTIC CONTINENT—VICTORIA LAND

The material originally positioned here is too large for reproduction in this reissue. A PDF can be downloaded from the web address given on page iv of this book, by clicking on 'Resources Available'.

At nine a.m., however, when I again saw it, it had become sufficiently well defined and clear in outline to enable me to get a sketch of it. It extended from S.E. by S. to S.W. by S.; very high, and enveloped in a mantle of snow, except the lower portions of the steep escarpment rising above the sea, and these were black, where not longitudinally streaked with snow; but the whole of the upper part of this vast mountain range was an entire glaciation beneath a white mantle of snow, relieved only at intervals by the dark apex of some hummock or projecting mountain peak, peering through the snow-clad mantle. The weather was all that could be desired for giving effect to such a magnificent panorama, as gradually unfolded itself like a dissolving view to our astonished eyes. The sky was a clear azure blue, with the most brilliant sunshine; the thermometer at 31°, with a fresh breeze from the westward. The refraction in the atmosphere caused the land to appear visible at a much greater distance, for we were all day standing in towards it. The northern side, which we were approaching, presented a very remarkable appearance: a cluster of white, angular-shaped hummocks or small peaks in the background, resembling a vast mass of crystallization, having a steep wall or escarpment of black rock like lava in the foreground next the sea, near which several large icebergs lay aground, and evidently had been separated from the barrier, for where the land trended to the south-east, a whole line of them were in process of formation, and off which a small island with several rocks are grouped, from which a narrow stream of ice extends out to seaward. We tacked ship during the first watch; and at eleven p.m. I got another sketch of the coast. Saw several birds about—a stormy petrel, a gigantic petrel, a white petrel, a pintado, and some penguins. The latitude at noon was 71° 14′ 45″, and longitude 171° 15′, consequently we are now beyond Captain Cook's farthest, and have

discovered a new land, of so extensive a range of coast-
line, attaining such an altitude, as to justify, from its
general aspect, the appellation of a " Southern Continent,"
in the highest latitude within the Antarctic Circle yet
known, and we have now but Weddell's track to get
beyond.

Tuesday, 12*th*.—At 9.50 a.m. I accompanied Captain
Ross and some others of the officers in the cutter to
take possession of the land, landing on an islet lying off
the mainland, which was christened " Possession Island."
Abernethy, Captain Ross's old follower, and our gunner
and ice-master, steered the boat as coxswain. It was a
long pull along shore, tossed about by the swell amongst
the ice, in a fruitless attempt to reach a projecting headland against a strong current setting us into the bight
amid a great ice-ripple, so that we were obliged to bear
up and run through an opening in the ice to leeward, a
perfect race, so rapid that had the water been shallow
enough to ground the boat, she would have been upset
instantly. The margin or ice-foot on which we at last
effected a landing took us upon a nearly level surface, a
guano-bed in fact, formed by a colony of penguins for
ages past. It had attained such a depth as to give an
elastic sensation under the feet, resembling a dried-up
peat-bog. It would afford valuable cargoes of guano for
whole fleets of ships for years to come, could they only
penetrate the vast packs of ice we have just forced our
way through at such risk, and which constitute an impassable barrier to ships as they are ordinarily constructed.

The penguins indeed, with their young all covered with
down, formed such a rookery here, that the whole place
and sea around seemed alive with them. In such countless myriads were they congregated, not only over the
incubation area below, but up the sides of the black lava
rocks in tier above tier to the very summit, which attained
the height of 300 feet, flanking the guano-bed on the

right, that it was like a thistle-bed to pass through, so thickly formed were their ranks, and without kicking them to right and left there was no getting through their dense legions. The old birds stoutly defended their young, attacking the intruder on their domain in front and rear with open mouth, sending forth at the same time such harsh notes of defiance, in which the whole colony united in concert, that we could scarcely hear each other speak so as to be understood. These sturdy, bold birds, standing erect on their tails, with the horny feathers of both head and neck ruffled in anger, their flipper-like wings extended from their sides, looked altogether the most ludicrous and grotesque objects imaginable.

In many places the young birds were grouped together in knots of a dozen or two, encircled by the old birds, forming a barrier around them.

Not a single specimen of an egg was to be found, the season of incubation having already passed. On taking the water they slide down the icy margin of the ice-foot on their tails and the soles of their feet, dashing into the sea with only their heads appearing above the surface, some leaping out like bonito in pursuit of flying fish; and I observed one bird make a most extraordinary leap upwards from the sea to the top of a perpendicular piece of ice, certainly not less than a fathom in height above the water, alighting on its feet like a cat. The perfume arising from this colony was certainly not of an Arabian sweetness, for even before the boat reached the shore the scent wafted off upon the waters was all but stifling. The population of this colony might be estimated by millions. Some of the old birds had their breasts stained red, as if from blood at first sight, but which I subsequently found arose from their resting on their breasts upon their excrement, which was frequently tinted of a red colour. I noticed several small heaps of pebbles about the size of a nest amongst the birds, which I could only account for by supposing they had been rejected from their stomachs, as I have often found the stomachs of the penguins, whose skins I have preserved, full of pebbles, as in the seals. The mature bird weighs about 10 lbs., the upper surface of the body being of a dark greyish ash colour, and approaching to black in some instances, with the under surface white ; beak and under the chin black ; legs of a pale flesh colour.

Several lestrises, and apparently a new species, had taken up their residence in the midst of the colony, and from their incessantly sweeping overhead on wing, and alighting amongst the penguins, would also appear to have their own young there. I shot one flying overheard, and knocked down an old penguin with my geological hammer, and put him in my haversack for the collection, with a few hastily collected specimens of the black lava rock. The only other bird I saw flying about the island was the gigantic petrel. I did not observe the faintest trace of vegetation, not so much as a lichen on the bare volcanic rocks, or even a seaweed on the shoal. But our stay was so brief—only some twenty five minutes—in consequence of the threatening aspect of the weather, that I had barely time to cross over the penguin rookery to the base of the black lava mount, which I was just about ascending when my recall was made.

After the flag was unfurled, hoisted, and the queen's health drunk in a glass of sherry by each of the party, three cheers were given on taking possession of the land, no less a domain than a continent, in all probability exceeding in magnitude the continents of either the Old or the New World, in the name of her Majesty. Its productions are indeed only ice and guano. We shoved off from this lone islet in the boat at 11.15 a.m , with the *Terror's* boat in company, containing her captain and some of her officers, who had landed after us.

To avoid the race on returning, we pulled round the

island, in by the bluff, black headland, passing several isolated rocks, some seven in all; one likened to a leaning tower, another of a rhomboidal form, and a massive black, circular rock, somewhat resembling a martello tower, with a hole through its base, vaulted over. On this side of the islet a perpendicular wall of basalt or lava faces the sea, and it has been worn smooth at its base, and to a considerable height above it by the action of the waves and the ice.

There is a beautiful cavern in the face of the rock, with a fine display of columnar basalt, quite a Fingal's Cave in miniature. The columns all most symmetrically vertical. In several places the rocks were perforated with caves. The summit of the island, in one bearing, resembles a bishop's mitre, and on this side two very remarkable projections of rock were compared to human forms, seated in front of a battlement, to which we at once gave the name of Victoria and Albert, as presiding over the new land we have been so fortunate to add to their dominions, now ranging from pole to pole. The former figure seemed draped in something like a veil, and the latter in a cloak. When we had rounded the Circular Rock, we got sight of the *Terror*, more than hull down, to windward, and on further pulling round a large berg forming on this side of the island, we opened the *Erebus*, but also at some distance from us. The weather now wore a very threatening aspect, black clouds rising in the horizon, with every appearance of blowing and thick weather coming on, rendering our position a somewhat perilous one. The " Mother Cary's chicken," ill-omened bird of the storm, and a lestris were incessantly flying over our heads, the latter so near it might have been knocked down by a stone or an oar. Doubtless we were as great a mystery to these lone denizens of the southern pole as Cook's ships were to the South Sea Islanders in the last century.

The *Erebus* soon discovered our position, and at once bore up, standing towards us, and the *Terror* shortly afterwards followed her example. We got on board at 12.45, much elated by our excursion, soon after which it came on so thick and foggy, with a fall of snow, that we wholly lost sight of the land, which would have been anything but pleasant for us had we been still in the boat.

In the evening saw large flights of the pintado, as many as from 200 to 300 in a flock, passing the ship with a white petrel or two. Air keen, with the thermometer at 30°, wind N.W., latitude at noon 71° 51′, longitude 170° 52′.

Wednesday, 13th, and Thursday, 14th.—Blowing a south-easterly gale, with cloudy and thick weather, and small snow at times, preventing any further communication with the land for the present. Ship rolling a great deal in a short, broken sea. Pintados flying around the ship in great numbers, with a white petrel or two. Saw two or three large whales blowing within a hundred yards of the ship. Employed to-day and yesterday skinning birds, penguins, white petrel, and lestris. Thermometer 29°, latitude at noon 71° 51′, longitude 172° 50′.

Friday, 15th.—Beating to windward off the islands all day. Strong breezes, weather remarkably fine, with a clear blue sky and bright sun. I took several sketches from the stern-boat, of this wonderful land. The southernmost headland in sight presented a very bold aspect. The lofty, magnificent-looking coast-line appeared to a great advantage this evening, standing out in such strong relief against the clear blue forming the background. The thickly grouped, angular-shaped, small peaks, or hummocks, clad in snow of the purest white, the whole resembling a vast mass of crystallization, but on such a huge and splendid scale, as nature's laboratory alone could produce. The highest peaks, majestically towering above a stratum of light, fleecy, white clouds, reflected the rays of the setting, or rather declining, sun—(for that luminary here now never sets throughout the whole twenty-four hours)—upon the facets of the angles on which it fell with a pure and silvery light, whilst the others were thrown into the deepest shade, at once displaying the continuity of outline of the mountain range, unfolding and bringing into view the minutest irregularity of the surface, whether eminences or depressions. Saw a large berg in shore, which had only separated from the land-ice this morning, and in capsizing had showed the rock and soil attached to it uppermost. We passed a thin stratum of widely scattered small bits of ice; a gigantic petrel and some pintados flying about the ship. Latitude 71° 56′, longitude 171° 51′, thermometer 31°, wind south-easterly.

Saturday, 16*th.*—Gained but little to windward of the islands since yesterday.

Sunday, 17*th.*—Still beating to windward. Saw this evening another headland to windward, extending out to the southward, which I had observed, though less distinctly, last night, having a trending-in of the coast between it and the bluff cape which has been so long in sight to windward. In the first watch I had a fine view of the mountain peaks. The angles taken gave to the highest peak an altitude of 8444 feet. In latitude 72° 10′, longitude 174° 23′, thermometer 28° Fahr.

Monday, 18*th.*—Cloudy, but fine day, with a moderate breeze and smooth water. Loom of the land barely visible, and not a particle of ice about us. Saw a white petrel, and many pintados. The beautiful crystal-shaped mountain peak has an altitude of 7867 feet, which, with the barn-shaped mountain, forms the most striking elevations of the whole of this wide mountain range. Latitude 72° 57′ 59″, longitude 176° 5′ 57″, thermometer 30°, wind southerly. On first making this land the

angles taken gave to the highest summit an altitude of
9096 feet. This afternoon I shot two pintados from the
deck, from near the fore-rigging, both to windward. The
first bird fell when the ship was on the starboard-tack,
striking the gunwale just abaft the main-rigging, and
bounding overboard. The second bird was flying before
me, when on the port-tack, hovering over the fore-
topsail yard, falling to leeward on the forecastle, and
was picked up by one of the crew, and brought to me.
It proved to be the young and immature *Procellaria
capensis*, and the first specimen I have hitherto been
able to obtain. I have been thus minute in what may
appear trifling details under the impression that it may
prove of service to others, who may hereafter be placed
under similar circumstances as myself, with a desire to
secure rare or new species of birds, with no boat with
which they could be picked up. During the first watch,
whilst on the port-tack, I shot a white petrel, which, in
falling, struck against the main-sail, and fell on deck to
leeward, the best specimen I have as yet succeeded
in thus obtaining. At eleven p.m. I saw the land very
distinctly on the lee-bow. The night fine, with a clear
blue sky. At midnight, when I left the deck, the sun
was about three degrees above the horizon. Our
observations gave latitude 76° south, longitude 145° 20′
east, placing the magnetic pole 500 miles south-west
of us.

Tuesday, 19th.—A lovely day for these regions, indeed
for any climate, the sun shining forth from a clear blue
sky, and we were becalmed at noon in smooth water. The
thermometer 32°, above which it rarely indeed rises within
the Antarctic circle. But the atmosphere seems as mild,
pleasant, and congenial as on the finest May day in
England. We heard the loud, harsh cry of the penguin
—" Qua, qua "—to windward, and a seal rose forty or fifty
yards on the starboard side of the ship. We are standing

in for the land, its magnificent mountain outline presented to the eye a wonderful scene of beauty, as the highest peak stood out in fine relief against the deep azure sky. As the day wore on, in the afternoon a fine northerly breeze sprang up, after having had a week of southerly winds to check us, off this newly-discovered land.

Both ships now crowded on all sail for the southward, in the 73° of latitude, under studding-sails, low and aloft, on both sides, from the two large lower studding-sails to topgallant and royals ; and under this tower of canvas the old ship only made five knots along the land. Several lestrises flying about us, and, during the first watch, a few of the immature petrels, and a white petrel. I shot and secured one of the former from the quarter-deck boat, on the port side, and it fell into the quarter-boat on the same side. When I went below at 12.45 the sun appeared like a beautiful, bright, white globe, about three or four degrees above the horizon, on the port side. An island was visible on the starboard-bow. Latitude at noon 72° 31′, longitude 172° 49′.

Wednesday, 20th.—Cloudy, overcast, gloomy day, with variable winds, and making but little progress to the southward. The land, for the most part, enveloped in clouds. We were abeam of a bold, black headland for most of the day. A stream of ice skirted the horizon. Thermometer 31°, latitude 73° 47′, longitude 171° 50′. Captain Ross to-day gave me some specimens of red coral and shells, brought up with the dredge-net, from a depth of 270 fathoms, when sounding yesterday at two p.m., in latitude 72° 31′, longitude 172° 7′.

Thursday, 21st.—This forenoon, the same dark, bluff headland seen yesterday appeared through the haze and clouds to-day, bearing W.N.W., with some white snowy peaks, having the appearance of land in the E.N.E. on weather beam. Weather gloomy, with a few flakes of small snow falling about noon. The pack seen from the mast-head to the southward and westward. During the first watch two lestrises passed the ship. In the calm the water had an undulating deep blue appearance, from the intensity of the tint of blue in the sky overhead. At midnight the sun looked unusually bright, being about three degrees above the horizon in the E.S.E., bearing S. 70° E. The land appearing like an island bore from S.S.W. to W.S.W.; latitude 74°, longitude 170° 43′, thermometer 30°.

Friday, 22nd.—Fine morning; the bluff headland still in sight; wind easterly, thermometer 31°, latitude 73° 56′, longitude 172° 20′. As night closed in we saw the last of the land at a long distance astern, the sun at the time shining brightly on it.

Saturday, 23rd.—Thick weather, with snow, blowing fresh, and a short head-sea. Our latitude at noon, by dead reckoning, made us to the southward of Weddell's farthest south, and consequently nearer to the South Pole than any other ship has hitherto attained; Weddell's farthest being in latitude 74° 15′, ours 74° 23′ D.R., in the longitude of 175° 35′ E. Captain Ross, on the occasion, spent the evening in the gun-room with us, and our toast was " Better luck still." Thermometer 30°, wind N.E. Captain Ross gave me some medusæ brought up in the dredge from a depth of 300 fathoms in the last middle watch.

Monday, 25th.—A fine clear day, but the air very chilly, the thermometer being down to 27° at noon, with southerly wind; land just visible on the lee-bow, with a very high peak faintly outlined on the weather-quarter. The pack was also visible, and at eight p.m., when·I went on deck, we were standing along the pack edge, which was backed by a mountain range. One lofty peak, named Mount Melbourne, not measured (magnetic dip 88° 10′) but supposed to be the highest land, evidently from its form had been a volcanic vent to this land. Its symmetrically and perfectly shaped crater at the summit, and altogether fine outline, could not have been surpassed even by Mount Etna or Vesuvius, as it reared its towering head above the mist which concealed most of the surrounding land, comparatively much lower, and flanking its conical peak on either side. Beyond, and to the right, the land again rose to a considerable altitude, and still farther to the N.N.W., another apparently in-sulated group of mountains appeared. About 8.20 we went about amongst some loose straggling ice. The main pack seemed to be formed of heavy massés and closely packed, more so than any we have hitherto met with, extending along a whole line of coast like a belt. I took a sketch of the land, and during the first watch I went up into the fore-top, to have a look at the moun-tains from an elevated position. The night was fine, with a clear blue sky and keen air; I shot three white petrel flying over the mast-head, two of them fell on board, which I secured, and the third but narrowly escaped me, a gust of wind just carrying it over the stern. The first one, which fell into the main-top, had a good-sized fish in its mouth, much resembling a pilchard, but its head was gone. The second one fell into the port quarter-boat, and I gave it to Captain Ross. Several more were flying about the ship. At midnight Captain Ross took an altitude of the sun, which was shining brightly about three degrees and a half above the horizon, and found the latitude to be 74° 45'. I turned in at one a.m.

Tuesday, 26th.—Cold, raw, gloomy day, with a short head-sea and southerly wind; thermometer 25°, latitude 75° 2' 38", longitude 169° 4', land very indistinctly seen.

Wednesday, 27th—Fine day; this forenoon an island appeared on the starboard-bow to windward, and the ship was hove-to off it at five p.m. The second cutter was lowered, and Captain Ross, with a party of officers, landed upon it, at the same time with a boat from the *Terror*, with her captain and some of her officers. An accident, however, happened to Dr. Hooker, the assistant-surgeon of the *Erebus*, who fell into the water in attempting to land from the boat, consequently he with others were not permitted to leave the boat at all, and Captain Ross himself landed with the *Terror's* officers. Possession of the island was taken in the customary way, when it was christened Franklin Island. Both boats returned to the *Erebus* at 8.40 p.m., and the *Terror's* boat did not leave with her party for their own ship until 1.30 a.m.

Franklin Island is purely volcanic, being composed of lava, and having a large berg forming at its southern extremity, with another large, square-topped island of ice lying at a short distance off it, aground; latitude 75° 48', longitude 168° 10, thermometer 24°, wind S.

It is with regret that I should have occasion here to allude to the circumstance of a very injudicious order of Captain Ross's in a voyage like ours; nor should I, but for its prejudicial effects in the results on the collection, as the sequel will show. This order forbade the ships being left without a medical officer on board of each, however near to each other or the land.

I do not hesitate to say that many highly interesting observations, both in natural history and geography and the collateral sciences, in a newly discovered land, in so high a latitude as we have had the good fortune to attain, and which may perchance never be again visited, have been lost to the world, through this ill-timed order. And notwithstanding my having had some personal influence with Captain Ross, both of us having served together as youths under our mutual old commander and friend, Sir Edward Parry, I could not induce him to cancel any order he had once placed in the order-book, so strong were his prejudices, and as a sequence so difficult to reason with. One medical officer left with the ships, as I pointed out to him, would be a sufficient guarantee in the remote chance of any accident happening on board during so short an interval of time.

Peak of Mount Erebus, 120 miles distant. (*See page* 176.)

CHAPTER XV.

The volcanic mountains " Erebus " and " Terror "—Great barrier of ice —The Barrier Bight—A fascinating scene—Twenty-four hours on deck—Seals and whales—Exploration of McMurdo Bay.

Thursday, January 28th, 1841.—We were startled by the most unexpected discovery, in this vast region of glaciation, of a stupendous volcanic mountain in a high state of activity. At ten a.m., upon going on deck, my attention was arrested by what appeared at the moment to be a fine snowdrift, driving from the summit of a lofty crater-shaped peak, rising from the centre of an island (apparently) on the starboard-bow.

As we made a nearer approach, however, this apparent snowdrift resolved itself into a dense column of black smoke, intermingled with flashes of red flame emerging from a magnificent volcanic vent, so near the South Pole, and in the very centre of a mighty mountain range encased in eternal ice and snow.

The peak itself, which rises to the altitude of 12,400 feet above the level of the sea, is situated in the lati-

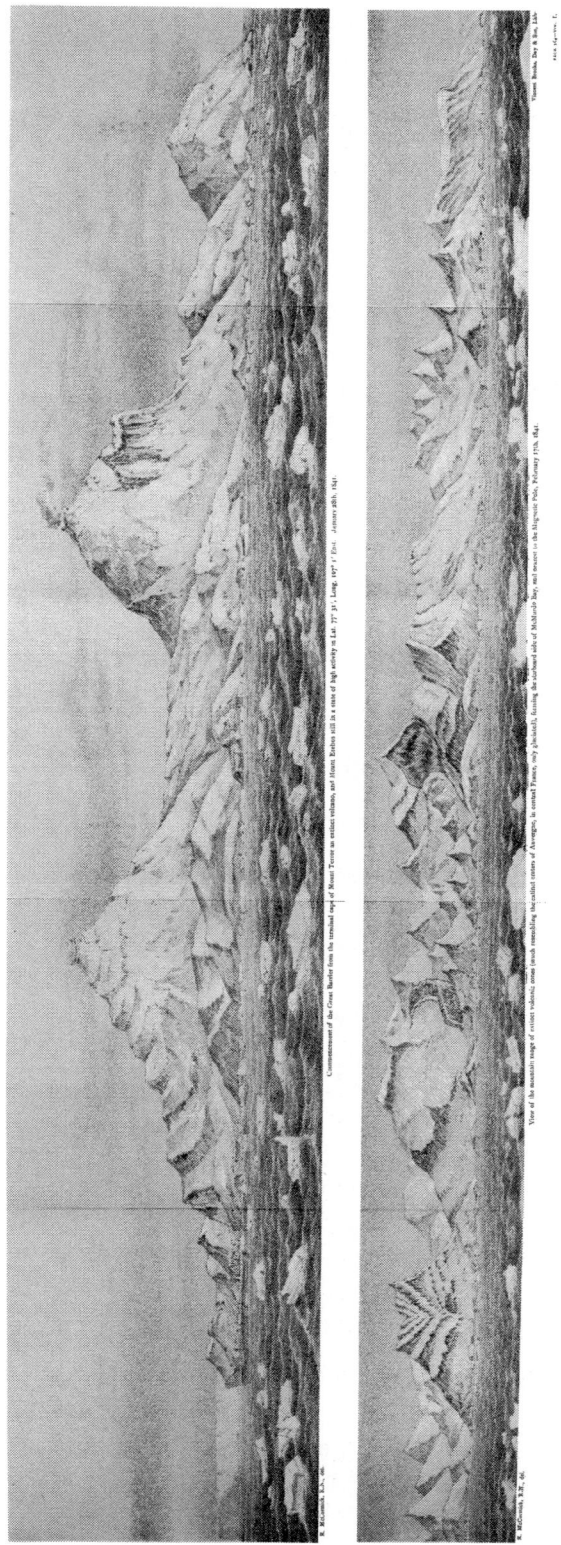

Commencement of the Great Barrier from the terminal cape of Mount Terror an extinct volcano, and Mount Erebus still in a state of high activity in Lat. 77° 31′, Long. 167° 1′ East. January 28th, 1841.

E. McCormick, R.N., del.

Vincent Brooks, Day & Son, Lith.

View of the mountain range of extinct volcano, steam jet ??? emitting, the central craters of Aberdeen, in named France, only glacially, forming the eastward side of McMurdo Bay, and nearest to the Magnetic Pole, February 17th, 1841.

E. McCormick, R.N., del.

The material originally positioned here is too large for reproduction in this reissue. A PDF can be downloaded from the web address given on page iv of this book, by clicking on 'Resources Available'.

tude of 77° 31′ S., and in longitude 167° 1′ E., and was named after our ship, Mount Erebus. Adjacent to it, and only separated by a saddle of ice-clad land on its east, arose a sister mountain to the height of 10,900 feet, but now extinct, though having the same general outline, also doubtless belched forth at no very distant period its volumes of smoke and flame. It received the name of Mount Terror, after that of our consort. Its sides were partially covered with snow, presenting the appearance of having been melted in many of the depressions on its sides, and again frozen into pools, glittering like molten metal in the sun's rays, and extending down the sides of the mountain, in a broken serpentine stream to the great wall of ice, which extends from its base, forming a point or cape. This sea-wall, having a perpendicular face and tabular summit, averages 150 feet in altitude, with caverns hollowed out by the constant action of the waves, producing a remarkable effect of light and shade along its whole margin which extends in a south-easterly and north-westerly direction, and along which our course lay to the southward, between it and the pack.

On the starboard beam another small island appeared, bearing south, which received the name of Beaufort Island, after our excellent and talented hydrographer to the Admiralty, Sir Francis Beaufort. There was also an appearance of land broad on the port bow. At five p.m. I went up to the crow's-nest, but could perceive no termination to the great ice-wall, which we have named the Great Southern Barrier, and barring our way to the pole. We are shaping a N.W. course along it, distant, perhaps, three or four leagues. A number of white petrel, and now and then a solitary lestris of predatory habits, like its congener the Skua gull of the north, have been the companions of our voyage for the last few days. Weather very fine and clear, thermometer 29°; wind S.W., latitude 76° 57′ ; longitude 169° 24′ 50″ E. ; ship under studding and topgallant sails.

Saturday, 30*th.*—Yesterday we ran parallel with the barrier, but to-day a change in the wind, and the weather becoming thick and gloomy, with a fall of small snow, compelled us to stand to the northward, and to relinquish, for the present, the following up the course of the barrier. There was neither land nor ice in sight to-day, thermometer 28°, latitude 77° 35′ D. R., longitude 181° 20′. We have followed this lofty barrier of steep and perpendicular ice-cliffs, varying in height from 100 or 150 to 200 and 300 feet, for upwards of 100 miles. The depth 410 fathoms ; the lead in sounding sank at least two feet in soft green mud, seemed to indicate that the outer edge of the barrier could not be attached to the bottom, but must be borne upwards by the water. The high land forming the background of the barrier, being the southernmost known land, was named after our worthy old arctic chief, Sir Edward Parry, the Parry Mountains.

Tuesday, *February 2nd.*—A beautiful day, with bright sunshine, quite congenial to our feelings. At noon the clear, blue sea was as smooth as a mirror from the calm reigning. Only a solitary berg just visible, through the light haze skirting the horizon on the port beam. In the evening we made the loose stream-ice, and our old acquaintance the fixed barrier again, appearing beyond the stream of ice or pack, like a long, low cloud skirting the horizon to the eastward. We entered the pack with light northerly winds, then bore up and got out of it again. A whale hove in sight and a gigantic petrel, with a white one or two. At the time we wore ship, 9.15 p.m., our latitude was 78° 3′ 6″ S. and longitude 187° E., the barrier being about fifteen miles distant, extending from about S. to N.N.E., skirted along its whole length by loose ice, preventing us from making any further attempt in this quarter. We got soundings in 260 fathoms, green mud and clay. The height of the barrier was here 160 feet, its thickness probably 1000 feet, extending 250 miles from the cape at the foot of Mount Terror; the face of the barrier here being in the latitude $78\frac{1}{4}°$ S.

Thursday, 4th.—Employed all day sailing through loose ice, working the ship between some very heavy masses in the narrow lanes of water. The pack is studded with several bergs. Saw three seals on the ice, and three or four of the largest kind of penguin, with several white petrel. This has been the coldest day we have yet experienced, but nevertheless fine, with a clear, blue sky, appearing through the white curdled clouds of the brightest and purest azure. As the day closed in, the whole scene presented a remarkably wild aspect, the bright "ice-blink," which lighted up the horizon in the direction of the barrier, which itself was not visible, was streaked longitudinally with the dark rays proceeding from the lower margin of a super-imposed bank of clouds, which rose thick and dark to windward, and threatened a heavy fall of snow, or thick, overcast weather. The thermometer at noon was only 22°, at nine p.m. it was 16°, and at midnight fell to 15°, the lowest that we have had it yet. A fine working southerly breeze blowing. Latitude 77° 0′ 17″ S., longitude 192° 18′ E. At eight o'clock last evening we hove overboard a cask containing a paper, with our latitude and longitude, when farthest south, signed by the captain and officers. It floated lightly and buoyantly astern, a very conspicuous object in this waste of waters and ice; in latitude 77° 0′ S., longitude 187° 24′ E.

Friday, 5th.—Sailing amongst loose ice mixed with some heavy masses. Three of the large kind of penguins were caught after having been wounded by small shot on a piece of ice from the boat, two weighing 66 lbs. each, and the third 57 lbs. Saw two whales spouting. At eight p.m. the ship was made fast to a piece of ice, and the cutter lowered to get some ice on board. At ten p.m. hoisted in cutter, and cast off from piece of ice, with about three tons on board, to replenish our tanks with water. At eleven the boat was again lowered and two very pretty seals with small heads and noses were taken from a piece of ice. The thermometer had fallen from 15° at noon to 13°, wind S., latitude 77° 10′ 32″ S., longitude 192° 48′ 10″ E.

Monday, 8th.—For the last two or three days we have been beating up for the pack-edge in an open sea, having only a berg or two in sight, with several whales. To-day, at five p.m., the barrier was seen from the mast-head during the first watch, as we sailed between the pack and the barrier.

As we ran for some 160 miles along the barrier, we discovered that a whole chain of table-topped bergs, shed from the barrier itself, had grounded on a bank about sixty miles distant from its edge and 200 miles from its origin at Cape Crozier, the bergs being from 150 to 200 feet in height.

Tuesday, 9th.—The wondrous scene nature has un-folded here, even beyond what might have been anti-cipated in this land of wonderment, has had the effect of riveting me to the deck for the last twenty-four hours, a volunteer and most willing sharer in the duties of every officer of the watches during that period. Last night, after sounding in 275 fathoms green mud, the barrier at midnight was seven miles distant, its ex-tremes bearing from S.E. to N. $\frac{1}{2}$ W., and the line of the pack-ice from S.W. $\frac{1}{2}$ S. to N.N.W. $\frac{1}{2}$ W Being myself most anxious to trace this mighty wall of ice continuously without a break so as to see all I could of it, I never turned in at all, but kept the deck throughout the night, a night never to be effaced from memory's tablet to the latest hour of existence; and well was I rewarded for the temporary sacrifice of a night's rest and sleep by the grand and sublime panorama which was unfolded to and arrested my gaze like some striking shifting scene on the stage, as the " noon-day " night of this high latitude wore on, and scene succeeded scene in nature's unrivalled display of her great Creator's works.

The night, so-called, although in fact day here, was indeed most favourable, being remarkably fine, the azure blue of the sky above was mottled over with curdled white, light cumuli, a mackerel sky in short. To windward the moon's pale, silvery disc every now and then emerged from beneath the clouds on the port beam, whilst the brighter rays from the glorious sun clearly indicated its position behind a bank of cirro-stratus on the starboard beam to leeward. We were sailing along a channel bounded on the starboard by the barrier, and on the port side by the heavy pack, passing through a quantity of young ice in streams varying in breadth, their outlines marked by a deeper shade of colour than the surrounding water. Each piece of ice assumed what we called the pancake ice, in form and size, having the margin slightly elevated and turned up, the pieces thickly packed together, some streams consisting of larger and more irregular-shaped masses, oblong, oval, and of irregular, hexagonal figures, from a foot to three or four feet in diameter, lined as if from several smaller ones having become cemented together.

Anxious to obtain what I believed to be a new species of the lestris, more especially from so deeply interesting a locality as the great ice-barrier in the highest latitude we have attained, I fired at one hovering over the starboard quarter about midnight, but having only wounded it, it escaped after tumbling over and over in its descent to nearly as low as the quarter-deck boat, then recovering itself somewhat, flew in a direct line for the barrier, with one leg hanging down broken. This

was a source of double vexation to me, first, in the loss
of so desirable an addition to the ornithological collection ; and lastly, in subjecting the poor bird perhaps to a
lingering death from any internal wound it might have
received.

For notwithstanding that my duties as ornithologist
compel me to take the lives of these most beautiful and
interesting creatures of all the works of their great
Creator, I never do so without a sharp sting of pain and
qualm of conscience, so fond am I of all the feathered race.
But as we have to sacrifice their lives for our food, we
cannot do otherwise than meet the claims of science in
the same spirit.   Beyond this, all wanton destruction of
these charming creatures is highly censurable.

Between midnight and one a.m. I succeeded in adding
two more of the elegant white petrel to my collection,
one falling dead on the quarter-deck, and the other on
the gun-room skylight, having fired at each when hovering in a very favourable direction over the mast-head.
A third I shot unfortunately proved a life uselessly taken,
as it fell overboard into the sea to leeward.   At 2.30 a.m.
I saw a seal swimming on the starboard quarter.   During
the middle watch the extreme point of the barrier was
obscured in mist, and the pack lying to the left of it had
an apparent opening between.   We passed inside of two
large bergs and the barrier.   A berg or two, with several
large pieces of ice like obelisks, studded the monotonous
surface of the pack.   The sun's rays reflected on the
facets of the curves in the barrier ahead gave to the
whole mass the appearance of a long range of white
buildings having the various abutments thrown into light
and shade.   Whales were numerous here, spouting in
all directions, and many white petrels.   At four a.m. a
whole line of finners were sending up jets of vapour
so high as to appear above the barrier.

At 5.40 a.m. the break in the barrier forming an inlet

R. McCormick, R.N., del.

Vincent Brooks, Day & Son, Lith·

The stupendous ice-cliffs forming the extraordinary bight in the great Antarctic Barrier, sketched as the " Erebus " tacked off its entrance.

or bight, perhaps, a quarter of a mile wide, and from a mile to two miles in depth, bounded on its starboard side by a very strikingly bold promontory of ice, for which we had been for some time standing in, and now went about when within a quarter of a mile of it, with a moderate breeze blowing. We had tacked none too soon, for its great height above our mast-heads, even at this distance, took the wind out of our sails as we hung in stays. Captain Ross, coming on deck at the moment, rated the lieutenant in charge of the watch for venturing so near before putting the ship about. But conscience forbids my letting him bear all the blame for this bit of daring, the temptation to have a nearer view into this extraordinary recess in the barrier having prompted it. When the good old ship's stern had been presented to the barrier, Captain Ross got soundings in 330 fathoms, and again at 6.15 a.m. in 318 fathoms, green mud as usual, whilst standing off. The only spot where the upper surface of the barrier could be seen was at the further extremity of the bight, the cliffs of ice forming the sides sloping down to a low angle, where they meet, and above which the upper plain surface rose like a smooth, snow-clad, swelling hill in the distant background. The enormous icicles suspended from the basal portions of the steep mural precipices forming the portal and promontory to this great inlet in the southern barrier had a most imposing and striking effect as the old *Erebus*, when nearest to it, turned her stern towards it, after getting her head round. The whole scene was one calculated to inspire no less awe and wonder than that of Mount Erebus itself, ejecting smoke and flame from the summit of its stupendous peak of thick-ribbed eternal ice and snow.

The extremes of the pack extended from S.E. to W.S.W., and the barrier from E.S.E. to W.N.W. At 10.15 a.m. we tacked off the edge of the pack, having passed through a patch of young ice, already beginning to assume a white appearance near a berg. The latitude at noon was 77° 56′ S., and longitude 190° 15′ E., thermometer 21°, wind E. About this time a thick fog came on, and the thermometer by two p.m. had sunk to 16° Fahr., rendering our return through a somewhat intricate channel no very easy matter. At 1.30 p.m. Captain Ross went on board the *Terror*, to consult with his colleague, Captain Crozier; and soon after his return the fog increased so much we lost sight of our consort. Between five and six p.m. signals were made to her by firing three muskets, three chamber and three large guns, shortening sail at the same time. The night, however, cleared up fine as we stood along the pack-edge on the starboard side.

As we coasted the line of the barrier we fell in with many whales, both finners and spermaceti. One large black one rose close under our bows, and actually dived under the hull of the ship; several others at the same time passed very close to us. A very successful whale-fishery might be carried on here; the whales are of the very largest size, especially the spermaceti, perhaps the most valuable of all; could but ships fitted in the ordinary way make their way through the vast and heavy intervening packs, a somewhat hazardous affair, however.

Monday, 15th.—Repassed Franklin Island on our return, in latitude 76° 2′, and longitude 170° 15′. Since the 10th the aspect of the weather has much changed, assuming a wintry appearance, the decks at times covered with snow, and the masts, spars, and running rigging encased in ice, which rattled down on the decks from aloft, hanging about often in long beads or bracelets, with the bows frosted over, giving the ship a very hoary aspect indeed. Whilst running along the pack-edge through streams of ice we passed several large bergs: one long, low one estimated at four miles in length. On Sunday last I recommended Captain Ross to have the hot-air stove lighted for a few hours, to ventilate the ship. The birds about us for the last few days have been white petrel, immature petrels, lestris, a dusky albatross or two, and a stormy petrel.

Tuesday, 16th.—Fine day, with light winds and a clear blue sky above; but the horizon being clouded, rendered the outline of the land indistinct ; as we stood up a deep gulf, with the sea covered with thin pancake ice, occasionally passing a fragment of white opaque ice, nearly a calm at times, I noticed several openings in the coast-line on the starboard side, large deep bays, as it were, bounded by very bold headlands.

Ahead of us once more appeared the grand volcanic mountain of the South Pole, Mount Erebus, with a band of clouds stretching across it. In the evening it was sending forth a dense volume of smoke, in the lower portion of which a red flame might be seen at intervals intermingled with it. It is not an island, as had been supposed on our first visit ; now clearly proved by our entering the gulf connected with the mainland. We passed a whole line of large whales, whose remarkably long, pointed, black fins bristled above the surface of the water, finners doubtless ; but so little of the outline of the back of the whale is seen above the water ; a group of three or four seals passed near us, and a number of white petrel, and a lestris or two, with some penguins. A man was sent up to the mizzen mast-head to-day, to clear the rigging of ice with a handspike ; when it fell and rattled about the decks like a shower of stones, so hard was it frozen. The thermometer was 32°, wind variable, latitude 76° 32', longitude 166° 12' E.

Wednesday, 17th.—About three a.m. the land at the head of the gulf was found to be continuous beyond all doubt, uniting Mounts Erebus and Terror with the mainland, forming a deep bay. The ship was consequently put about, and her course directed out of the gulf again. During the day we stood across the bay towards the land on the starboard side, and in the direction of the Magnetic Pole, which appears to be situated in the midst of the lofty mountains about 120 or 160 miles inland of us. We passed through a very large quantity of young ice, the pieces of which it is composed having greatly increased in both breadth and thickness since we entered this large bay. I took a sketch of the land, which is comparatively of moderate height, and inclined to an angular form, margined by a line of bay ice, white, smooth, and snow-clad.

The opening between this land and Mount Erebus, forming the entrance to the bay, occupies three points of the compass ; there is a small island off the mount. At 4.15 p.m. Captain Crozier and two of his officers came on board, remaining till 6.30 a.m., owing to the difficulty in getting a boat to the *Terror*, the young ice with which the whole surface of the bay was covered thickly as far as the eye could reach all round the visible horizon, and through which we were passing all night, greatly interrupting all communication between the ships.

The pancake ice, from its original pentagonal or hexagonal form, becomes irregularly rounded, turned up at the margin, and the surface crossed by similar-shaped curved edges, as if some three or four had been cemented together ; in this way the inshore pack is rapidly formed.

I remained on deck, a not unusual circumstance with me, all night; and at 10.30 p m. I saw the sun set for the first time; the moon on the wane, forming a slender crescent. During the first watch Mount Erebus presented a splendid spectacle, sending upwards a tall, dense column of smoke, tinted red on the right side, and extending out in that direction in an oblique line of pale red along the sky, a smaller vent appearing on the first small eminence to the right of the crater; ultimately a bank of dark clouds of a deep neutral tint colour, surmounted by a reddish flush with foam-like edges screened the mountain; the volume of flame-tinted smoke curling just above this stratum of clouds to the right disappeared altogether about one p.m.

At midnight I shot two white petrel; the first fell upon the ice on the starboard quarter, and the second on board on the port gunwale, behind the quarter-deck boat by the mizzen-shrouds; saw one stormy petrel, and counted twenty seals lying on the young pack-ice. During the night passed many penguins sitting on the pieces of pancake ice, generally solitary, but occasionally two together, frequently pluming themselves, one reclining on its breast and the other standing near it. These were the smaller kind, having dark upper and white lower surfaces. A small seal, apparently the fur-seal, passed under the port quarter on a piece of ice. There were many white petrel about, so that the otherwise still solitude of the scene was thus broken in upon by the variety of animal life around us. During the morning watch we weathered the pack-edge, a dark sky to windward; thermometer standing at 25°.

Thursday, 18*th.*—At 7.30 a.m., after my long night's vigils on deck, I gladly descended to breakfast, the most social meal of the twenty-four hours. This day we made the nearest approach to the Magnetic Pole, in the early part, running along the pack-edge, through the young ice, and towards the close of the day in an open sea, clear of ice, and very fine weather, with moderate breezes. A small whale swam round the ship with a seal. We found the dip the greatest we have yet had, being 88° 44′, and the variation 92° 53′. During the first watch I shot six white petrel, of which I secured three for the collection, and singularly the third bird, when wounded, having its leg hanging down, took a flight to windward, then, suddenly tacking in its course, swept round and fell dead on the quarter-deck, as if in the last moments of its ebbing life it yielded up its beautiful form to the altar of science. Of the others, the first fell between the capstan and the gun-room skylight; the second into the catharpens of the main-top; a solitary stormy petrel following in the wake of the ship. When I left the deck at midnight Mount Erebus presented a beautiful appearance astern, rearing its lofty peak above the surface of the waters in solitary grandeur; the surrounding land having become submerged beneath the horizon by the great distance, being not less than 120 miles off, the peak so completely isolated as to present all the appearance of an island. The sky was so clear that the volume of smoke from its crater was most distinctly visible, curling upwards in the blue ether.

Remarkable rock, like a Roman gateway.

CHAPTER XVI.

Remarkable rock, like a Roman gateway—Singular black dyke on
mountain summit—Off Possession Island—Cape McCormick—
Balleny's Island—Frozen fresh meat—The aurora australis—On
the return track—Sighting Bruin Light.

Friday, 19*th.*—Fine day, running along the land to the
northward, which was but indistinctly visible, from the
cloudy state of the horizon on the port bow. The polar
flags were got on deck for an airing, as it now appears
very improbable that they will ever fly nearer the
Magnetic Pole, at least for this season. During the first
watch I saw several bergs and islets on the port quarter,
one a very remarkable perforated rock, resembling an
old Roman gateway between two towers; another like
the leaning tower of Pisa. At 9.30 p.m. the sun set
most beautifully astern, lighting up the horizon with a
dazzling flush of red between the land and a bank of
dark clouds, which extended along the horizon. The
sea was open and clear of ice. Thermometer 24°, wind
easterly, latitude 75° 3′, and longitude 168° 45′, dip
87° 49′. Dark nights are now fast closing in again, as
the sun now sets for upwards of four hours, which,
together with the rapid formation of young ice within the
last few days, and the general aspect of the weather,
indicate the near approach of winter.

Saturday, 20th.—Blowing a fresh gale, running along the land at the rate of seven knots. It is the same island which appeared on our outward voyage so much like a bold promontory of the mainland. In the evening we were off Possession Island.

Sunday, 21st.—Cloudy and unsettled weather; going before a strong breeze. The land on the port side high and covered with snow, the lofty peaks enveloped in clouds. We doubled a bold, black cape, christened Cape McCormick by Captain Ross, off which several large heavy bergs were lying, and I took a sketch of it. Latitude 71° 5′, longitude 169° 58′. Land about thirty miles off. In the evening I saw a remarkable black dyke intersecting the summit of an inland hill and lofty cone.

Wednesday, 24th.—Thermometer down to 25°, wind westerly, latitude 70° 14′ 39″, longitude 167° 34′ 30″. Going along the land all day, which I took a sketch of soon after tacking, at five p.m., when it bore south-east astern. A number of heavy large bergs skirted the coast-line, and the summits of the snow-clad mountains were half concealed by a girdle of light clouds. The ship was under double-reefed top-sails and close-hauled. A strong, fiery breeze blowing, with a clear blue sky, and bright sun, was very bracing and elastic. A shoal of spotted whales passed on our port quarter. At eight p.m. the sun set behind a narrow bank of clouds, very bright and dazzling.

Thursday, 25th.—Fine clear day. Beating up for the extreme point of the land, off which a line of packed ice was attached, and bergs lying. Westward of the Admiralty range of mountains and the icy barrier extending from Cape North, in a westerly direction, the land trends, apparently to the southward again, and maintaining its great altitude, a south-westerly course between Victoria Land and Balleny's and D'Urville's Islands might lead to a nearer approach to the

Magnetic Pole. In latitude 65°, and longitude 162° E. Whilst drifting down towards the pack, large whales were numerous along the edge of it. About eighty large bergs seen from the mast-head, forming a chain. After some anxious hours we were within some half a mile of the bergs, on which the surf was breaking and roaring, looking out for Wilkes' Land, in the latitude of 64°, and longitude of 164°. At 7.10 p.m. we tacked off the land. Five small islands were seen, and I took a sketch of a very remarkable peak, bearing S.E., strikingly resembling the rock called Lot in Fairyland, St. Helena. Thermometer 20°, wind westerly, latitude 70° 6′, longitude 167° 27′. At eight p.m. extremes of land bore from S. ½ E. to E. by S. ¼ S.

Friday, 26th.—Overcast and gloomy weather, snow and mist, with much swell, no land or ice in sight. We have had to-day a greater number and variety of birds flying in the wake of the ship than we have seen since entering the Antarctic circle—white petrel, immature petrel, ash-backed, a stormy, and a gigantic petrel, a sooty albatross, lestris, and several of the pretty blue petrel. Had the hot-air stove lighted to air the ship. Latitude 69° 52′, longitude 167° 53′.

Saturday, 27th.—Gloomy weather, with overcast sky and a strong breeze. At three p.m. I shot a pintado, or immature petrel, flying over the ship, which I secured by its falling in the port-waist. The sky this morning presented a most extraordinary wild aspect, curdled cumuli, mottled over with a fiery red, and the horizon astern was suffused with a glaring red mist.

Tuesday, March 2nd.—In latitude 68° 27′, and longitude 168° 6′ 18″, at three p.m. Saw an appearance of land on the starboard bow, which at eight p.m. proved to be Balleny's Island, bearing S.W. by W. to W. by S., appearing like two small islands, and very high land, with a lofty peak on the right, on the weather port bow.

A few ash-backed and numerous pintados about the ship, flying generally very low, skimming the surface of the sea to leeward, and passing under the stern to windward, but not attaining a sufficient altitude to justify me in shooting them, as the odds would be very much indeed against their falling in-board. So our feathered friends were permitted, doubtless to their great satisfaction, to follow their vocations instead of becoming, at the sacrifice of their lives, immortalized in some of our museums of natural history. Some stars appeared for the first time this evening, looking very bright. Our consort the *Terror* looked like a ship of glass, so embossed all over in ice was she, her side presenting a long, broad band of white.

Thursday, 4th.—Blowing a gale of wind, accompanied by a heavy sea. Weather overcast with snow and mist, surrounded on all sides by bergs and heavy stream ice. Dolphin-striker carried away. Latitude 66° 43′ 38″, longitude 165° 45′.

Saturday, 6th.—The finest day we have experienced for some time past, mild to the feelings, with a clear blue sky and sunshine, light winds from the eastward, and calms. Thermometer 31°, latitude 65° 50′ 31″, longitude 164° 45′ 13″. Passed several bergs, one singular-shaped mass strikingly resembling a village church, with a short, peaked spire rising from a round tower. A solitary white petrel passed the ship. These well-named snow-birds appear to be confined to within the limits of the Antarctic circle, we having first of all fallen in with them in the latitude of 66° S., on the outward-bound voyage, and they are now disappearing again in about the same latitude, the blue petrel now superseding them, with great numbers of the pintado, and an occasional sooty albatross and stormy petrel; an ash-backed petrel now and then mixing with them.

Sunday, 7th.—The articles of war read, followed by

prayers. Mild, gloomy day. Thermometer 32°, latitude 65° 30', longitude 162° 15'. I counted sixteen or seventeen bergs to leeward, on the port side, some of them very large ones; one in particular was both large and lofty. We have passed over the spot where the American Antarctic Expedition have laid down land, but we found only an open sea, with not the vestige of a bit of land in sight. We had some fresh beef roasted for dinner to-day, which had been killed in December last, and suspended under the mizzen-top in a bread-bag ever since. It turned out most excellent, being more tender and juicy than when first slaughtered. I opened a jar of Tasmania honey on the occasion which had been given me by my kind friends the Gregsons on my taking leave of them, under their hospitable roof at Risdon. I also broached a bottle of whisky.

Friday, 12th.—Fine day, with strong breezes. At three p.m. I shot and succeeded in securing another specimen of the pintado I have named the *Procellaria antarctica*, which fell on board between the companion and main-mast to windward (port beam). There was a larger proportion of blue petrel about to-day. Passed a fine, large, white-looking berg, with smooth sides and not a crack in it, looking as fresh as if only just sliced off the barrier. Another long, low berg appeared some distance to leeward. In the evening several large finners passed to leeward, sending up large jets of spray, with a loud spouting noise, the upper surface of their backs alone visible, leaving a track of foam in their wake. Latitude 64° 12', longitude 161° 28', thermometer 31°.

Friday, 19th.—Weather overcast and misty, going eight knots and a half before a favouring breeze. I shot three Antarctic petrel, but only succeeded in securing one of them unfortunately, which fell into the stern-boat. Latitude 64° 21' 20", longitude 148° 45'.

Saturday, 20th.—All last night the ship was sailing

through young ice, at times receiving some heavy thumps from larger pieces, as she ran along the edge of the heavy pack before an easterly breeze, with studding-sails set, in latitude 65° 10′ D.R., longitude 143° 21′ D.R. We are now in the longitude of Tasmania. The Antarctic petrel numerous to-day ; I shot four, and succeeded in reclaiming three of them, one falling on the weather, and one on the lee side of the deck, and the third into the stern-boat. I sat up during the night skinning five petrel for the ornithological collection.

Sunday, 21*st.*—Weather overcast and gloomy, with snow at noon. Had divisions and divine service. At dinner fresh roast beef, which had been kept for the last three months, and was excellent.

Tuesday, 23*rd.*—Last night, between the hours of ten and midnight, during the first watch, I saw for the first time the aurora australis. It formed an arch extending from north-west to north-east, from which pale yellow or straw-coloured rays shot upwards to the height of some twenty or thirty degrees, converging towards the zenith. It proceeded from a bank of dark clouds (13° altitude) on starboard side, and to windward. A diffused, pale light crested the bank of clouds from which the rays of the aurora shot upwards. The night was beautifully clear, and the stars numerous. Over the starboard quarter to windward the Southern Cross shone conspicuous near the zenith. To-day the weather was overcast and misty, with a fall of snow. A large berg on the port quarter. This evening I again saw the aurora faintly in the south-east quarter. The crow's-nest was taken down this morning ; and during the night we crossed the line of " no variation," the dip being 83° 57′.

Wednesday, 24*th.*—Thick weather, with much snow ; wind increased to a hard gale, ship going before it, and rolling heavily, her decks covered with sludge from the falling of lumps of snow from the rigging and sails aloft. The only birds seen were a few blue petrel, a sooty albatross, and a bird with long, pointed wings, skimming the surface of the sea at a distance, like the sheerwater, which in all probability it was.

Friday, 26*th*.—This evening I saw the first large albatross (*Diomedia exulens*) since we first entered the Antarctic circle, in latitude 59° 23′ 38″, longitude 130° 9′ 6″.

Saturday, 27*th*.—During the first watch I saw some brilliant displays of the aurora australis, forming an arch of white light astern, at an altitude of 30°, extending from south-east to north-west, and another arch ahead, like light, white, fleecy clouds, forming an arch of 20°, a faint band of white light extending from the east to the west horizon across the zenith. The ship's head at the time being N. by W. ; latitude 58° 2′ 7″, longitude 128° 40′, thermometer 35°, wind E.N.E to W. Weather misty and disagreeable, with sleet and snow.

Friday, April 2*nd*.—This forenoon the first cutter was lowered, in which Captain Ross went away to sound with 5000 fathoms of line on the reef, and got bottom in 1540 fathoms, in latitude 51° 10′ 6″, longitude 136° 55′ 45″. A cask, containing a paper signed by the captain and officers, was thrown overboard.

Sunday, 4*th*.—No divisions or divine service. At 6.30 p.m. a heavy sea carried away the lee quarter-boat (second cutter) from the port quarter. Strong breezes and gloomy weather, with much sea on. Numbers of the small black-backed albatrosses about the ship, with a few *D. exulens* and blue petrel.

Tuesday, 6*th*.—In latitude 44° 0′ 32″, and longitude 145° 57′; thermometer 39°, wind W. ; a beautiful day, mild and genial after the boisterous weather we have experienced on our return from the Antarctic circle, going six and seven knots in smooth water, with all studding-sails set, low and aloft ; accompanied by numerous albatrosses, both large and small, and the blue petrel.

In the afternoon the land appeared like two hummocks on the port bow. Night moonlight. At 6.30 p.m. saw Bruin Light on the port bow, and a lunar rainbow astern, standing up Storm Bay.

Remarkable black trap dyke, intersecting a lofty mountain peak.
(*See page* 178.)

CHAPTER XVII.

Summary of our attempt to reach the South Pole—Physical features of the Antarctic continent—Climate—Geology—General sketch of changes in the earth's surface.

THIS our first voyage south, on the meridian of the Aucklands, has been rendered memorable by the discovery of an Antarctic continent, to be added to the terrestrial globe, capping the Pole to a vast extent. The eastern coast of this, on the meridian of New Zealand, we traced from 70° to 79° of latitude, when the great icy barrier, commencing from the terminal cape of Mount Terror, in the longitude of 170°, extended in a continuous perpendicular wall of ice, varying from 100 to 200 feet and upwards in altitude above the level of the sea for some 250 miles in length, arrested all further progress to the southward. Beyond, and at the back of this stupendous glacier, the distant peaks of land were just visible, named the Parry Mountains, after our distinguished chief, Admiral Sir Edward Parry, of Arctic celebrity. This mighty barrier could not have been less in thickness than 1000 feet, and off which we got soundings in 260 fathoms, where the height of the face of it, in latitude 78¼°, was 168 feet.

The gulf or bay connecting Mount Erebus with the mainland was named after the first lieutenant of the *Terror*, McMurdo Bay, a compliment well deserved by that worthy and meritorious officer. An island lying off it and the volcanic mounts was named, after the talented and distinguished hydrographer to the Admiralty, Admiral Sir Francis Beaufort, Beaufort Island; and another island, not quite so far south, was named Franklin Island, after the Governor of Tasmania, Sir John Franklin, whose melancholy fate in after-years became so painfully associated with the mysterious loss of our old ships the *Erebus* and *Terror* in the opposite hemisphere, after having buffeted and escaped the bergs and perils of the Antarctic seas.

My worthy fellow-boating companion in the exploration of Kerguelen's Land, Lieutenant Gerrans Phillips of the *Terror*, had a cape named after him in latitude 73° S.; and Captain Ross did me the honour of associating my own name with a striking bold, black promontory, abreast of Possession Island, on which we landed to take possession, in the name of her most gracious Majesty Queen Victoria, of this great southern continent, Cape McCormick, in the latitude of 71° 55′ S., and longitude of 170° 58′ E., forming the terminal coast-line extremity of the ridge descending from Mount Whewell, situated about midway between the two most remarkable peaks of this magnificent mountain range, Mounts Sabine and Herschell, towering above all; the former on the north side, having somewhat the shape in outline of the opera-house at the Havana; and the latter on the south, resembling a stupendous crystal of quartz, projecting above the surface of this vast mass of glaciation, to which the angles taken gave the height of 7867 feet, and to Mount Sabine an altitude of 8444 feet. The highest summit on first making the land rose to 9096 feet. The extreme height of this grand mountain range, culminating in the two magnificent volcanic peaks, attaining elevations of above 12,000 and 10,000 feet. Mount Erebus, the highest, in a high state of activity, belching forth dense volumes of smoke, at intervals with flame intermingled, from its lofty crater, emerging from a colossal pyramid of eternal ice and snow; and Mount Terror, now extinct,—were named after the two ships. From the terminal ridge of the latter, the Great Barrier extends in an easterly direction. The remainder of the coast-line received the names of the other officers of the expedition, and the mountains of the inland range those of the official authorities at the Admiralty and other public and scientific men as far as Cape North, where we first sighted the land 100 miles at least distant. The voyage extending over five months at sea.

The climate of this Antarctic region of our planet, with its vast accumulation or cupola of ice at the pole, keeps the temperature, even in what goes by the name of the summer season here, down to the freezing-point of 32°, for it rarely ever exceeds this a fraction, on its most congenial days—vividly recalling to the imagination some faint idea of the condition Europe and our own islands must have presented in the remote past, during the great glacial epoch, with the winter in aphelion.

The soundings in high south latitudes indicate a general rise of the ocean bed, as we approach the southern pole, and the deep azure blue colour of the cracks and fissures in the Antarctic ice is due to the blue rays only being reflected, whilst the others are absorbed. The Antarctic bergs differ from the Arctic ones, in their general formation; being detached from the outer margin of the barrier, the ice is much purer and freer from earthy

matter than are the Arctic bergs, for the most part
formed in deep ravines or fiords along the coast; consequently the greater transparency of the southern ice
may give to it the intensity of blue which so distinguished it from the northern ice.

It may be worth while here to make a few observations on the changes and vicissitudes of climate which
the surface of our earth has undergone since it was first
launched into the boundless realms of space, in so far
as it may bear upon the eventful history of the successive
creations of organic beings, animal and vegetable, which
the ever-changing conditions of land and water have
brought about and unfolded to us in the strata of the
crust of the earth, in which they have been so wonder-
fully preserved.

The glacial epoch, however, was not due to terrestrial
changes alone so much as the result of cosmical causes
dependent on the eccentricity of the earth's orbit, and
the precession of the equinoxes, combined with changes
in the obliquity of the ecliptic, and nutation, a small
gyratory, nodding, conical movement or oscillation of
the earth's axis, a retrograde movement along the ecliptic from the east to west, and very slow motion of the
pole of the heavens among the stars, round the pole of
the ecliptic, occupying no less than nearly 26,000 years
in completing the tour of the ecliptic; during which the
sun loses one day in the year on the stars, by its direct
motion in longitude, and the equinox gains a day during
the same period on them by its retrogradation. This,
indeed, of late, has been a subject of controversy which
I leave to others to settle.

The pole-star, a star of the second magnitude in
the constellation of the lesser bear, is at the present
time somewhat more than a degree from the pole and
will continue to approach till within half a degree of it.
Its movement through space is not more than one mile

and a half a second, but at such a vast distance from us that its light takes half a century to reach the earth, whilst that splendid star Vega, the pole-star of the future, next to Sirius, one of the brightest of the first magnitude stars in the northern hemisphere, moves at the rate of thirteen miles a second, its light taking twenty-one years to reach us.

The precession of the equinoxes was observed some 2000 years ago by Hipparchus, who first compiled a catalogue of the stars. It is brought about by the diurnal motion of the rotation of the earth on its axis, and with which the phenomenon of nutation, a minor movement due to the moon's attraction, is intimately associated. The earth revolves at the rate of 1000 miles an hour, and at the same time moves in its orbit through space at the rate of 70,000 miles in the same space of time. Moreover, the friction of the tides would appear to have some influence on this diurnal rotation of our planet. The pole being merely the vanishing-point of the earth's axis, the retrograde motion of the latter is consequently a conical one.

The changes of climate from variations in the eccentricity of the earth's orbit must have been considerable, deviating as it does from a circular orbit, the sun, whose mean distance from the earth is in round numbers some 92,000,000 miles; but when the eccentricity is at its superior limit ·07775, will be no less than 99,000,000 of miles distant, when the earth is in the aphelion of its orbit; when in perihelion it is only 85,000,000 of miles, so that our planet is really 14,000,000 miles farther from the sun when in the aphelion than in the perihelion of its orbit. The difference between the perihelion and aphelion distances of the sun at the present time amounts in round numbers to only 3,000,000 miles. The eccentricity of the earth's orbit, it would appear, has been diminishing for ages past.

Although astronomers will not admit that the earth has changed its axis of diurnal rotation as opposed to the known law of a spheroid rotating round its shortest diameter, and that the latitudes would not remain constant as they now are, nevertheless it may be movable as regards space, arising from the pole of the heavens opposite the earth's axis describing a circle around the pole of the ecliptic whose obliquity is 23° 28'; the decrease in the obliquity of the ecliptic amounts to 45' per century only, so almost imperceptibly slow is the movement. Had the earth's axis been perpendicular, instead of being inclined to the plane of its orbit at an angle of 23° 30', the day and night would have been equal all the year round, but this angle not being constant, oscillates to at least a degree and a half.

The earth is an oblate spheroid whose equatorial circumference is in round numbers 25,000 miles, and the difference between the polar and equatorial diameters is twenty-six miles and a half. The siderial exceeds the tropical year by about twenty minutes.

Our planet in its path round the sun has in the past doubtless been subjected to extraordinary vicissitudes of temperature, however brought about, commencing with its own heated nucleus, changes in its position with reference to the sun, possible variations in the amount of heat received from the sun itself, from changes in the relative distribution of land and water, together with perhaps irregularities in the temperature of space itself, which has been estimated at 239°, or, according to La Place, 100° below zero. In stating 239° as the temperature of space, it must be understood that of stellar space, and not the absolute zero without the heat of the stars, which would be 222° below that of stellar space.

From the constant low temperature of the sea-bottom, everywhere cold and dark, with a constant temperature under every latitude of about the freezing-point, it would appear to derive no increase of heat from the earth's interior. Yet it has been found, in descending the deepest mines, the temperature increases in the ratio of one degree for every sixty or seventy feet below the surface, and during the carboniferous epoch it has been supposed that the internal heat of the globe affected its surface so much as to produce a subtropical climate from pole to pole. The thickness of the crust of the earth has been variously estimated, some supposing it to be about thirty miles; and the nature of the nucleus, whether fluid or solid, remains still an open question. All we know is that the lowermost rock in the crust is granite, this passing into gneiss, and above these lies the Laurentian formation, made up of the disintegrated granite and gneiss, syenite being a subsequent eruption to the granite, and differing from it in structure by having the mica replaced by hornblende, and being often found in large spherical masses, resembling cannon-balls, remarkable specimens of which I found in New Zealand.

Above these Plutonic rocks we have the argillaceous schists and sandstones, and the various other sedimentary strata of the limestone and sandstone formations, through which the igneous rocks have been erupted by volcanic action, or upheaved the strata from the original horizontal position at various periods.

The deepest mine does not exceed half a mile in depth, and no boring a mile. The highest mountain on the surface of the globe does not exceed five miles in altitude, and the greatest depth of the ocean does not amount to more than this. Mount Everest, the highest elevation of the Himalaya, reaches, in round numbers, 29,000 feet, and the deepest soundings in the Pacific Ocean 4475 fathoms. The deepest depression of the Caspian Sea is 3000 feet, and that of the Dead Sea 1300 feet below the ocean. One half of the weight of the rocks constituting the earth's crust is oxygen, and the most abundant rock is silica.　Chalk is a fine calcareous flour derived from foraminifera and other marine organisms, as are also layers of flint.　Mountain-ranges form the crests of the great wave into which the crust of the earth has been upheaved by subterranean agency.

After these few brief observations on the structure of our planet and its relations to space, a glance at the glacial period or ice age of the post-tertiary or pleistocene time, when our own islands and Europe covered with a mantle of ice and snow must have presented a very similar aspect to the Antarctic regions of our own time, may possibly afford some explanation of the problem as to how those fossil remains of both animals and plants have become entombed in such high latitudes, north and south, as Greenland in the one hemisphere, and Kerguelen's Land in the other; forms, when living, so organized that they could only exist in a tropical clime, and their remains found in so perfect a condition, the corals, for instance, that they must have lived and died on the spot where imbedded in the rocks.

The glacial epoch, as we have seen, has been the sequence of cosmical causes in combination with terrestrial changes brought about by the precession of the equinoxes about every 10,000 years, and alternately in the two hemispheres, that is to say, when one hemisphere is glaciated under a mantle of ice and snow, the other has a warm and temperate climate, and *vice versâ*, when the cold has reached the maximum in the one hemisphere the warmth would attain its maximum in the other. When the eccentricity is at its superior limit ·07775 and the winter occurring in aphelion, the earth would be in round numbers some 8,000,000 miles farther from the sun than at the present time, with the eccentricity at the inferior limit of 0168 and the distance somewhat above 90,000,000 miles, with the winters at present nearly eight days shorter than the summers, whilst, with the eccentricity at its superior limit, the winter solstice being in aphelion, the winters would then be no less than thirty-six days longer.

During this remarkable cycle of intense cold there would appear to have been inter-glacial periods or recurrent changes of climate, accompanied by oscillations of the sea-level, and consequent submergences and emergences of the land, giving rise to the submerged forests and raised beaches, and to which the coal formation owes its origin.

Thus the warm interglacial periods in the Arctic regions produced the miocene flora of Spitzbergen, Greenland, and North Devon, forming the shores of Barrow Strait and the Wellington Channel, as indicated by their fossil remains of animal and vegetable life that could only have existed in tropical or sub-tropical seas and lands, some 12,000 years in the past perhaps.

How often and often have I pondered and meditated over those relics of ages long gone by ; corals and other forms of oceanic life, by their organization only adapted to exist in the seas of the tropics, their delicate structures as perfect as when living, which I have myself found embedded in the Arctic table-land limestone formation of North Devon, and the entombed forests of coniferous trees I found in the volcanic land of Kerguelen, in the Antarctic seas, where now not a vestige of arborescent form of vegetation exists. As all these forms have come under my observation, I have wondered how they could ever have existed there ; for whether admitting that the heat derived either from the causes above stated, or from the greater temperature of the crust of the earth not then cooled down to its present condition, yet whatever might be the amount of heat required, and so obtained, still there is this difficulty to encounter, the absence of the sun's light during the six months that orb remains

beneath the horizon at the pole in the winter of the
Arctic regions. The difficult question which here meets
us to answer is, could these dicotyledonous, hardwood
forest-trees, found fossil in Greenland and Spitzbergen,
such as sequoia, salisburia, numerous conifers, besides
oaks, maples, beeches, poplars, planes, magnolias, &c.,
have existed without the sun's rays in a land of darkness for one half the year, had the earth's axis been the
same then as now? A problem, I believe, only to be
solved by the admission of some adequate change in the
relative position of earth and sun, however brought about.
A transference of the ice-cap, or accumulation of ice at
the pole from one hemisphere to the other, it has been
supposed, might result in a displacement of the earth's
centre of gravity.

During the warm interglacial period of the miocene
flora, in mid-tertiary times, was also a period of the maximum of intensity of volcanic action on the crust of our
globe. Lyell supposes Dover and Calais were united by
an isthmus in miocene times. The concentric rings in
the coniferous trees found in the coal measures would
appear to indicate the existence of seasons at that period.
Volcanic action in Auvergne broke through the granitic
plateau in miocene times.

Before bringing to a close this glance at the changes
our planet has gone through in the past, a few brief
remarks on the effects of denudation on its surface may
render this deeply interesting subject more complete.
The vast and startling periods of time assigned by most
geologists to the silent operations of nature in bringing
these changes about, would seem to be founded on rather
questionable calculations; for from recent observations,
both on bone caves and raised beaches, the inference to
be drawn is that denudation has gone on far more rapidly
than has been generally admitted. With reference to
the bone caves, the time required for the deposition of

the travertine, forming the stalagmites and stalactites, from the carbonate of lime held in solution, may be greatly influenced by an excess of carbonic acid, increasing the dissolving power of the rain-water as it percolates through the limestone, thus accumulating a greater amount of deposition ; and the calculation made from a long series of observations on the raised beaches of the North American lakes, gives only some 6000 or 7000 years for the deposition of all the several beaches, so that thousands may be nearer the mark than the millions of years so often assigned to the operations of nature.

As these pages were passing through the press, my friend Mr. Carruthers, of the Botanical Department of the British Museum, South Kensington, called my attention to the last number of the *Tertiarflora Australiens* for 1883, sent to him, in which Professor Dr. C. von Ettingshausen had done me the honour to name some fossils, which I found in the Lindisferne travertin limestone deposit near Hobart Town, when there in H.M.S. *Erebus* with the Antarctic Expedition. The willow named after myself, and the cinnamon after Hobart Town, with the third specimen, *Echitonium obscurum*, will be found at the end of the next chapter, from drawings kindly made for me by Mr. Carruthers.

CHAPTER XVIII.

Quail-shooting at Risdon—Lindisferne limestone fossils—Levées—The
Erebus and *Terror* ball—Visiting—The bushranger defeated.

Wednesday, April 7th, 7.45 *a.m.*—Beating up above the
lighthouse on the Derwent, with a strong breeze; firing
several guns for a pilot. About noon Sir John Franklin
came in a cutter to meet us, saluting us with three
cheers, which we returned off Hobart Town. The squalls
were so heavy as to heel the ship over to a considerable
angle; and at 3.45 p.m. we let go our anchor off the
paddock.

Thursday, 8th.—At seven p.m. I dined at the mess
of the 51st, and slept at the barracks. At eleven a.m.
I mounted one of the officers' horses and rode out by
the new town road to Risdon. A delightful day with a
refreshing breeze; I reached my friend Gregson's at one,
and returned on board at six p.m. When at Risdon I
found my paroquet alive, and greatly improved in his
vocal powers.

Wednesday, 14*th.*—Mr. Gregson and Captain Forman
dined on board with me; the former remained on board
all night. The following morning I met Judge Montagu,
just landed from his yacht at the wharf. At seven p.m.
I dined at the army mess, meeting Gregson there; I sat
next to the Inspector-General of Hospitals, Dr. Clark,
and was called upon for a speech on the toast of the
Naval Medical Department being given.

Monday, 19*th*.—I started in Captain Forman's gig-boat for Risdon on a quail-shooting excursion, with two or three others, and accompanied by two dogs. After breakfasting with Mr. Gregson, he accompanied us round his grounds about noon. The first covey of birds was flushed in the stubble-field adjacent to the house ; several shots were fired, but no birds fell ; and, as this covey is a pet one of my friend's, his regulations forbid any one firing a second time who had missed his bird. I did not fire myself till we had proceeded for about two miles, when I put up the first bird without the aid of the dogs, and bagged it. The next bird I shot rose to one of the dogs, and Mr. Gregson said it was the largest quail he had ever seen here. On our return to the house by the home-field, I took the shot due to me at the pet covey, and killed my bird to keep as a memento of the pleasant day's sport, and in mercy to my kind host's pets did not avail myself of the second shot I was entitled to from not missing a shot ; so that the covey came off well, losing only one of its mess. The birds were altogether only seven brace bagged. Gregson, the best shot of the party, killed two brace, Captain Forman one brace, Lieutenant Breton, R.N., one bird, Lieutenant H. three birds ; the fourteenth was a disputed bird. Besides my three quail, I shot two miners for the collection, both on the wing returning. We dined at six p.m., had music and dancing in the evening, and left Risdon at midnight.

Friday, 23*rd*.—Mr. Blackett's yacht, the *Albatross*, anchored in shore of us, and her owner having been a shipmate of mine in H.M.S. *Tyne* in bygone years, I called on board to see him, and asked him to dine with me on board the *Erebus*, which he did on ·the following Sunday.

Saturday, May 1*st*.—At six p.m. dined at Government House with Sir John Franklin.

Monday, 3*rd, seven p.m.*—Went to the Hobart Town

Theatre to see the new play got up in honour of the expedition, and called the " Antarctic Expedition." It was but rather indifferently got up, and not much better acted. It was concluded at ten, and the after-piece, the " Robber of the Rhine," commenced at 10.30, and concluded about midnight. We had a front box, curtained round, to ourselves. Yesterday I saw my paroquet in Gregson's garden.

Thursday, 6th.—Dined at Government House with Sir John Franklin at six p.m.; and on the following day, at 8.30 p.m., I went to a large party at Stoke, given by Mrs. Spode, and left at two a.m.

Thursday, 13th.—Our ball committee met on board for the purpose of making arrangements and sending out invitation cards to the inhabitants of the vicinity, from whom we had received so much hospitality during our sojourn at this beautiful island. I was elected honorary secretary for the *Erebus*, and my old friend and boating colleague, Lieutenant Phillips, to the same office for his own ship, the *Terror*.

Tuesday, 18th.—At ten a.m. attended the ball committee, and received a specimen of the ornithorhynchus from Macquarie Plains as a present, but unfortunately it died in the transit, and I preserved its skin.

Thursday, 20th.—I attended the committee at the Observatory, and lunched there. Went to the *Courier* office, and saw the paper printed ; and at six p.m. dined at the mess of the 51st regiment with Dr. Clark.

Friday, 21st.—At 5.30 p.m. I dined at the Rev. Mr. Lillie's, and had music in the evening. To-day both ships were warped in shore alongside of each other in preparation for the ball.

Sunday, 23rd.—Divine service ; and a ball-room committee assembled afterwards in Captain Ross's cabin, when I signed some invitations to personal friends, to whom I specially wished to send cards for the ball, we having arranged for a certain number to be specially invited officially, and the surplus cards were divided amongst the officers of both ships, to enable them to pay a compliment to friends from whom they may have individually received attention during their stay.

Monday, 24th.—I went on shore at two p.m., and attended a *levée* at Government House. It was a very confused, bustling affair, even for such ceremonials, wedging through a narrow doorway between two very small apartments, from the waiting-room to the governor's closet ; and at nine p.m. I attended the governor's ball, and left at 3.40 a.m. On the following day I accompanied the Gregsons to " Fry's School," and to hear the band of the 51st regiment, after which we all dined together at their friends, the Crombies, a lawyer of Hobart Town. The next day, at two p.m., I called at Mr. Crombie's, and took the family on board the *Seahorse,* a beautifully fitted steamer, just arrived from England, to run between Hobart Town and Sydney.

Friday, 28th.—At eleven a.m. I walked across the paddock to Risdon, which I reached at one, and left at eleven p.m. The evening had been wet, but it afterwards cleared up ; in returning along the new town road I was up to the knees in mud, so bad is the road after rain. Mr. Gregson had not returnea.

Saturday, 29th.—Fine day. Upon the receipt of a note from Risdon, I started at 4.20 p.m. across the paddock ; sky lowering, with threatening black clouds. We dined at six, and retired at midnight. On the following morning, Sunday, 30th, we arose at 8.30 a.m,, and after breakfast Gregson accompanied me to examine a limestone quarry at Lindisferne. Our way lay through some pleasant woods ; and a lovely morning. The limestone is of a yellowish or buff colour, a fossil univalve shell occurs in the upper part, beneath which are found trunks and branches of trees, with the impressions of leaves. The deposit is perhaps sixty to seventy feet in depth, and apparently a tertiary formation. Gregson dined on board with me in the evening.

THE "EREBUS" AND "TERROR" BALL.—We have been favoured by a fine evening for our ball, which commenced at eight p.m., on Tuesday, the 1st of June. The approach to the ships was through a canvas-covered way, forming an arcade, lined with flags intermingled with branches of the "wattle," in its full yellow bloom, and other plants, the whole supported on a bridge of boats, and of sufficient breadth for two persons to walk abreast along it. A lamp-post was placed on each side of the entrance, so ornamented with native plants, as to resemble the mouth of a grotto, between which and the road through the paddock Sir John Franklin had got constructed a branch road, to enable the carriages to pass down the hill to the very entrance of this tunnel-like approach to the ball-room, which was formed by the upper deck of the *Erebus*, the innermost ship, whilst the *Terror*, outside of us, secured head and stern, with a bridge connecting the gangway, was allotted for the supper-table.

Our ball-room was covered in by a canvas awning, lined throughout with flags, and decorated with the various native plants, branches of the beautiful orange-yellow wattle, ferns, &c. The band of the 51st Regiment occupied an orchestra, covered with dark cloth rising to some feet above the deck, and ornamented with shrubs and flowers, in front of which was suspended a portrait of our Queen, encircled in a garland of flowers. Just abaft the main-mast rose a second orchestra, for the Hobart Town quadrille band, in the midst of a labyrinth of foliage. The capstan, also, supported a pile of Flora's productions, and in the centre of the flags, forming a screen between the forecastle and the waist (where lemonade was served as a refreshment to the dancers) was a floral device, representing the letters "V.R." The tops of the sky-lights were converted into ottomans covered with flags, and benches covered with scarlet cloth were ranged all round the sides of the ship, as seats for those not engaged in the dance. The whole was brilliantly lighted up by chandeliers obtained from the shore, and with lamps placed at intervals around the sides, the effect of which was very much heightened from the approach by the tunnel having been barely lighted sufficiently to enable the guests to find their way, so that, after wending along a gloomy, narrow passage for some sixty or seventy yards, a flood of light all at once burst upon them on stepping from the gangway upon the quarter-deck, and here the captains and officers were standing to receive their guests.

Captain Ross's cabin and the gun-room of the *Erebus* were assigned as dressing-rooms for the ladies, and were supplied with mirrors and most of the etceteras of a lady's toilet, down to hair-pins, eau-de-Cologne, and other perfumes. The descent to the lower deck was by the main hatchway, the steps covered with red baize, having a circular awning of flags decorated with flowers of the wattle, and rosettes made of bunting by the sailors. The ring-bolts had been removed from the decks, and everything that could possibly leave more space.

The governor, Sir John Franklin, and his suite arrived soon after eight p.m., and by nine o'clock the deck presented a very gay and animated scene, upwards of 300 guests must have been present during the evening. Each officer being a steward, and having the option of inviting ten personal friends, was well able to secure attention to all, and the more especially as each had a station assigned him in charge of a division at the supper-table, for which lots had been drawn, as the most equitable plan to avoid all partiality. I was very fortunate in this lottery, for, had the choice been offered me, I should most un-hesitatingly have selected the one which fell to my lot. It was the small table over the gun-room companion, just abaft the main-mast, with the capstan immediately behind me ; a snug, isolated berth, with just sufficient room to accommodate my own little party of seven, consisting of my friend Mr. Gregson, Mrs. Gregson, and Miss Gregson, Dr. Clark, the Inspector-General of Hospitals, Mr. Crombie, the lawyer, and two other friends of theirs. Supper was served at eleven p.m., and after some squeezing and pressure in the passage, through the narrow gangway between the *Erebus* and *Terror*, all found their seats at the table, the governor and his suite, with the two captains, occupying the after-part of the quarter-deck table, which was terminated by a small table athwart-ships for that purpose.

As usual on such occasions many toasts were drunk, and speeches perpetrated, accompanied by loud cheering and emptying of wine-glasses. McMurdo, the senior lieutenant of the *Terror*, with whom rested the arrangement of the supper-table, had, with his customary good taste, left nothing to be desired in its decoration, and certainly threw us in the *Erebus* into the shade, the difference being just in the same degree as existed in the minds of the two senior lieutenants themselves, with whom these preparations rested. The sides were lined with black and scarlet cloth, having candles placed at intervals backed by mirrors, for which purpose the looking-glasses intended as presents to the natives of any lands we might visit were taken from their frames, small bouquets were attached to these, and the effect was very pleasing. The chandeliers were tastefully formed of bright steel bayonets, which had a far more ship-shape appearance than our hired commonplace glass ones from the shore. The productions of Flora were most tastefully arranged in small bouquets variously grouped. The supper-tables bore on them poultry, dressed in various ways, pies, pastries, cakes, and jellies, with fruits ; of wines, port, sherry, and hock, and an abundance of cham-pagne. On returning to the ball-room dancing continued until daylight. My friends the Gregsons, who had arrived at nine p.m., I escorted to their carriage at the paddock at four a.m. The whole affair passed off well, every one seemed highly gratified with their entertain-ment, and even the elements were propitious, as our guests had a fine evening for their arrival, and a morning not less so for their departure.

The *Erebus* and *Terror* Ball will doubtless long be remembered by the Tasmanians as a memorable event in the history of their very beautiful island; and most assuredly the boundless hospitality which every member of the expedition received at their hands will be as long remembered on their part as a no less interesting epoch in their own wandering lives.

The decks of the old ice-and-weather-beaten ships never before responded to the elastic step of so much female loveliness and beauty, as this small island of the Antipodes mustered on the occasion; and at that division of the supper-table over which as a steward I presided, it fell to my lot to have the honour of having seated next myself at table the " belle " of the island, Miss Elizabeth Gregson, of Risdon, a beautiful and accomplished girl of eighteen, who, with her father and mother (as already stated) and a mutual friend or two, constituted the small group at my table.

Friday, 4*th.*—Captain Forman's boat took me to Risdon for a kangaroo hunt on Grass Tree Hill; Mr. Gregson with his fine pack of hounds, and we with our guns. Not a very successful kangaroo chase, as the dogs having started one, took to the hill and got over it. I shot a guinea-fowl on the grounds of my friend's house, and a ground thrush. We dined at six p.m., and on the following morning, Saturday the 5th, after breakfast, I strolled round the grounds with my host's children, and shot an island crow, a thrush, and four paroquets. Dined at six p.m., and it being a moonlight evening, my friend's eldest son accompanied me to the bottom of the hill to shoot opossums, but we were not successful in meeting with any.

Monday, 7th.—After breakfast I started for Hobart Town accompanied by Gregson, and at six p.m. we both dined on board the *Terror*, my friend remaining on board for the night.

Friday, 18th.—Rainy day. After breakfast I set some of the Kerguelen Land cabbage seed in my friend's garden, and started on my return to the ship at 2.15 p.m., and at 6.30 p.m. dined at the mess of the 51st, and left at eleven p.m. Sir John Franklin was present, with Captain Ross. Thirty-three in all sat down to table, and Captain Forman called upon me for a speech on the occasion.

Saturday, 19th.—I called on Lady Franklin, who had been absent on an excursion to New Zealand, and only returned to-day in H.M.S. *Favourite,* of eighteen guns.

Sunday, 27th.—Fine day, with a fresh breeze. After divisions I started for Risdon. Dined at Gregson's at three p.m., where I met old Major de Gillan, a Peninsular officer. Had sacred music in the evening, and the house being full of guests, I slept on the sofa in the parlour.

Monday, 28th.—After breakfast I walked with the young ladies down to the creek, and whilst crossing it in their boat one of the party, in her over-haste to get out of the boat on its reaching the opposite bank, was very near going overboard, and certainly would not have escaped a cold bath but for the presence of mind and promptitude of her young friend, Miss Elizabeth Gregson, who, relieving me at the oar, kept the boat steady, whilst I sprang to the rescue of her friend. After this little adventure we walked on to the ferry, and looked over the new house in the course of construction there

R. McCormick, R.N., del.

Vincent Brooks, Day & Son, Lith.

Mount Wellington and River Derwent, Tasmania.
Sketched from Risdon.

for Mr. Gregson. On our return to the house, after joining the family at lunch, I started at four p.m. for the ship, accompanied by the Rev. Mr. Gell, who subsequently married Miss Franklin.

Thursday, July 1st, 10.30 a.m.—Walked to Risdon, and at three p.m. drove in Gregson's carriage to Major de Gillan's, and reached his picturesque cottage residence at 4.15 p.m., where I dined, and started on my return at 6.45 p.m. It is situated about a mile from the Richmond Road, and approached by a winding avenue. Many birds were flying about the trees, especially paroquets, and large flocks of sheep grazing. Reached Risdon at eight p.m. A fine moonlight night, and keen air. Found my friend Gregson but just returned from Hobart Town.

It proved very fortunate in the sequel, that I was induced to remain under my friend's roof for the night, as it was in all probability the means of saving my host's highly valued old family plate. I slept in the room usually allotted me in my many visits here, which is situated on the ground-floor in the opposite wing of the building to the one in which the family have their sleeping-apartments. The windows opened upon a veranda outside, under which the carriage stands. About five o'clock in the morning I was suddenly aroused from a sound sleep by a harsh grating sound, which I at first thought arose from the grating of the door on the hinges, by the servants opening it to remove my boots. But on glancing in the direction of the window, on the left side of the bed, I discovered that it was half open, and the burly figure of a man in a blouse outside, shadowed in strong relief upon the white window-blind. Although barely awake from that heavy sleep induced by weariness and over-physical exertion, it at once occurred to me that this nocturnal visitor could be no other than one of those bushrangers at that time by no means an extinct race in the colony, and who by some means or other had ascertained that this identical room was the depository of my friend's chest of plate.

Anxious to secure the marauder, I remained still and quiet to watch his movements, fully expecting that he would get in through the open window, when I was on the alert to seize him before he could get a footing on the floor inside. However, on his attention being for the moment arrested in his effort to raise the sash higher, and which did not readily yield, I seized the interval to slip out of bed and grope for the boot-jack at my bedside, the only available weapon at hand. This movement of mine, quietly as it was made, caught his ear, for he suddenly withdrew the half of his body, already within the window, and retreated. I quickly followed him out of the window in my night-dress, head foremost on the pavement outside, and grasped the iron-shod cross-piece of the carriage trace, the nearest thing at hand. Thus armed, I gave chase to the fugitive, pointing it at him, and threatening to shoot him unless he at once hove-to ; but, supposing it to be a gun, this idle threat only apparently accelerated his speed, and after a fruitless search round the garden and lawn, I, in returning, examined all the doors by walking round the house and trying them ; and, finding them all secure, I again turned in, and rested undisturbed till morning.

At breakfast I made no allusion to the incident of the early morning, for fear of alarming the female part of the family ; but Mrs. Gregson pointedly asked me if I was not disturbed by noises about the house in the night, and moreover said that she had found one of the latches of the doors raised this morning. Mr. Gregson himself seemed much surprised, and said that during the many years he had resided here he had never had any attempt made upon his house before.

After breakfast was over I took him to my bedroom, where his own eyes convinced him that such at least was

R. McCormick, R.N., del.

Vincent Brooks, Day & Son, Lith.

Risdon, Tasmania,
The residence of T. Gregson, Esq.

the case last night; for the large screw, the old-fashioned way of fastening the window, had been forced through the wooden frame, fracturing it; and this was the noise which awakened me. Mr. Gregson's eldest son John then recalled to mind that he had observed a very suspicious-looking character, resembling a bushranger in appearance, lurking about the premises on the preceding day, and speaking to some of their own workmen.

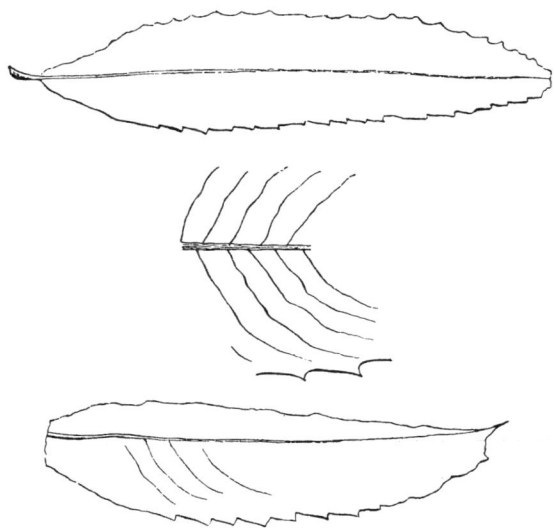

Salix McCormickii.

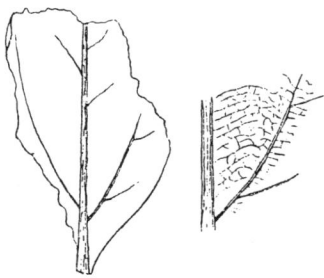 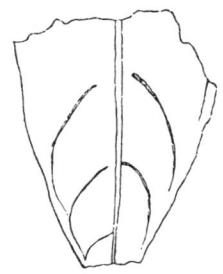

Cinnamomum Hobartianum.   Echitonium Obscurum.
(*See page* 199.)

Pomaré's Pah, Bay of Islands.   (*See page* 216.)

CHAPTER XIX.

Leave for Sydney—Visits—Trip to Paramatta—New Zealand—
Description—Visit to the chief Pomaré—His farm.

Tuesday, 6th.—Did not sail as was intended. Day remarkably fine. Walking out to Risdon, I met my friend
Gregson at the Ferry. Lunched with the family, as did
Dr. Hooker, who arrived after me ; started on return at
3.40 p.m. Gregson, with Mr. Crombie and the new
Surveyor-General, Mr. Power, took a farewell dinner
with me on board. Lieutenant Kay, of the *Terror*, was
the bearer of a book from Lady Franklin as a parting
present to myself.

Wednesday July 7th.—Weighed anchor at 6.30 a.m.,
and at 6.45 a.m. Sir John Franklin and suite came on
board from a brig. Manned yards and cheered him,
which was returned. Gregson slept on board last night,
and breakfasted with Captain Ross, and Sir John

Franklin this morning. At 10.30 a.m. the governor took his farewell leave of us in Storm Bay, with a mutual exchange of cheers. My old friend Gregson left at 11.10 a.m., and joined the governor in the brig. She bore down and cheered the *Terror*, and in passing us did the same. We had fine weather and a fair wind round Cape Pillar, but afterwards foul. The night was very fine. We had Lieutenants Otway and Carey, of the 51st, passengers with us to Sydney; and Captain Forman in the *Terror*. We were a week on the passage. Anchoring off Garden Island at 4.30 p.m. on Wednesday, 14th. The health-officer, Arthur Savage, Surgeon, R.N., came on board. We passed the entrance to the notorious Botany Bay, having a forbidding aspect; a heavy surf breaking over the reef at its entrance. Barren-looking cliffs of horizontal sandstone, interspersed with low sand-hills and sandy beaches. A heavy surf also breaks over a reef on the starboard side of the entrance to Port Jackson. The lighthouse and signal-station with a flag-staff appears on a promontory about 300 feet in height. On passing the entrance, which is about a mile in width, and turning sharply to the left, a noble bay opens, or rather an arm of the sea. The town of Sydney, the masts of the shipping, with the picturesque village of Wooloomooloo, appear at the head of it, five or six miles up. The surrounding hills are low and scrubby-looking.

Thursday, 15*th.*—I landed at Sydney for the first time at the government jetty. Wrote my name in the governor's visiting-book, and paid a visit to the club, where I met Captain King, R.N. Walked through the town to the race-course, returning across the government domain to the rocks called Lady Macquarie's Chair, and got on board at five p.m.

Sunday, 18*th.*—Fine day, but the last two or three days heavy rain, with thunder and lightning. I attended

morning service at St. James's Cathedral; Bishop Brough-
ton preached the sermon.   Walked to Wooloomooloo,
afterwards called on Dr. Savage.   On the following day
I met another brother-officer, Thomas Bell, at the club,
he had been out with Parry in one of his earlier voyages
to the north, and was now a settler out here.

Tuesday, 20th.—Fine day; went on shore and visited
the Botanical Garden.   On the following day heavy
rain; when I dined with Captain Ross to meet Dr. Bell,
Captain King, Mr. McLeay, and Captain Crozier.

Sunday, August 1st.—Paid a visit to the Roman
Catholic Chapel, service from eleven to one; being high-
mass day, it was very fully attended.   There were no
pews, we were seated on a bench, about the third or
fourth from the altar, a highly ornamented structure,
with the steps and platform richly covered with carpet-
ing and scarlet cloth.   Twelve tapers burning, the windows
of richly-coloured glass.   Seven priests officiated at the
altar, and the chief priest was distinguished from the rest
by the richness of the embroidery on his robes.   I visited
the cemetery afterwards, and called in Jamaica Street, on
an old brother officer, Sir John Jameson, a retired physician of the fleet, whom I had never before met.   He
invited me to dine with him on the following day, at
6.30 p.m., when fourteen in all, including some naval and
military, officers sat down to table.   Cards, singing, and
music followed, and I left at 12.30.   Old Sir John, on
whom the marks of age were becoming but too plainly
visible, told me that he knew my father, when he was
surgeon of H.M.S. *Defence,* with the Baltic fleet in the
year 1811, and endeavoured to dissuade him from returning home in her with the crippled ship *St. George,* of
ninety-eight guns, in company; as he said that from
his own position as physician to the Baltic fleet, it was
in his power to have effected an exchange for him into
some other ship, but that he could not be induced

to leave his own ship, and consequently perished with her.

Tuesday, 3rd.—After calling on Sir John Jameson, I dined with the governor, Sir George Gipps, at seven p.m. Twenty-three sat down to table at Government House; Captains Ross and Crozier present. Left at eleven p.m.

Wednesday, 4th.—At nine a.m. started from the wharf in the steamer, on an excursion to Paramatta. In passing up this narrow arm of the bay or river, a number of small bays or coves follow in quick succession on either side; beyond Kissing Point, about nine miles up, we passed some fine orange and lemon plantations, the trees weighed down with their golden fruit.

We reached Paramatta at 11.40 a.m., situated on a plain about a mile distant from the wharf, and through it the main road to Windsor in the interior passes. Passed a timber waggon drawn by twelve oxen. We visited the penitentiary, containing some 900 women and 400 girls and boys, employed chiefly as laundresses and needlewomen. At one p.m., on returning to the town, we lunched at the " Red Cow," a very pretty and quite rural-looking little inn, embosomed in trees and flowers, having a verandah hung round with cages, containing parrots and canaries. In the garden in front was a curlew, and the majestic-looking "native companion," or gigantic crane of the country, with its ash-coloured plumage and reddish head. Returned by the same steamer at four p.m., but did not reach Sydney until some time after dark, landing at 6.45 p.m.

Thursday, 5th.—Captain Sullivan, appointed to the *Favourite* from India, arrived last night, and takes a passage with us to join his ship at New Zealand. At 11.15 a.m. we sailed with a fresh breeze and fine weather, but afterwards stormy, with rain and foul winds.

Monday, 16th.—Mainland in sight; and during the

first watch last night we were off Cape Maria, Van Die-man. To-day we weathered the North Cape, two miles distant.

Tuesday, 17*th.*—Beating up for the Bay of Islands, and on the following day at nine a.m. we were off the bay.

There are two rocks on the starboard side, the outer one called Tiki-Tiki, with Cape Brett on the port side of the entrance, and somewhat less than two miles up the bay on the right, we passed the entrance to the Kiddi-Kiddi, or Gravelly-Gravelly, from the nature of its bottom. The New Zealanders, or Maoris, have a habit of repeating many of their proper names in duplex. Lying off it is the island of Motouroa, with several rocks; three or four miles still higher up is the Waitangui River, on the same side, and meaning in the Maori language, the "Crying, or Weeping Waters," from a waterfall in its course. Between the two streams is a low black ledge of lava, intervening between the argillaceous cliffs of which the shores of the bay are chiefly constituted. A broken-down, crater-shaped hill, but a short distance inland, sufficiently explains the source from whence this lava current has flowed at some period or other. At the entrance to the Waitangui lies the small island of Motoumane.

On the opposite or left side of the Bay of Islands is Paroa Bay, the scene of the murder of the French Captain Marion by the natives, separated only from Korora-riki by a narrow peninsula a quarter of a mile across; and just round Point Wayhihi, skirting a small bay with a shingly beach, is situated the settlement of Kororarika itself; in the native language—always euphonious—termed the valley or bay of Sweet Penguins; consisting of some score of houses, chiefly stores, ranged in a line along the beach, with perhaps double that number scattered about in a straggling manner at the back.

The native pah occupies the centre of the beach, formed of a few low huts, in the centre of which is the chief's, the whole being enclosed within a fence of stakes twelve feet and upwards in height. Above this is the new church, and to the right of it the Roman Catholic chapel; the signal-station crowns the hill above the town.

Nearly opposite to Kororarika, on the right side of the Bay of Islands from the entrance, which is here about two miles wide, is the missionary station of Paihia, or Paheha, the residence of the Rev. Mr. Williams, the head of the Mission, and formerly a lieutenant in the Royal Navy. The station consists of three or four good houses, having pretty flower-gardens in front, and skirted by a fine, white, sandy beach, in front of which lies the small island of Moutou Rangui, covered with trees, and situated seven miles up the bay. Just above Kororarika a deep cove runs up, called Kiddi Cove, terminated by a rock, Torré-Torré, connected with the mainland by a narrow neck, which is dry at low water, but may be crossed over in a boat at flood-tide.

At 10.30 a.m. we anchored in the Kava-Kava, or Bitter-Bitter, about two miles above Kororarika, and nine miles from the entrance to the bay opposite Tavumay beach, a sandy flat on the right, backed by a mangrove swamp, through which a creek or narrow river runs. This is said to be the spot where Marion was eaten by the cannibals, after murdering him in Paroa Bay.

About two miles above our anchorage, Pomaré's (the chief of the Bay of Islands) pah appeared, cresting a hill or ridge, where the Kava-Kava gives off the Waikadi branch to the left. On the left side of the bay, nearly midway between us and Pomaré's pah, is Russelton, on which is situated the prettiest house in the bay, in which Captain Hobson, the governor, lived when here. The barracks is built on the point. We found the American corvette *Yorktown*, with a whaler and several small vessels at anchor off Kororarika. The weather was so thick, blowing hard, coming in sudden squalls and violent gusts, as to obscure the distant hills, the rain falling heavily.

Monday, 23rd.—I landed this morning for the first time at Pahia, in Captain Ross's galley, accompanied by Captain Sullivan. The day was fine; saw a native chief, and called at Colenso's printing-office; walked over the hills at the back of Pahia, which are clothed with what is here called the tea-scrub, a fragrant, aromatic plant, bearing a pretty white bloom. Returned on board at one; found that Mr. Williams had been on board during my absence. In the afternoon I again landed by the Observatory, on Tavumay Beach, and shot two of the tui or parson bird, a beautiful bird, about the size of our starling having a plumage displaying the same metallic lustre and of the same species as I met with at the Auckland Islands; I also shot a small dark warbler; returning on board at five p.m.

Tuesday, 24th.—Landed at eleven a.m. at the Observatory; weather overcast with drizzling rain. I strolled along the beach to Pahia to call on Mr. Williams; I found him at home. He is a fine, dignified-looking man, inclined to be stout, with the frank, independent manner of the ex-naval officer. I was much pleased with him, and we had a long conversation about the land of his adoption; he told me that he had crossed New Zealand from Port Nicholson to the Bay of Islands, through the then unknown interior, in two months; he saw the central volcano with the hot springs and lakes, one thirty miles across, containing fish; and also a river; some of the highest hills were covered with perpetual snow. He had to cross a desert in the interior, which alone occupied him a week, and he had to carry his provisions for this time. He met with several species of birds unknown in the parts of the island hitherto visited. The rapidity with which the general diffusion of the Scriptures had taken place throughout the various tribes in the interior, he states as most remarkable, and effected entirely by converted natives from the Bay of Islands. Native Prayer Books were general amongst them; and, as a substitute for the church bell, they jingled together pieces of metal at their meetings for public worship.

Auckland, the capital, it appears contains the best land in its vicinity. Port Nicholson is subject to much blowing weather and sudden and heavy squalls. The Cowdie Pine is limited to the northern portion of the island; the southern island he states to be mountainous, and but sparsely inhabited. On White Island there is an active volcano. He showed me his track chart across the island, with the ridges of mountains running parallel S.W. and N.E.

Just before I left him, the widow of a chief came into the room—a well-mannered and very decently-dressed woman. The weather clearing up, I returned along the beach, shot a kingfisher, crossed the Observatory River, up the Mangrove Creek, and over the inland hills, along the native paths which intersect and wind round them in every direction, concealed in the distance by the thickets of tea-scrub which overhangs them. Returning by the river, I got on board at five p.m. to dinner. Mr. Williams was a lieutenant of the *Endymion* frigate when she captured the American frigate *President.*

Wednesday, 25th.—The tide surveyor called alongside in his boat, for me to go with him to see Pomaré's pah at ten a.m. We landed on a small shingly and sandy beach, flanked on the right by two gin-shops, chiefly frequented by the whalers, and called the "Sailor's Return" and the "Eagle Inn," having the shutters and verandah painted green. A narrow path, with three or four peach-trees in their pink bloom, led up the face of the hill to the pah. The huts of the natives commenced at the summit of the hill, and were scattered down the opposite side in divisions, separated by fences of stakes, amounting perhaps to fifty or sixty in all, each very low, with doors just high enough to crawl in at on hands and knees. The chief Pomaré's hut was situated at the bottom of the hill, only just above the beach, forming one of three placed in a row.

Pomaré himself was absent, planting potatoes at his farm, about six or seven miles up the River Kurrito. His hut was more ornamented than the rest; but this was the chief difference. The door was higher, being nearly five feet in height by two feet in width, sur-mounted by a red carved board and a group of dark feathers. The door was also red, and latched with only a bit of cord. On the left side was a small square wicket or window, having a shutter of the same colour, and over it was suspended a French coloured print, " L'Hiver." The roof was formed by two boards, meet-ing at an angle, having a porch in front, surmounted by two carved figures, a male and female. A plank across the basement formed the threshold. The whole structure was thatched with reeds, those on the inside of the porch were crossed by red and yellow bands, and the plank of the ceiling wreathed red and white.

On entering the interior a carved image presents itself in the centre of the floor, on which some scattered embers, the ashes of the last fire, remained. On either side were the sleeping-places, simply portions of the bare ground floor parted off by narrow ledges of wood. The only furniture consisted of an old black and yellow, long-fibred mat, an old musket, pair of canoe paddles, and a spear. Near the hut is a storehouse, elevated on posts from the ground. The whole pah and huts are encircled by a railing of stakes, irregular in height, ten to twelve feet high, and bound together with grass, a signal-staff crowning the hill. The only natives we saw were an old Rangitera—the term here for the rank of gentleman— coming next to the chief, his wife and daughter, an old woman, and a few children, the tribe being absent with their chief. We embarked again at noon, landed at Russelton, and again at the American Cousul's wharf on the Kororarika side of the bay, a little below the anchorage of the ships, calling at a pretty cottage and garden of an old resident. Started again at 1.30 p.m. Made sail on the whale-boat, which, carrying a very large lug-sail, as much as she could well stand under without a reef in the strong breeze, and sudden violent squalls that often struck her, with her lee gunwale a-wash with the sea, I all but anticipated a capsize, and was not sorry, though no novice in boating, when we at last effected a landing at Kororarika. Ten minutes had scarcely elapsed when I saw from the Custom House one of the native canoes upset in the surf opposite, during one of those heavy gusts of wind accompanied by rain. Paid a visit to the pah here; but old Rivers, the native chief, was absent. It was enclosed with the usual stakes, ten to twelve feet high. I rambled over the hill to Paroa Bay, and started on our return at 4.15 p.m., sailing a portion of the way, and reached the ship at five p.m.

Thursday, September 2nd.—H.M.S. *Favourite* arrived in the bay. Went on board.

Tuesday, 7th.—At 9.40 a.m. I left the ship in the gig for Pomaré's farm, up the Kurrito River. Blowing a gale, and threatening rain. On our way, calling at the " Sailor's Return," we met Pomaré's general, or fighting-man, as he is termed; a stout, resolute-looking fellow, much tattooed, which added a fierceness to his expression. The banks of the Kurrito, the left branch of the Kava-Kava river, are low, flat, and muddy; in places very shallow. We had the flood-tide up and the ebb down; the stream narrowing as we ascended. On the right a succession of mangrove swamps, mingled with rushes, ferns, and abundance of flax, growing luxuriantly in the low and swampy underwood and rushes. Beyond the low banks of the stream appeared a receding range of hills of moderate height, having a smooth, undulating outline, clad with short fern. On the left bank the argillaceous hills are clothed with dense woods of lofty trees and underwood, sloping down to the banks of the river. As we approached Pomaré's village the river became so narrow, as it winds through flax and rushes, that we were compelled to shorten in the oars to get the boat along, the stream being only the boat's length across.

We landed soon after noon, amid heavy rain, at the village, consisting of about a score of huts, scattered up the sides of the sloping bank. On reaching the first hut we found Pomaré himself squatted on his hams in the customary native mode in the outside porch. He was smoking his pipe, and enveloped in the folds of a blanket of a light-blue colour, which doubtless had been dyed to suit the taste of the wearer. Under this appeared a red guernsey-frock ; his feet were bare, a green jade earring was suspended from one ear. His aspect is that of a tall, powerful man, about six feet in height, and apparently between thirty and forty years of age ; but from his being much tattooed, and, like the rest of his country-men, his head covered with a profusion of coarse, bushy, jet-black hair, renders it a somewhat difficult matter in judging of the age of a New Zealander. He received us in the cold, calm, taciturn manner natural to these savages on a first introduction. Without rising from the posture in which we found him, he simply made a sign with his hand for us to come under the porch of his hut out of the rain, at the same time shaking hands with each of us, a ceremony rarely omitted by these people. Seated on his left was his wife, seeming to be about his own age. She had long, black hair, lips tattooed of a blue colour, and wearing earrings of blue stone, round in shape. She was dressed in an old blue gown, over which was thrown the native mat of light-coloured flax; she also was smoking her pipe, and at the same time suckling her infant, about two months old.

The favourite daughter was a very lively-looking girl, about eleven or twelve years of age, with jet-black eyes, and hair as black, worn short, a shark's tooth and red sealing-wax suspended from one ear; her dress an old print frock, with a black outer covering, and a dark, thick, long-fibred mat worn over her bronzed shoulders as a tippet. Her feet were naked. She was all life and vivacity, a pretty, intelligent, young savage. My haversack of specimens having attracted her attention whilst I was talking with her father, she darted off, and, quick as thought, caught a passing mouse, so keen was her bright, black eye, so rapid her movements; when, returning to me with a gratified expression, she added this diminutive representative of the mammalia to my haversack of specimens. Grouped around were his other wives and children, both boys and girls, with an aged native or two, some squatted at one end of the porch, and others, lounging outside; there might have been thirty or forty altogether.

On returning to the village we had a lunch of sandwiches and ale at the boat, whilst the natives were making their repast on potatoes and cockles, to which fare we added some of our sandwiches; but they were more anxious for rum, both men and women, nay, even the children showed a decided propensity for the "fire-water." This habit, with others equalling demoralizing, have been introduced among them by the whalers and other Europeans of a low order who have settled amongst and around them.

At two p.m., after I had given the chief some percussion-caps for his gun, we sent the boat round to a tree, a hundred or two yards lower down, as the tide was fast ebbing, and the consequence shallow water. We then walked across a swampy marsh to it, and finally started at 2.30 p.m. On our return I gathered a beautiful white clematis here, and saw several ducks, but they were far too wary and shy to permit us to get within shot of them. At 3.15 p.m. we reached the confluence of the Kurrito with the Kava-Kava amid heavy rain all the way, continuing without the slightest intermission until our return after dark.

Thursday, 9th.—Fine day. This afternoon Mr. Williams, accompanied by his wife and Miss Williams, with one of his sons, came on board to see the ship. I gave them some Antarctic specimens, and accompanied them on shore to the Observatory, where Captain Ross showed them the various magnetic instruments.

Monday, 13th.—Fine day. I left the ship at 7.40 a.m. in the gig, on an excursion up the Kava-Kava River, with the ebb-tide against us. We passed a quantity of timber logs floating down stream. Landed at several of the sawyers' huts on the banks, which presented the same luxuriant vegetation as on the Kurrito the other day. As the river became narrower, we had no small labour in getting the boat through, passing two rapids having islets in mid-channel. About noon we passed the Kava-Kava pah, about 200 yards from the left bank, and landed. A rich, alluvial, level tract extends along each side of the river, studded with huts and potato patches, enlivened by numerous clumps of peach-trees in full bloom. As we continued our course for four or five miles, all the way through dense woods of lofty pines, amongst which the tall, tapering spars of the famed Koudies were very conspicuous, rising straight as an arrow, the stream made numerous fine curves as it meandered through the flat, marshy ground fringed with underwood, in many places overhanging the stream so as to form a natural arcade. Ascending to where the river divides, we followed the branch to the right, which at a short distance farther on again divides, and becomes so narrow that the boat could only be got along by poling.

As it was now getting late in the day, I made up my mind to return, first landing on the potato plantation of a fine, venerable-looking old chief, much tattooed, who was busily engaged in planting his crop of potatoes ; his hut was completely embosomed in the densest wood. I saw more birds here than I have met with anywhere else, and shot several tui. At 3.30 p.m. we commenced our return, with the freshets running about three knots in our favour. In returning I landed at the Kava-Kava pah, consisting of fifty or sixty huts, enclosed within the usual stockade, twelve to fourteen feet high. A rudely built church, somewhat resembling a barn in appearance and dimensions, stands in the centre, and here the Rev. Mr. Williams preaches a sermon in Maori every Sunday. In the potato patch, through which a stream runs on its way to the river, the natives were planting the tubers. The chief, a young man, not much tattooed for a chief, came out and shook hands with me. I think he has been christened Matthew by the missionaries, under whose care all the natives of this pah have placed themselves.

The ebb-tide being in our favour, we started immediately. At 5.15 p m. I landed at the entrance to the Kurrito, for the boat's crew to get their supper, which they cooked over a fire burning outside of a native hut, on which cockles were roasting ; the natives were returning with bundles of wood from the forest. Whilst the boat's crew were at supper I went up the hill in search of pigeons, but did not meet with any ; only shot a small paroquet. Before I got back to the boat it was dark, and so dense was the wood and impenetrable the long, tangled fern, in many places higher than my head, that it was with no little difficulty I made my way back to the boat. At 6.40 p.m. we shoved off, passing some canoes and a light at a sawyer's hut, with the barking of a dog. On rounding a point the light at Pomaré's pah showed itself. At eight p.m. we got on board. Our farthest distance up the river must have been twelve to fourteen miles from the ship, and the pah up the Kava-Kava not less than eight or nine miles.

Tuesday, 14*th.*—At 10.30 a.m. I landed at the cove on the west side of the bay, and in the ravine over the hill I shot a shag, a pigeon, and an owl. In returning at high water my course along the beach was cut off, and I had to cross over the hill by a native path, through the tea-scrub and fern, to Mangrove Point, where I had to wait for a boat for an hour and a half, after firing at least a dozen times for one.

CHAPTER XX.

Trip to Waimate—The missionary settlement—Ascent of the mountain —Descend the crater—The sulphur-springs—Pomaré visits the ship.

Tuesday, 21*st.*—At eleven a.m. I landed at Pahia in the gig, to meet the Rev. Mr. Taylor, who, through the kind offices of Mr. Williams, had arrived for the purpose of accompanying me back to his home, to show me that station and the country in its vicinity. A group of the natives were assembled round Mr. Williams's house. We commenced our journey on horseback. The wind happened to be in a rainy quarter; the weather looked overcast and threatening. We ascended the hill above the creek at the lower end of Pahia. Our road at first lay over hill and hollow, by fern and tea-scrub, wood-crested summits, and along a narrow path, through which both horse and rider had to force their way. At 12.45 we discovered that we had taken the wrong path, by a few huts that appeared on our right, and had to retrace our steps, but only for about a hundred yards or so, descending a hill more to the left into a flat valley. About one p.m. passed by a Maori village of a few huts, surrounded by patches of potatoes, kumeras, and other vegetables. Crossed a rivulet running through it, and wended our way up a hill, from the summit of which the aspect of the country appeared quite changed to fern-clad hills, destitute of wood, and presenting a most bleak and barren appearance, arising, it is said, from the woods having formerly been burnt down by the natives, who, in clearing an acre to plant a few potatoes, often destroyed a whole forest of the finest timber.

At two p.m. we entered a path winding through dense brushwood, six to eight feet high, along a flat valley. A great deal of flax was growing amongst the rushes, in a sedgy, marshy ground. Some peach-trees, with their pink bloom, relieved the scene. At three p.m. we forded the river Waitangui, about ten yards across; and on coming to four cross paths, over a fern-clad table land, we followed one to the left, in the opposite direction to Waimate, to examine some blocks of limestone cropping out of the bed of the river, taking us some five or six miles out of our way. Reached this spot at 3.45 p.m., where the river winds through wood and tall fern, between steep banks where it is ten yards wide, and a strong current running.

Two blocks of highly crystalline marble show themselves four or five feet above the surface of the river, and a third projecting from the bank, having an E.N.E. and W.S.W. bearing, the course of the river itself being S.W. and N.E. I rode into the stream for the purpose of obtaining specimens from the blocks in the centre, to compare with the limestone formations about the Bay of Islands, but my horse losing his footing in a deep hole of water, alongside one of the blocks, was very near being carried down stream by the force of the current, and I could not induce him to approach the other block. However, being resolved to succeed in getting a specimen, after having diverged so much from our course for that purpose, I tried Mr. Taylor's horse, and, mounting him, once more rode into the stream, and after a little buffeting between horse and current, I got alongside the second block; but, between reining in a restive horse, and using my geological hammer at the same time, I only succeeded in chipping off a very small fragment, so hard was the marble, but this served my purpose. A pair of ducks rose from the banks of the river.

At 4.45 p.m. we continued on our way over a fern-clad and undulating country. When within a quarter of a mile of the settlement, the path led through a narrow belt of wood, having a most symmetrical and beautiful tree-fern, with a circular top, like an umbrella growing isolated on the right-hand side. On emerging from this wood, Waimate opened upon us all at once, reposing on a level spot encircled by hills of moderate height. The first building was the blacksmith's, with a hut or two, and a little farther on, on the left of the road, we entered a white gate, from which a straight carriage-road brought us to Mr. Taylor's house, after riding about 100 yards. We alighted at 5.45 p.m. The house fronts the N.W., and is the centre one of three, all constructed on the same scale, and the only good houses here, having verandahs and large flower-gardens in front, and separated from a paddock by a railing. On the right is a pond for the ducks, with a water-mill and clump of trees. At about 200 yards to the left of Mr. Taylor's houses, rises the light tapering spire of the neat little church in progress of building, beyond which is another clump of trees.

The whole scene, with its enclosures of clover, and parterres of English flowers, reminded me more of some quiet, peaceful hamlet in England, than a missionary settlement in the wilds of New Zealand, which barely a dozen years ago was covered by an almost impenetrable fern. After changing my drenched clothes in a small bedroom on the left side of the verandah, having two windows looking into the garden, which was allotted me as a sleeping-room, we all sat down to tea, and turned in at eleven p.m.

Wednesday, 22nd.—Arose at seven a.m., and before breakfast took a stroll round the garden, which was full of English flowers, so welcome to the eye of the wanderer.

We breakfasted at 7.30, after attending prayers. Both the walls and ceilings of the dining-room are of Kauri pine, which has a smooth, polished, and dark appearance. The tables were all covered with the native mats.

At 10.15 a.m. Mr. Taylor accompanied me on horseback on an excursion to a crater-shaped hill, about ten miles distant; the morning was gloomy. On passing through the wood by which we approached Waimate yesterday, our course lay to the right, over an undulating, fern-clad land; but in consequence of an impassable wood, we had to make a considerable *détour*, passing a clump of trees on the left, railed in, where wood-cutting was going on. At 11.40 a.m. we alighted at the Rev. Mr. Williams's farm, managed by his son, about six or seven miles from Waimate—a pretty little Maori-built cottage, having a thatched roof and reeded sides, with glazed windows, and enclosed in a garden, in which the wattle was growing in profusion. It is divided by the road from the huts occupied by his people, and where two or three of the native girls were busily occupied with their household duties; very good-looking lasses they were. Two of the Maories were ploughing in an adjacent field with four oxen yoked to the plough. The back yard was full of poultry. I took a glass of ale with young Williams in his picturesque little cottage, which, small as it is, he has made very comfortable; a very fancifully ornamented rifle hung over the sofa.

We started again at 12.15 on horseback, accompanied by him. At first across level, highly-cultivated fields, and extending over many acres, well enclosed by wood railings—a fine tract of rich soil, well cleared and productive. After passing a hut on the right, the door guarded by a dog on either side, we rounded a hill, through long fern, to a small arm of a lake, skirted with rushes and underwood, in a valley below the crater hill.

Here I shot a very pretty little grebe ; and at one p.m. we commenced the ascent of the mountain.

At 1.45 p.m., after tying our steeds to the bushes, we began the descent through the dense and tangled thickets with which the whole interior of the bowl is clothed, to the bottom of the crater, a depth of 300 feet, which we reached at two p.m. ; fragments of scoria and lava were scattered about amongst the rank vegetation and under-wood at the bottom. The death-like silence and solitude of the scene was only broken by the lively and varied musical notes of the agile tui or parson-bird, in the top-most branches of the tallest trees. I reluctantly shot two of them for the collection, and one of them hav-ing been only wounded by the discharge of the first barrel, after falling on the ground, instantly began climb-ing, in nautical phrase, hand-over-hand, up the branches of a tree, with the most astonishing celerity, till he was checked in his career by a discharge from my second barrel.

We made the ascent from this circular bowl of wood at another point which was found still more intricate, when, having reached the summit, we remounted our horses at 3.15 p.m. This old extinct crater, although of small dimensions, not exceeding 300 feet in width at the brim, is, without exception, the most perfect and sym-metrical model of a volcanic vent or cone, in an extinct state, that I have ever seen. On one side is a deep ravine, bare of wood, through which the lava-current once found an exit. Between it and the margin of the crater is a saddle or ridge formed from the scoria and ashes, but now clothed with fern, and through which, winding round the opposite side of the hill, we reached the sulphur springs, to the north, at four p.m.

These springs are situated in some small swamps and pools, overgrown with rushes on a level surface. The gas, with which the water is impregnated, is incessantly bubbling to the surface. The grass around is encrusted over with deposits of yellow sulphur. After drinking some of the water, and filling a bottle with it, we remounted our steeds, and reached Mr. Williams's farm at 4.30 p.m., where we took our leave of him.

The remains of former native pahs on the summits of the highest hills, present such a ridged appearance as to be easily taken for trappean terraces in the distance. At five p.m. we forded a river, and I had a shot at a brace of ducks which rose from its sedgy banks, but missed them, and I have not yet succeeded in obtaining specimens of the duck of this country. We arrived at Waimate at six p.m., dined at seven p.m., and attended the evening lecture from 7.30 to 8.20 p.m., where eighteen missionary youths attended, and Mrs. Taylor with her maids, both English and Maori. After tea we looked over some native shells in a cabinet in the drawing-room, and I turned in at 10.30 p.m.

Thursday, 23rd.—After breakfast, Mr. Taylor accompanied me to a Kauri pine-wood, about three miles distant, most of the largest trees having been cut down, and the largest living tree which I measured was twenty-four feet in circumference, from seventy to eighty feet in height, straight as an arrow, and without a branch, till very near the top; a quantity of resin was exuding from the bark of this tree. A dead one standing about the same height measured twenty-eight feet in circumference, and a section of a trunk lying on the ground was five feet in diameter at the top, and nine feet a little lower down. We commenced our return at 11.30; crossed a stream over the trunk of a large tree which had fallen across it; reached Waimate at 12.30, and after partaking of an early dinner, at 1.30 p.m. I took leave of my hospitable friends; at 2.40, mounting my horse, accompanied by one of Mr. Taylor's Maories on foot, to carry my traps, started on my return to the ship, which I reached at eight p.m.

Saturday, 25th.—Gloomy day, blowing fresh, in squalls. In the afternoon, as our dingy was returning from our sheep-station, where Abernethy, our gunner, was in charge of the sheep, she was capsized whilst under sail, in one of those sudden squalls so frequent here, and one of the two marines in her at the time was drowned; poor Barker, one of our best men, not much over twenty years of age, who had been one of my own boat's crew in the Kerguelen's Land boating expedition. The other marine was taken from the keel of the boat, to which he had clung till a boat, crossing the river at the time, rescued him from his perilous position, and took him to the barracks, to which I at once proceeded in the cutter to see him; he was in a state of great exhaustion. Afterwards I crossed the bay to the pah, to which the dingy had been towed, but could get no tidings whatever of the other poor fellow.

Tuesday, 28th.—About noon, Pomaré, the chief of the Bay of Islands, paid a formal visit to the ship, accompanied by twenty-four of his tribe, in his large canoe; a flag was displayed in the bows, and the model of a war-canoe lay at the bottom of the boat. She was paddled alongside by the women, when the whole party, with the exception of two or three left in charge of the boat, came on board.

Pomaré, on reaching the quarter-deck, took up his position in a very stately attitude by the companion-hatch. On Captain Ross coming to receive him, he asked in very good English if he was the captain. He was dressed in his state robes, a cap with broad gold lace round it, showy-coloured flax mat negligently thrown over his shoulders, beneath which a buff-waistcoat with gilt buttons appeared, as also his shirt-sleeves; his trousers of scarlet cloth, with a black band running down the side seams, and a pair of high shoes, completed his costume. His chief wife had her long black hair bound up within a polished metallic fillet with a number of bead necklaces round her neck, having a black and yellow flax mat over her shoulders.   The youngest daughter (whom I had seen when up the Kurrito River, at her father's potato ground, where she caught me a mouse for my haversack) wore earrings of a mass of white down from the albatross suspended in striking contrast with her copper-coloured shoulders, over which the ordinary flax mat was carelessly thrown.   Like the rest of her tribe, she had a pipe in her mouth—smoking is a general habit even amongst the youngest children in New Zealand.   The chief and his wife descended with Captain Ross to his cabin, where they had wine and trinkets given to them ; and Pomaré himself got a rifle as a present.   The rest of the suite assembled in the gun-room, where they partook of wine and biscuit or grog, according to their tastes.   They soon cleared our shelves of all the newspapers ; and from my own cabin I distributed amongst them skeins of thread of various colours, which seemed to be in great request amongst them, especially by Queen Pomaré, as she styled herself.   When shown the portrait of our own queen suspended over the gun-room table, she remarked, " Ah ! ah ! queene, queene ! all de same as me ! "   They remained on board about two hours, highly amused with all they saw.   From us they went on board the *Terror*.

Wednesday, 29*th.*—Fine day.   At eleven a.m. I landed at Mangrove Creek, and walked up the ravine, over Kingfisher Hill, through a sedgy swamp at the extremity of a mangrove creek, after making the circuit of the wooded ravine, till I fell upon a native path, where I shot a specimen of the beautiful pigeon of the island, on the very edge of the ravine, which, from its falling amongst the trees and long fern beneath, cost me an hour's loss of time in searching for it, but as I ultimately found it, it was well worth the trouble.   Following a path in the direction of the observatory, I reached it at six p.m., after shooting two tui.

Sunday, October 3rd.—After divine service on board, I landed at Pahia, and called on the Rev. Mr. Williams, who presented me with a copy of the New Testament in the Maori language, the " first " copy printed, and the selfsame copy which Mr. Williams himself had been in the habit of using for years past in his pulpit, when preaching to his native congregation in their own language. He also gave me an invitation to dine with him on Tuesday next, to meet Mr. Fitzgerald, the Government Resident here. After partaking of some wine and cake with the family, he gave me a passage on board in his own galley, on his way to the barracks, where he preaches this afternoon.

Kiddi-Kiddi Falls, New Zealand. (*See page* 236.)

CHAPTER XXI.

Up the Kava-Kava River—Excursions—The Kiddi-Kiddi River and Falls— Public discussion between Protestant and Roman Catholic missionaries.

Monday, October 4th.—Left the ship at 5.30 a.m. in the gig, on an excursion to the Waiomio marble formation, Captain Ross having given me his own boat's crew for the purpose.   The morning was moist and overcast, with rain at intervals.   At six a.m. I entered the main branch of the Kava-Kava, and left the boat at the second sawyer's hut on the left bank.   Then, ascending a ridge, followed a path to the pah, through which we passed at 7.30 a.m., a mile from the boat.   From thence we continued our route along ridges of tea-scrub and long fern, bare of trees.   At 8.30 a.m. crossed a rivulet, having a small cascade, or fall, of ten feet over basalt.   At 8.35 reached the first three blocks of the Waiomio limestone, nine or ten feet in height, on the side of a low hill, amongst tall fern.

On descending from this to a valley on the opposite side, and just as I was about clambering up the highest of the groups of rocks, some forty feet in height, my steps were arrested by the sudden apparition of an aged Maori chief, who hailed me, and endeavoured to make me understand that I was approaching " tabooed " ground, consecrated as a burial-place, and the natives are still very sensitive in all these matters with reference to the " taboo." I at once desisted from advancing a step farther, and accompanied him to his village at Waiomio, just below, situated in a fine valley skirted by a meandering river, which takes a south-east and north-west course through it.

It was 9.30 a.m. when we reached the village, consisting of about thirty huts, and at the chief's wigwam I saw the most aged-looking native I have yet met with, and, from the way in which the chief introduced him to me, I concluded he was his father. I ascended a coarse, bluff, buff-coloured mass of limestone, having a patch of potato-ground at its base, and from the summit of a fern-clad hill, strewed over with fragments of greenstone, I had a general view of the range of limestone rocks arising from the declivities of the valley in isolated masses, from ten to forty feet in height, horizontally pointed, with sharply-defined angular edges, irregularly pointed at the top, and presenting a castellated appearance, the whole intermingled with trees and underwood. These groups of limestone form an irregular and interrupted circle, the general bearing of which have a W.S.W. and E.N.E. direction, the apparently magnesian limestone being nearer north and south. The general aspect of the country around consists of barren and fern-clad ridges, with here and there a clump of wood merging from some ravine. The distance from the river, where we left the boat, about four miles. At 11.30 a.m. we commenced our return by another path, as I make it a rule never to retrace my steps, and thus waste time, if a new route by any chance presents itself. Our way now lay along the edge of a wooded ravine, through fern and tea-scrub. We heard the report of guns in the village, doubtless the chief trying some of the percussion caps I had given him. After proceeding some two miles, we descended a deep, circular bowl or hollow, clothed with fern and underwood, and about 300 feet in depth, having a narrow river running through it, winding along in a very tortuous course, nearly forming the figure of 8 in its doublings. A few natives had built their huts here, amid potato plantations. On crossing the stream, and emerging on the opposite side, I came upon the only outlet to this singular glen in the mountains, due, evidently, to some ancient crater, of which it is the only remnant left. I now followed the mazy windings of the stream downwards, in the direction of its course towards the main branch of the Kava-Kava, to within about two miles of its confluence with that river, just above the pah. So tortuous is its course between wooded hills on either side, one curve doubling on another, that we had every few yards to ford the stream. We passed many huts and native villages, scattered along its banks. The natives were busy planting their potatoes. In one spot I tasted some kumeras an old native woman was roasting over a fire in the open air, and found them delicious, as baked in this underground oven in heated stones covered over with soil. Saw only one pigeon, a hawk, and a few tui.

Following a native path over fern-clad table-land for about a mile, we reached a good cross-path, having a view of both the Kava-Kava and Kurrito rivers, reaching our boat, concealed by a headland on the left, at two p.m. Passing down the river, we were off Pomaré's pah at four p.m., and proceeded up the Waikaddi with the flood-tide. I shot a gannet from an old native woman's canoe, who paddled me nearer to the mud-flats than our own boat could approach, from the shallowness of the banks. After drinking some wine which I gave her for her trouble, and with which she seemed much pleased, she returned to a slave-girl she had left on the mud-flat picking up cockles, their favourite food.

The Waikaddi, as it passes by the left of the pah, is a wide river for about eight miles, afterwards, winding round some flats to the right, it becomes suddenly reduced to a narrow stream, terminating about half a mile higher up among mangrove bushes. It now being six p.m., and having the wind, as well as a strong flood-tide against us all the way back, we did not get on board the ship till nine p.m.

Tuesday, 5th.—Showery day. At three p.m. I left the ship for Pahia, to dine with Mr. Williams, where I met at dinner Mr. and Mrs. Fitzgerald, with their two children. After dinner looked over an herbarium of native plants, seaweeds, fossils, and drawings, in the drawing-room, where we had tea at nine p.m. Mr. Williams's boat put me on board, landing the Fitzgeralds at Russelton on the way, and where, on the following day, I dined with them at 4.30 p.m., and left at nine p.m.

Tuesday, 19th.—At five a.m. I left the ship in one of our boats, the morning overcast and threatening rain, and when off Pahia we encountered a pelting shower. But at eight a.m. the weather cleared up, and I made sail up the river, which is wide, passing between islets and rocks on the port shore. At nine passed a creek on the left, a little beyond which the river divides into two branches, becoming narrower. I followed the branch to the left, winding in zigzag curves. Saw a hawk on the bank, low hills on each side. At 9.30 p.m. passed a pipe-clay formation on the right bank, about twelve feet in height, direction south-west and north-east. A little higher up a fine mass of columnar greenstone of hexagonal form crops out. At 9.40 reached Mr. Kemp's station, a pretty place, with a neat church and farmlike-looking house at the top of the river. Met Mr. Taylor here, and walked up the hill above the house to the spot which had witnessed so many of the murders and atrocities perpetrated by the notorious chief Shongi, who once resided near this, and died a few years ago. At 11.30 Mr. Kemp gave me as a guide to the Falls of the Kiddi-Kiddi a native lad. On crossing the river, and over a fern-clad table-land, by a narrow track, a distant group of what appears to the eye merely underwood, rising a little above the level of the plain, alone indicates the ravine in which they are situated, when the Falls burst upon you very suddenly. On reaching the edge of the ravine, however, what appeared to be bushes in the distance now assumed the form of tall trees, growing out of the acclivities of the ravine, with their tops alone showing above it.

The fall is over a perpendicular wall of rock—basalt— on the right, about eighty feet in height, and the width of the stream perhaps fifty feet. It descends into a deep and finely-wooded ravine, strewed over with fragments of rock, the river below the falls being about 100 feet across. I waded through it about 100 yards above the falls, and walked along the opposite bank to the mouth of the cave at the back of the falls, over the ferns amongst the underwood and trees.

The cave is nearly 100 feet wide, about forty feet in height, and much the same in depth; the roof encrusted over, and the floor strewed with ochreous clays, amongst which small herbaceous plants and ferns were growing.

The striking scene before my eyes, formed by the curtain of falling water intermingled with vapour and mist in front of me, with the rushing sound of the falling water, as it descended in a wide sheet from the steep precipice above, amply repaid me for any amount of labour entailed in reaching so remarkable a position for observing it. On the other side of the cave I succeeded in getting over a narrow ledge of rocks up the bank, which saved me from retracing my steps—always objectionable to me—for a long way round, and having to cross the river again ; but I found it very slippery, with barely any foothold even for the toes, so that a false step would have plunged me into the foaming stream beneath me. At 12.30 I had reached the summit of the bank, and started on my return at 12.45, after making a hasty sketch of the falls, as shown in the accompanying engraving. I reached Mr. Kemp's station at 1.40 p.m., and with a fresh breeze shoved off in my boat down the river. At two p.m. I got some specimens of the columnar basalt, or rather greenstone, and some of the pipe-clay. It being ebb-tide, I landed on the mud flats, which are extensive, and dry at low water; saw a heron or bittern, and a brace or two of ducks, but all were too wary to get within shot of them. I shot one small sandpiper out of a flock. The weather now becoming wet and squally, I made sail on the boat, and, beating down, cleared the river at six p.m. Landed just round the headland on the right, near a cormorant rookery, built on the tops of the trees ; there were a number of nests, chiefly confined to two trees ; some of the birds were sitting, and I also heard the cry of young birds. The parent birds were hovering overhead in great numbers, in much excitement and alarm. I shot four for specimens, and might have killed any number had I needed them. I shoved off again at 6.45 p.m., in a dark and stormy night, eight miles still between me and the ship, but we now made sail with both the wind and tide in our favour. Saw the Kororarika lights in passing. I found the same argillaceous character of the hills, with greenstone along the banks of the Kiddi-Kiddi (with the addition of the pipe-clay) as on the shores of the bay itself. I got safe on board the ship again at 8.30 p.m., after a somewhat boisterous passage.

Thursday, 21*st.*—Fine day.   At ten a.m. I landed at Russelton on a shooting excursion.   Shot a bittern in the mangrove swamp, three small sedge-birds, a lark, and two warblers.   On the following day I was fully employed skinning and preserving my birds, ten in all.   Captain Leviche, of the *Herione*, French frigate, recently arrived, came on board, accompanied by his first lieutenant, to call on Captain Ross; and Captain Ross on the following day returned his call, and was saluted with nine guns, which number we returned.

Sunday, 24*th.*—H.M.S. *Favourite* arrived at nine a.m., and I dined on board of her.

Tuesday, 26*th.*—Fine, warm day.   At eight a.m. I breakfasted at Mr. Fitzgerald's, and accompanied him and his wife to Kororarika, to attend a conference held there between the Protestant and Roman Catholic Missions. At ten a.m. we found both rival parties assembled under a temporary shed, erected for the purpose with some planks, near the native pah, encircled by a large group of natives. Having escorted Mrs. Fitzgerald and another lady to Mr. Burroughs's (the clergyman) house up the hill, and left them under the verandah with Miss Williams, I returned to the beach and remained at the conference till three p.m., when it was adjourned till the following day. All three of the captains arrived just as it was over, and we repaired to the clergyman's house for lunch, after which I took a passage with him in his boat, accompanied by Mr. Taylor, to Pahia.   From thence I walked along the beach to the observatory, and got on board at 6.15 p.m.   This meeting appears to have been brought about to try the strength of the two contending parties by discussing certain points of religion in the native language, thus affording the New Zealanders an opportunity of judging for themselves which of the two creeds it might be most desirable to become converts to.

Mr. Williams, as the head of the Church Missionary Society, was the principal speaker, and, from the attentive manner in which the Maories listened to him, most unquestionably had the best part of it and the greatest influence with them. The point which gave rise to the warmest part of the controversy was that of the worshipping of images, in which the Roman Catholics clearly went to the wall. There were three of their priests present, two of them the chief speakers. Their attendants occupied another table in the rear, containing books and plates for reference, with pens, ink, and paper. On the opposite side, seated with writing materials before him, was Mr. Williams, chairs and benches being placed between them for the accommodation of visitors. Mr. Fitzgerald took his seat between the contending parties as president. Each party was allowed a quarter of an hour for his speech. The priests made a great fuss about the answers and signatures to some questions.

A son of the talented and amusing author of Waterton's " Wanderings in South America " was present ; a staunch Catholic, and evidently, from his personal appearance, a more eccentric character even than his father, the great ornithologist himself. The motley group of Maories assembled around formed not the least interesting part of the programme, in their many and varied grotesque costumes. In the front rank were several old chiefs, squatted on their haunches on the bare ground, some with gold-laced caps, others in round hats or straw ones, and not a few having only their natural covering of thick, coarse, bushy, black hair. One would appear wrapped within the folds of a flax mat of native manufacture, another with a common blanket carelessly thrown over his brawny shoulders. Pea-jackets also, with red guernsey frocks, and even camlet cloaks, formed part of their costumes ; coloured shirts being in general use as under garments, with scarlet merino and silk stocks. Amongst the female part of the audience prints and ginghams were the fashion ; and there was one pretty Maori girl, about fifteen or sixteen years of age, whose graceful and sylph-like figure did ample justice to her dressmaker, whoever she was. The whole scene was of so novel a nature, enhanced, too, by such lovely weather, as one is not likely to have an opportunity of witnessing again. The French frigate sailed to-day.

Wednesday, 27*th*.—Rainy day. I landed on the observatory shore shooting, but only succeeded in getting a tui and a fantail. The conference at Kororarika was brought to a conclusion this afternoon, after which Messrs. Williams and Taylor dined with me on board at 6.30 p.m.

From Thursday, 28th, to Saturday, 30th, I have been very busily employed in skinning and preserving birds, bittern, cormorant, &c., and arranging my specimens of natural history. The natives coming on board furnished me with many of the vernacular names in their own euphonious language.

Monday, November 1*st*.—The Fitzgeralds sailed for Auckland, and the two captains started on an excursion to the Waimate.

Thursday, 4*th*.—The *Albatross* yacht arrived, her owner, Mr. Blackett, bringing me the first number of the journal of the Tasmanian Natural History Society, kindly sent me from Hobart Town by the Governor, Sir John Franklin.

CHAPTER XXII.

Trip inland, and my night out in the woods—Shooting birds—Sail from
the Bay of Islands after three months stay.

Tuesday, 9th, at eleven a.m.—Finding that the beautiful
pigeon of the island was so scarce in the immediate
vicinity of the anchorage, I determined on making my
way through the dense woods and ravines as far as I could
accomplish in the day direct inland from the ships, with
the view of obtaining specimens for the Government
ornithological collection, to be transmitted from this place
to England.

Starting from the observatory above the small river, and
over the hill at the head of the mangrove swamp, I
followed the native paths through the tea-scrub, winding
along the summits or sides of the hills and ridges,
sometimes through short or tall fern, now and then penetrating some densely-wooded summit, through which it is
frequently very difficult to follow the faint traces of a
path bewildered amid the rank vegetation and tall fern
growing beneath the lofty timber-trees. These native
paths are found passing over most of the hills in the
vicinity of the bay, and branching off in every direction,
but seldom descending into the deep and thickly-wooded
ravines, which are so closely interwoven with the long
lianes and parasitical plants, fern, and underwood, as to
render them all but impermeable even to the native, and
to keep a path open is out of the question where the

growth of vegetation is so rapid.   Here I shot two small,
dark-coloured species of sylvia, a bird of such quiet,
unobtrusive habits, as only to be found in the bottoms of
the most thickly-wooded ravines, in the silence and solitude of which they hop from twig to twig, like young
robins in their habits.

In the farthest ravine I reached, certainly not more
than some five or six miles in a direct line, as the crow
flies, from the bay, though at the very least double that
distance in the course I had to follow—winding round the
edges of some ravines, and threading through the dense
mazes in the depths of others—I ultimately lost every
trace of a path, which, for some time previously, had
become all but obliterated.   I shot a new bird in this
ravine—at least, one we had not met with before, belonging to the *corvidæ* family, and about the size of a jay, of
a dark slate colour, with wattles of an azure blue on its
throat—and lost some time in searching for its mate,
and also in the search for a pigeon, out of two which I
had shot, falling into the long fern.   So difficult a matter
is it here to find your game without the aid of a dog,
that the bird in question was cold and stiff when at last
I picked it up.

It being already 4.30 p.m., the night fast closing in,
with an intricate, almost trackless course back to the
ship, before me, I commenced my return; but missing
the track between two wooded ravines, I took the
wrong side of one of them, and eventually got so out of
my course to the right, that I became benighted before I
could recover the track in the direction of the ships,
of which about dusk I got a momentary glance from the
summit of a hill bounded by an impassable ravine, cutting me entirely off from them.   After a fruitless attempt
to turn the hill, the darkness increased so much that I
could no longer see my way.

I gave up all hope of reaching the ship, and at once

prepared to rough it out in the bush for the night. Striking up to the right, through a great deal of tall fern, I descended a very deep and densely wooded ravine in search of water, so exceedingly thirsty had I become from the extreme heat and toil I had gone through during the day. When about halfway down, groping in the dark, I laid my ear to the ground, and the sound of running water was distinctly heard. I discovered a tiny rill, oozing over the rocks at the bottom, as it murmured along, and from it I quenched my thirst with a draught of nature's refreshing beverage ; this, with a bit of ship's biscuit I had in my pocket, formed my frugal supper ; for having intended being on board to a late dinner, I had had nothing else since breakfast, and was rather lightly clothed (in a thin duck shooting-jacket) to bivouac out in the open air.

The bottom of this ravine was brilliantly illuminated by numerous phosphorescent particles, arising from the decomposition of the decaying wood, glittering like so many glow-worms or fire-flies, affording light enough to enable me to see the time by my watch, which indicated eleven p.m. I now descended to the opposite side, in search of a suitable place in which to take up my quarters for the night, and on the ridge above, by the side of a wooded hill, found a small, open glade, between the trees, having a large mangrove swamp below and in front, extending onwards to Pomaré Bay, the pah lights being visible. I felled the dead slender trunk of a tree, around which some withered clematis had entwined, which was standing near ; separating the dry clematis from the dead wood, I rolled it round me as a covering for the night, this, with some long fern, constituted my couch, on which I resigned myself to sleep, worn out and wearied. The green fern growing around was all wet with the dew.

The night proved mild and starlight, the silence being broken at intervals by the harsh notes of a small owl, much resembling, both in size and colour, our own little passerine owl; the Ruru-Ruru of the natives, or *Strix Nova Zealandiæ* of ornithologists. This bird had selected the topmost branches of a tree, immediately over my head, for his station, from which he incessantly sent forth his monotonous cry of " more-pork, more-pork," throughout the long and wearisome night. About three a.m. I heard the crowing of the cocks and the barking of the dogs in a native village not far distant, but separated from me by the mangrove swamp, over which it would have been next to impossible to make my way in the darkness of the night; though not more, perhaps, than a mile off on the shores of the bay. Before the sun appeared above the horizon, as the small hours wore on, I heard the musical notes of that pretty mocking-bird of this country, the tui, or parson-bird, and another small bird belonging to the fly-catchers, the earliest risers here. A light shower now fell, and the sky became much overcast. I arose at 4.30 p.m., and heard the voices of the natives in their village at some distance. But the rain now falling in heavy showers, delayed my departure till 5.30 p.m., when, freeing myself from my tunic of entwining clematis, and taking up my gun from beneath me, I ascended the hill in my rear. Saw two pigeons amongst the trees, one of which I shot. I next made my way over a fern-clad hill to the left, following a native path which seemed to lead in the direction of the ship, descending a hill to the bay, by a sawyer's hut. The ridges ran E. and W., the lateral spurs at right angles, N. and S. generally. I shot a fantail and a paroquet returning. Got on board at eleven a.m.

Wednesday, 17th.—I landed at Pahia; very fine day. I called and took leave of Mr. Williams's family. Miss Williams had kindly obtained for my herbarium the first blossoms of the Pohutokava (*Metrosideros tormentosa*), a large forest-tree, attaining a height of from sixty to seventy feet, the timber of which is so hard and durable as to obtain for it the name of the New Zealand oak ; the leaves, which are of a shining, deep green, change to a bright scarlet before falling. The beautiful rich coral-like crimson of the flowers contrasts strikingly with the normal deep-green, clustering foliage. The tree from which Miss Williams succeeded in procuring the blossoms rises from the embankment overhanging the shores of the bay in the vicinity of Pahia, just before reaching Mr. Williams's residence, and was only just beginning to unfold its blossoms.

Thursday, 18*th*.—Very fine, warm day. Mr. Williams's two sons came on board to take leave of me, bringing me a large bouquet of flowers ; and soon after they had left, their father came on board and lunched in the gun-room, at which Captain Ross joined us. The *Jupiter* arriving from Auckland this morning, I embarked on board of her my three cases of specimens of natural history for England, when she sailed in the afternoon for Sydney.

Sunday, 21*st*.—Fine day. I went on shore to Pahia, and took my final leave of my kind friends the Williams family, who had given me every aid in their power in my natural history pursuits. On the following day I again landed for the last time in New Zealand, at the observatory, at nine a.m. Fine, warm, sunny weather. I strolled along the beach as far as the Waitungui River, and, returning over the hills at the back of Pahia, obtained an almanack in the New Zealand language from Colenso's printing-office at Pahia. I also gathered a few more of the blossoms from the pohutokava tree, which were rather difficult to get at, from the height of the branches above the argillaceous cliffs overhanging the bay. Returned on board at 5.20 p.m.

On Saturday last a murder was committed by the natives at Paroa Bay, and the house set on fire ; an

Englishwoman and a man-servant, with a native child, were the victims. As a rising amongst the natives was apprehended last night, the *Favourite's* boat about midnight came alongside of us, manned and armed, and under the command of her first lieutenant, for orders before proceeding to the town; but, on finding all quiet, returned on board.

Tuesday, November 23rd.—At 4.45 a.m. we sailed from the Bay of Islands in company with H.M.S. *Favourite*, after a sojourn of somewhat more than three months. At eight a.m. we were fairly outside of the harbour and at sea. Our decks had all the appearance of a farm-yard, from the sea stock, consisting of oxen, sheep, goats, pigs, and poultry, and each quarter was festooned with a line of pumpkins.

The entrance to the Bay of Islands is about eleven miles in width, and the entire length of the two islands may be estimated at about 800 miles, and averaging 100 miles in breadth, distant from Australia about 1200 miles. The mountains attain a height of some 14,000 feet; the Keri-Keri Falls descend from a height of ninety feet. The mean temperature of the climate at midsummer (February) is 66° Fahr., and in June, midwinter, 48°. Waimate is twelve miles from the Bay of Islands; Mount Campbell, near the North Cape, rises abruptly from the sea to 800 feet, having greenstone on the south and sandstone on the north side.

CHAPTER XXIII.

Second attempt to reach the South Pole—In the ice-pack—Our Christmas fare—Large penguins—A ball-room cut in the ice—Our Antarctic hotel—A white petrel falls a victim to the New Year.

Tuesday, November 23rd, 1841.—At noon we found ourselves in lat. 35° 14' S., and long. 174° 39' E., thermometer 65°, wind S.W., a fine, fresh, and fair wind for Chatham Island. Captain Sullivan came on board and took leave of us, and on his returning to his ship, the *Favourite* manned her rigging and cheered us, which we duly returned, and stood to the southward, whilst she shaped her course for Auckland; saw many gannets and stormy petrel about.

Friday, 26th.—Having entered west longitude, by crossing the meridian of 180°, we consequently make eight days in the week, and a repetition of this day in to-morrow's log. The gale of last night has subsided, but the wind coming round unfair for Chatham Island, prevents us from anchoring there as originally intended.

On Monday, 13th of December, in lat. 55° 18', long 149° 20', W., we crossed the circle of uniform temperature of the ocean, the temperature not varying one degree throughout the entire depth.

Thursday, 16th.—Nothing worth recording has transpired for some days past. The birds about us have been *Diomedia exulens*, or the wandering albatross, with the sooty and the black-backed kinds, many young immature birds, blue and black petrel numerous, a few penguins, and a group of black and white porpoises. Several soundings have been made for the temperature of the water. At six a.m. I saw the first berg at some distance in the horizon, but smaller and far less imposing in its appearance than the one we first fell in with last year. At 8.30 a.m. we passed a larger one. The temperature of the air at noon was 48° and that of the sea 33°. Only a few straggling pieces of ice about; weather hazy, chilly, gloomy, and overcast; saw some Cape pigeon, blue petrel, and *Diomedia fuliginosa*, but the common albatross has now altogether left us. The crow's-nest was got up to-day.

Saturday, 18*th.*—Early this morning we entered the pack; the ice was generally loose, but having some heavy pieces amongst it, with several large bergs. Working through lanes of water all day, with a fair wind for the south. I saw the first white petrel to-day, with an Antarctic and Cape petrel or two, a gigantic petrel was flying about the ship. Saw several whales spouting in the distance; one finner passed close to the ship, and another dived down under her bottom from the bows.

Our deck all day has presented a sad scene of slaughter, the six sheep remaining, with most of our pigs, were killed, and suspended over the quarter-boats. The last ox was killed yesterday, and I preserved the horns. Our fodder running short cost the poor animals their lives at an earlier date than would otherwise have been the case. This afternoon I went up to the crow's-nest for the first time since it was got up, to have a look at the ice; saw a large berg, which I took a sketch of after coming down from aloft. At 8.30 p.m. I shot the first white petrel as it was flying over the main-truck, when it fell on the keel of the boat amidships. At 9.45 p.m. I shot the first Antarctic petrel, which fell into the weather-chains, and I gave it to Captain Ross, who takes much interest in collecting birds, and skins them himself. The day was very fine, with a beautiful sunset behind a berg, bearing S. ½ W. I heard the note of the white petrel, a kind of murmuring cackle, as two were chasing each other. At eleven p.m. the thermometer was 28°, lat. 62° 50', long. 147° 25'.

Monday, 20th.—Sailing through loose ice, several whales blowing near the ship. Three seals caught on the ice, the last one having the stomach full of shrimps. Captain Ross went away in the boat and got soundings in 1700 fathoms. I shot another white petrel during the first watch, which I secured by its falling into the quarter-boat, or rather upon the quarter deck.

Tuesday, 21st.—Sailing amongst heavy pack-ice, passing many seals, and a silver-grey one was caught. Whilst I was sketching the ice, a seal asleep on a piece of ice passed on the weather quarter, when, having my gun charged with small shot lying alongside of me, I could not resist the impulse to fire at him, to see how he would act, for at the distance he was it could not injure him in the least, but simply tickle his tough hide and frighten him. Indeed I never saw an animal look so astonished as he did, on being roused from his siesta by the shower of No. 4 shot falling around him. He lost no time in shuffling his unwieldly form towards the edge of the ice, at the same time elevating his head with open jaws, and staring about him. Thunder-storms are somewhat rare phenomena in high latitudes, or he might have thought that he had been struck by lightning. The latitude to-day was 64° 50', long. 153° 23'.

Wednesday, 22nd.—This morning we found the sea nearly clear of ice, and were congratulating ourselves that we had passed through the whole of the pack, when about noon we entered within another margin, which became heavier as we proceeded on, till about midnight, when we were completely hampered amongst heavy masses. The day was otherwise fine, and we passed a number of seals on the ice; three were caught, an old and two young ones, and another young one of a darker colour, and I measured and weighed them all.

At six p.m. a penguin of the largest kind was seen on the ice, and the starboard quarter-boat being lowered, I went in pursuit of him, accompanied by Abernethy, our gunner, and the junior mate. Being the first of the large species that we have seen, and new to us, I resolved to give him no chance of escape, by shooting him through the centre of the body with a ball from my old double-barrel; yet on landing upon the piece of ice to secure him, he displayed as much strength and energy as if he had only been struck by a few grains of small shot, so powerful is the structure and tenacity of life in these magnificent birds. I had to put an end to his sufferings on getting him on board; he weighed sixty-four pounds. I shot a white petrel on the same piece of ice. As we were returning to the ship two more large penguins appeared on the ice at some distance. At nine p.m. I again went away in the same cutter after two small penguins, of which I shot one and a white petrel, in lat. 65° 20′, long. 154° 19′.

Thursday, 23rd.—Weather gloomy and overcast. At eleven a.m., accompanied by Abernethy, I went away in the starboard quarter-boat in the pursuit of a large penguin on a piece of ice. He took the water from this, and got upon an adjacent piece, which I pulled round to, leaving the gunner and part of the boat's crew behind to intercept him in the event of his retreating to the same spot again; which he did after a chase across the ice, and was ultimately caught by the gunner on the very same piece of ice on which we at first saw him; his weight sixty-four pounds and a half. At 1.40 p.m. we made fast to a floe-piece, and took on board twelve tons from the hummocks on it, to complete our water. At 7.45 p.m., just as we were about casting off from it, I

Edwin Wilson, del.

Vincent Brooks, Day & Son, Lith.

Kerguelen's Land Penguin
(Pygoscelis tæniata).

Emperor Penguin
(Aptenodytes forsteri).

Possession Island Penguin
(Eudyptes adelix).

PAGE 250—VOL. I.

shot a white petrel, which fell on it. At 8.3ᴏ p.m. I
went in pursuit of a large penguin, accompanied by Dr.
Hooker and the junior mate. He gave us a long chase
over the piece of ice, making off on his breast along the
surface of the snow, propelling himself with his flipper-
like wings and his feet with astonishing rapidity, whilst
we sank up to the knees at every step. I came up with
him first, and, as I stopped his way with a stick, the
mate got hold of him, and he was finally escorted down
to the boat between myself and one of the boat's crew,
one having hold of each flipper, as depicted in Captain
Ross's narrative of the voyage, in the heading of a
chapter. He weighed sixty-one pounds and a half. We
had to make a considerable circuit to the piece of ice,
forcing the boat through very narrow channels by breaking away the ice. Passed two seals swimming in the
water as we returned on board at 9.30 p.m. Two whalers
passed the ship. Another penguin, weighing fifty-four
pounds and a half, was subsequently caught, but I was
not present. Lat. at noon was 65° 59′, long. 155° 44′,
thermometer 28°, wind E.S.E.

Friday, 24th.—Employed all day in superintending
the preservation of the skeleton of the silver-grey seal,
weighing 414 pounds, exclusive of the blood lost. The
stomach was empty. Our decks full of ice caused a
feeling of chilliness. At one p.m. I went in the cutter,
with the gunner and mate, in pursuit of another large
penguin, which weighed seventy pounds and a half. In
the afternoon we were tacking about in an open pool of
water, off a large berg, which I took a sketch of. We
passed our Christmas Eve in lat. 65° 58′, long. 155° 54′,
with the thermometer 31°, in the midshipmen's berth, where
Captain Ross and all the gun room officers assembled,
and were regaled with punch, cake, and snap-dragons.

Saturday, 25th, Christmas Day.—We had divine
service, but no sermon; and at 3.30 p.m. Captain Ross

and the gentlemen from the midshipmen's berth dined in the gun-room, and, although surrounded by ice, and having been some time at sea, we managed to provide a very fair dinner on the occasion, roast goose and plenty of fresh meat. The weather in the earlier part of the day was overcast and gloomy, and our decks, unusually cold and chilly, from the quantity of ice piled abaft, and from which we derive our supply of fresh water. The weather, however, in the afternoon cleared up fine, the ship tacking about in an open pool of water. The *Terror* appeared beset behind a most remarkable berg, having two cupola-shaped hummocks on its summit, which we christened the " Christmas berg." I took two sketches of it, giving one to Captain Ross. There were several white petrel about. In the evening I went up to the crow's-nest, and found the ice more open. So ended our Christmas Day within the pack; the thermometer at 27°, wind N.E.

Monday, 27th.—Snow fell last night; day gloomy; beating about in an open pool of water. It took me four hours in skinning and preserving the large penguin I shot the other day. At six p.m. three more were caught, one weighing sixty-four pounds, which Captain Ross himself skinned; and another of fifty-three pounds he put in pickle; and the third one, weighing fifty-seven pounds, I preserved in a cask of pickle, to present to my own college, for a skeleton for the Hunterian Museum.

Tuesday, 28th.—Still beating about in the pool of water, with foggy weather. Skinned the second penguin, which also occupied me for four hours, from the benumbed condition of the fingers in this cold climate. At three p.m. another was caught. At six o'clock still another, and I noticed a small flock of tern on a piece of ice ahead.

Wednesday, 29th.—Foggy weather. I skinned the penguin caught yesterday at three p.m. by Abernethy in four hours. Pebbles and half-digested fish were found in the stomachs of all.

Thursday, 30th.—Hot-air stove lighted to-day, which filled my cabin with smoke. Six p.m. made fast to the ice.

Friday, 31st.—This being the last day of the old year, great preparations have been in progress all day upon the piece of ice forming a fender between the two ships, one being made fast on either side, admitting a free communication between them, for welcoming in the new year and seeing the old one out. For this purpose a quadrangular space has been excavated in the ice for a dance, albeit a somewhat novel kind of ball-room. On this an elevated chair of the same material has been constructed for the accommodation of both captains; adjacent to this crystal ball-room a refreshment-room has also been cut out, with a table carved in the centre for the bottles of wine and grog-glasses for the use of the dancers. The whole of this sculptured ice almost rivals in hardness and whiteness the finest Carrara marble.

Our worthy boatswain undertook to act the part of Boniface on the occasion, and went through the character in a most admirable manner; but not rejoicing, as he thought, in a sufficiently portly person himself for duly supporting the new character he had assumed, he made up the deficiency by stuffing a pillow beneath his waistcoat, and in this guise, very much like a cropper pigeon, strutted about with his hands in his shooting-jacket pockets, an apron round his waist, a bunch of keys dangling in front, inexpressibles buckled at the knees, which, with a round cap jauntily tipped on the crown of his head, completed his rig. Two of the younger seamen, acting as his waiters, handed genuine ices on a tray all round.

In front of this Antarctic hotel a sign-board was fixed to a pole, having "Pilgrims of the Rhine" chalked on

one side, and " Pioneers of Science " on the other.
These devices were of the landlord's own suggestion.
A flag was unfolded to the breeze, guarded by a cannon
with a pile of shot alongside of it, all shaped out of the
frozen snow, with steps cut through the ice down to the
table.

Near the *Terror's* gangway a female figure in a sitting
posture was formed out of the snow, her head ornamented with a profusion of ringlets, and surmounted by
a card, on which was the word " Haïde," though possibly
bearing small resemblance to the beautiful Greek girl,
the creation of Byron's imagination.   In front of the
Erebus's gangway was the bust of a male figure in a
foraging cap.

As the ships' bells struck the nautical number eight,
announcing the midnight hour, the frolic began ; the
new year being ushered in by three hearty cheers from
stentorian vocal organs, which strangely broke in upon
the deathlike silence of the solitude reigning around.   At
this very moment an ill-fated white petrel (*Procellaria
nivea*), like a phantom, white as the surrounding snow
itself in its purity, appeared as a herald borne on the
waves of ether, to announce the first hour of the morn in
the birthday of a new year.   Poor little, confiding creature, it was destined to pay with its life for the curiosity
that fated it to become another victim to the claims of
science, for I happened to have my gun at hand, whilst
reclining on a solitary hummock of ice, not far from the
scene, and meditating on all that was enacting around
me, when the temptation proved too great for me to
resist the impulse to fire at it, and as it whirled round
and round, fell upon the ice in its descent, just as Captain
Ross himself appeared on the scene.   I presented it to
him as the first victim to science in the new-born year ;
and, as such, he skinned and preserved it himself.   Just
as I was about leaving the ice to turn in at three a.m., a

second bird, whether its mate or not I cannot tell, hovered over, as its predecessor had done, and shared the same fate, falling to my gun, as a memento for myself of so novel an advent of a new year on the confines of the Antarctic circle. Dancing and singing wound up the whole, in which both captains joined. The day had been overcast and gloomy, but unusually calm and mild, some fine snow even thawed as it fell. The ice close all around. A barrier berg and another were seen in the north-east. The other birds seen were two or three gigantic petrel, a penguin or two, and one solitary stormy petrel.

Saturday, January 1st, 1842 —New Year's Day has been ushered in by fine weather. The state of the ice the same as yesterday, with several bergs in sight. This forenoon a supply of Government clothing was issued to the officers and ship's company gratis, consisting of a box-cloth jacket and trousers, red guernsey frock, two comforters, Welsh wig, pair of boat hose, five skeins of thread of different colours, ten needles, and a sailor's clasp-knife.

At four p.m. all the officers from both gun-room and berth dined in the cabin with Captain Ross. The table displayed such a bill of fare as was scarcely to have been anticipated within the Antarctic circle, which, not a little singular, we have crossed to-day, on the same day as last year, in longitude 156° 28′ W., therefore some 1400 miles to the eastward of last year. We were regaled with roast beef and mutton, roast goose, and mince-pies, delicious preserves, gooseberry and cherry tarts, &c., were amongst the dainties. Lieutenant Phillips and myself this evening went up to the *Erebus's* crow's-nest, or rather the main-topmast cross-trees, to have a better view of the scene on the deck beneath us. A fine lestris was hovering over the *Terror's* mast-heads at the time. I turned in at three a.m.

CHAPTER XXIV.

Closely beset in the pack—Episode of the dying petrel—Animal
instinct—Twelfth-night—The Christmas Berg.

Monday, 3rd.—A fine day. The ice so closely packed
around us that it enabled me to walk in various directions
from the ship for at the least half a mile. My first shot
was at a gigantic petrel, breaking his wing, yet he gave
me such a chase over the ice before I could capture him,
that I had to stop his further progress with the contents
of my second barrel, and even then had some difficulty
in securing my prize, having to reach him with the aid of
a boat-hook from a sludgy piece of ice treacherous to
trust the feet upon. I am sorry to have to record here,
as I do reluctantly and with remorse, that I was the
cause of an instance of devotion and affection in the
animal creation, which, however interesting to the naturalist as a study of animal life, was most painful to witness.
I happened to fire at a white petrel as it flew past me,
when it fell on a treacherous part of the floe. I lost it,
but its mate, flying in company with it at the time,
instantly alighted near the wounded bird, and placing its
own beak in juxtaposition with the dying creature's,
began a painful lamentation over its dying companion,
curving its own neck over the prostrate form, and giving
expression to a plaintive, murmuring, cackling note,
which it continued for some minutes, apparently evincing
the greatest amazement at the sudden change which had

come over its dying mate; but finding that all its atten-
tions were unheeded, then, as if acted upon by some
impulse or instinctive feeling impressing on its percep-
tion that this was death, and it could be of no further use,
it took wing and flew straight away. Whilst I was
endeavouring to get round the sludge-ice to pick the other
up, though ineffectually (for, as may be imagined, after this
little episode the bird would have had a double value for
my collection), the poor thing feebly raised its head,
after slowly staggering along the ice for a few paces,
apparently its last effort.

How little do we know of animal life and mind! for
mind they have unquestionably, call it instinct or what
you will. They are constructed on a very similar type to
their lordly master, man, consequently must necessarily
be endowed to a certain extent with his faculties. The
same brain and nerves, which in common with him render
them sentient. beings, must also endow them with the
power of thought; the main difference between them
arises from the want of language which he possesses as a
medium of communication.

What indeed is man, in his savage, untutored state,
but for the language he inherits, the superior of the lower
animals? Many of these much-wronged and under-
estimated beings should make him blush, exhibiting, as
they often do, moral attributes of a high order, with
quite as much intelligence. Those whose opportunities
have afforded them the means of comparing man in his
primitive, savage condition with the most sagacious of
our own domesticated animals, the friends of man, like
the dog and others, will, I feel sure, bear me out in
this, although they may not be prepared to go so far with
me as to claim for these often-persecuted creatures a
" hereafter," of which, however, we certainly know nothing
to the contrary. We can all readily enough comprehend
the perishable nature of all material things, though the

atoms themselves remain ; but cannot so easily reconcile to our reason the destructibility of a living principle, whose very essence is immaterial, any more than we can doubt the wisdom and benevolence of a great Creator of all, by supposing that He has had so little of considera- tion for creatures as much the work of His own hands as man himself, consigned, as many of them are, to unmiti- gated sufferings here, without any compensation or reparation in the future. We trust that their case is not so hopeless. Shrouded in mystery as this wonderful principle of life may be, it is at least as inseparable from space, time and eternity, as imperishable and ever- lasting, as all matter and material things are evanescent and transitory.

Wednesday, 5th.—A very fine day; bright sun in a clear blue sky. Abernethy caught at one p.m. to-day a very fine large penguin, weighing seventy-five pounds.

Thursday, 6th.—We are drifting seven or eight miles a day to the northward. We cast off from the ice, leaving a cask on it containing a paper with all our signatures and the latitude and longitude. A very pretty, darkly- mottled young seal was caught this afternoon, having a deep, old wound on its side, which had burst open afresh; the pain arising from this must have been excruciating, for the poor creature was rendered so irritable and fierce, that on reaching the deck it bit at everything that came in its way, deck and ropes, so I put an end to its sufferings by a thrust of the sword. A thermometer placed in the wound gave a temperature of 100° Fahr.

This being " Twelfth Night," all the officers partook of " Twelfth Cake," with cherry brandy, in Captain Ross's cabin. On a signal being made to the *Terror*, Captain Crozier came on board and joined us. It fell to my lot to cut up the huge cake, and the junior mate, who sat next to me, distributed the enclosed painted figures and enigmas

R. McCormick, R.N., del.

Vincent Brooks, Day & Son, Lith.

The " Discovery " Ships drifting down on a berg when fast to a floe, with a fender of ice between them in Lat. 66°, Long, 158°.
January 18th, 1842.

all round, creating considerable amusement in our novel situation. At eleven p.m. we passed our old acquaintance the " Christmas Berg " on the port side ; and a very large finner whale, seventy or eighty feet in length, passed very near the ship ; the weather becoming thick with snow. Wind E.S.E. ; thermometer 32°.

Friday, 7th.—The large penguin of seventy-five pounds took me five hours to skin to-day ; it was a female, and had no less than three pounds of fish in its stomach.

Monday, 17th.—For the last ten days we have continued beset in the pack, sometimes boring through, and at others beating about in pools of water, with interchanges of visits between the two ships. Our Christmas Berg frequently in sight. Flocks of tern and white petrel, with now and then a gigantic one, were seen, and scarcely a day passed without a seal being caught, mostly from nine to eleven feet in length—one was harpooned, it weighed 644 pounds, and was nine feet in length. On the 8th a penguin was caught, weighing sixty-four pounds. But the *Terror* carries off the palm for having caught the largest of all, weighing seventy-eight pounds. Two finner whales, seventy to eighty feet in length, passed close to the ship, having a large fin above the water abaft, and humps forward. Both ships are now made fast with hawsers to one of those heavy, table-topped pieces of ice, rising from ten to twelve feet above the sea, with perpendicular sides—which, indeed, have become a new feature in our Antarctic landscape— the ships having a low piece of ice, forming a fender, between their sides. To-day I took a sketch of the *Terror*, from the stern boat. Between two and three p.m., having shot half a dozen of the brown and white petrel, on the port quarter-boat being lowered for me to pick them up, to leeward, I had scarcely got into the boat, when a silver-grey seal was seen on the ice, which the junior mate and myself immediately went in pursuit of ; and we secured our prize, by each of us planting a harpoon in him, and then we towed him alongside of the ship, but, unfortunately, during the interim, most of my birds had drifted away amongst the ice and were lost, having only bagged two of them ; but the seal was the most valued desideratum. The pieces of ice to which we were made fast ahead, rising and surging in the long and heavy swell, occasioned by a strong breeze blowing, imperilled our stern, which had a narrow escape from coming in collision with the large, hard, table-topped pieces this forenoon.

Tuesday, 18*th.*—Some fine rain fell during the day, with a gloomy atmosphere, ending in a heavy, long swell from the W.S.W., with smart showers and a thick fog. Towards night it blew very hard, and throughout the night the ship thumped and surged so heavily, occasioning such a strain on the bow hawsers, that both were carried away, and we consequently parted company with the table-topped piece, leaving our ice-anchor on it, both still holding on to the fender of ice between them. This took place at four a.m. I found, upon going on deck, that we were fast drifting down upon a large berg on our port quarter, the white summit of which was barely visible through the dense fog enveloping it, which seemed to add in appearance to the vast magnitude of this stupendous mass of frozen water, as it rose and fell in the long, heavy swell, presenting the wildest and most threatening aspect that the most fertile imagination could conjure up. We only just escaped a fearful collision with it, by making a press of all possible sail on both ships, so as to increase their rate of drift, passing not more than fifty or sixty yards to leeward of its steep, hard, washed, blue-looking and perpendicular sides, which in places were festooned with enormous icicles. It towered to the height of not less than 200 feet above the sea. I took a hasty sketch of it as we passed. A pair of brown and white petrel, nestled close together, were apparently slumbering on its summit, quite undisturbed and undismayed by the elemental war raging around them. Captain Ross was on board of the *Terror* at the time. This immense berg left an open pool in its wake, and not far distant from us were two smaller bergs. The ice opened more during the first watch, and some heavy masses passed us, one but just clearing the *Terror's* bows.

Wednesday, 19*th.*—The ship thumped heavily throughout the night against the margin of the piece of ice, shaking every timber of her framework, and producing such a strain on the hawsers, as she surged in the heavy swell, that we parted from one, and stranded the other; so that at 2.30 a.m. we were adrift, and had to make sail under the top-sails, firing guns and muskets all the morning, as fog-signals to our consort, to enable her to keep company with us in the dense fog concealing everything around. About nine a.m. passed within a quarter of a mile of a large berg. Observed a strange, blackish-looking petrel, and at seven p.m. we passed close to the piece of ice with the two carcasses of seals we left upon it, when we cast off from it this morning, Rain in the first watch; lat. 66° 18′, long. 158° 38′; thermometer 38°, and wind N.W.

CHAPTER XXV.

A heavy gale—Waves of solid ice—Seals, whales, and penguins—Eight
hundred miles of the pack passed through—In the open again—
Shot a lestris—A beautiful view of the great ice barrier—Highest
latitude reached.

Thursday, January 20th, was a day not soon to be
forgotten, while memory has the power of recalling
vivid impressions of the past. We were destined to witness one of the most extraordinary scenes perhaps that
ever occurred in the annals of navigation. We encountered a heavy gale of wind, little short of a West
India hurricane in its force, whilst beset in this vast and
close pack of ice. It was a heaving sea, with a long
swell, unprecedented in the Arctic seas. Each moun-
tain-wave was crested, not by spray and foam, but bore
on its summit huge masses of solid ice, hard as adamant,
intermingled with brash and *débris*, resulting from the
tremendous collision of ice with ice, in the combined
tumult of waters, both fluid and solid; and notwithstanding the enormous pressure of the ice borne on their
surface, some of these waves ran so high as frequently to
render the *Terror's* main-topsail yard barely visible above
them, when she fell into the trough between two of them,
scarcely half a mile ahead of us. Both ships had been
rolling heavily all through the preceding night, coming
so violently in collision with the ice, as to shake their
whole framework in such a way as to render it doubtful
whether their timbers, strongly put together as they were,

could much longer resist the fearful strain on them. The swell appeared to come from the W.N.W., and the ships' drift to S. by E. The *Terror* was under her main-topsail on the cap. We were limited to the main-trysail, and fore-staysail, backing and filling as requisite, to clear the heavier pieces of ice, or by lowering the fore-staysail and squaring the main-yard, to drop astern of them. Then, again, forging ahead by dropping the fore-sail, &c., the main-topsail hanging loose upon the cap. Barometer at two p.m. 28° 49'. We passed perilously close to some enormous hard masses, having white, table-topped summits ten to twelve feet above the surface of the sea, having a horizontal hollow line, in their perpen-dicular sides, reflecting a beautiful cobalt-blue colour, and vertically streaked with an appendage of white pendent icicles, apparently resting on older ice as a basis, having a pale, yellowish-brown colour at the water's edge, divided by short pillars. Beneath the surface of the water, large tongues of ice, having a convex upper surface, and smooth, blue appearance, hard as the granite rock itself, stretched out far beyond, on which the roaring surf broke. Were a ship's bottom—her weakest part—to strike on this, no human power could possibly preserve her from instant destruction in a sea like this, with such a hurri-cane raging around. We, indeed, passed in very close proximity to one mass, of a rounded, hard, washed, blue appearance, pitching, as it were, bows under, like a ship going down, in the turmoil of waters raging around.

Fortunately for us there were none of the large bergs in our line of drift, and only two far to leeward. Two poor seals were quietly sleeping on a piece of ice ahead, apparently, if not unconscious, indifferent to the turbulent scene of the elements around them. A solitary black and brown, and a white petrel or two, were now and then seen hovering overhead in the height of the gale. The sky itself presented one uniform, lurid, leaden colour ; the wind was from the N.N.W., and the barometer falling all day ; snow in large flakes fell at intervals, and in the afternoon the weather became thicker with fog. At 12.30 we drifted into a lane of open water. During the last dog-watch the wind shifted round to the westward, and the gale and swell both became much abated.

We had our rudder injured, and on exchanging signals with the *Terror*, learnt that hers was in a much worse condition than our own ; made the signal to rendezvous at the Falkland Islands, in the event of parting company. The *Terror*, as she rose on a sea, showed her copper sheathing, very bright and polished from the scrubbing it has sustained in her late collision with the ice.

At seven p.m. we passed a very beautiful young seal of the large dark kind, reposing on a piece of ice not ten yards from the port side of the ship. He was four or five feet in length, blackish-brown above, hair short and thick, crisp-looking underneath, grey, mottled with black, both on flanks and flippers. The poor animal seemed much astonished at his close proximity to the ship, looking round him with a bewildered expression, which was soon converted into fear and dread by the laughing and noise on deck, and he at once set about crawling off the ice, propelling himself along on his chest, without making any use of his flippers, progressing by curving in of his spine, thus shortening in his body as a caterpillar would do, the hinder or tail flipper being vertically closed, and passively stretched out on the ice. On rolling off the ice into the sludge, he then made use of his fore flippers in endeavouring to get upon another piece of ice, but being unsuccessful, he rolled over on his back and disappeared.

Thursday, 27th.—The weather for the last week since the gale has been gloomy, overcast, and threatening, much heavy ice about, with lanes of open water, and snow at intervals. We have had the carpenters of both ships at work, in making an entire new rudder for the *Terror*, and repairing our own and other damages from the gale in the pack. About nine p.m. to-day, seeing two large penguins, apparently new species, on a piece of ice ahead, I was very naturally desirous of securing them for the government collection, and asked for a boat to go and capture them; but, unluckily for me, Captain Ross being on board the *Terror* at the time, our automaton first lieutenant, whose prestige, if he has any at all, is more for holy-stoning decks in his morning watch than in the paths of science, did not deem them worth the trouble of lowering a boat for. Fortunately for the *Terror's* credit, his brother-officer in that ship, Lieutenant McMurdo, thought differently, and had a boat manned, and a chase on the ice. Both the birds were secured, when they turned out to be the young of the great penguin, still in their grey, immature plumage, and as such a highly interesting addition to the ornithological collection. One weighed thirty-seven and the other thirty-five pounds. Just before we cast off from the piece of ice to which both ships had been made fast, a large sheet of copper sheathing was removed from our bows, torn off in her recent struggle with the ice in the late gale, and I had a piece of it cut out to preserve as a memento. Lat. 67° 39', long. 155° 59', thermometer 34°, wind W.S.W. A finner whale crossed our bows the following morning.

Tuesday, February 1*st.*—The weather cleared up, with the first sight of the blue sky for some time past. Sailing in open water, but ice closely packed around, with the exception of to the southward and eastward; wind light. Saw three gigantic petrel nestled together on a piece of ice, asleep. The latitude at noon was 67° 18', long. 158° 12', thermometer 30°.

During the first watch we were sailing through very loose, small ice, with a very dark water-sky to windward, indicating open water in that direction; indeed, a faint streak in the horizon perceptibly marked out the line of water, which, taken with the free, quick, and undulating movement of the swell in short rolling waves, following each other in rapid succession, so unlike the long, broad, heavy-moving swell in the pack, produced there by the weight of the superincumbent mass of ice, left no doubt that we were now on the confines or margin of the pack. To leeward there appeared over the pack and a chain of large bergs a very bright ice-blink, so that before dawn we shall be in all probability clear of the pack.

Wednesday, 2nd.—We found ourselves at 2.30 a.m. fairly out of this enormous pack, in which we had been beset since the 18th of last December, no less a period than forty-six days. The breadth of the belt of ice we have passed through may be fairly estimated at 800 miles. We entered it in the latitude of 62° 50', and in the longitude of 147° 25'; and our observations at noon to-day gave 67° 57' lat., and 160° 3' long. We are now about 450 miles from the place where we entered the pack. The average drift to the southward was some ten miles a day.

Upon going on deck this morning I saw only one solitary fragment of ice in the horizon, on the starboard bow, bearing S.W. by W. Being once more in such a vast expanse of open water seemed quite a novel spectacle to us, and through which we are going at the rate of four knots , making a S. by W. course, with a fine fresh westerly breeze and a swell from W.S.W.; weather very fine, with a few light fleecy clouds, floating along a clear blue sky. Towards the close of the day the pack was again seen, and at 6.30 p.m. we tacked off its edge, which extended from the north in an easterly direction to S.S.E., flanked by two long flat bergs; the ships sailing amongst straggling bits of ice, one stormy petrel in sight. At 11.30 p.m. I saw the first star, for some time past, bearing W. half S. twenty degrees above the horizon, on the weather-bow. I shot two brown and white petrel, securing both by their falling on board. We carried away our chain-bobstays against a piece of ice. Thermometer 30°, wind W.S.W.

Thursday, 3rd.—In a fine open sea till 4.15 p.m., when we tacked within a mile of the pack-edge, which was here flanked by seven large, table-topped bergs. It appeared very close and heavy, extending from N.N.E. to S.E. Saw a blue petrel in the first watch. The *Terror* had a fire break out in her main-hatchway on Sunday last, but it was speedily got under without any serious consequences.

Thursday, 10th.—The finest day we have had for some time past; for during the last week we had the thickest fog, and heavy falls of snow, with frequent squalls, during the voyage. Whilst sailing along the pack-edge, off which we tacked at three p.m., at 10.30 p.m. I shot another brown and white petrel, which fell on the deck. The sun set at 9.30 last night.

Saturday, 12th.—Early part of the day very fine, going south. In the afternoon it came on thick and foggy, with a considerable fall of snow. We passed the highest latitude of Cook to-day, in the longitude of 179° 13′ E.; many birds about us—the sooty and the black-backed albatrosses, blue petrel, the brown and white, a Cape pigeon, a gigantic petrel, a stormy petrel, and one solitary white petrel, and a light-coloured one, having a black patch on its head, I have not before seen; most probably an immature bird in change of plumage.

Tuesday, 15th.—Blowing a gale of wind all day, with misty, thick weather: ship rolling heavily with her rigging thickly coated with ice. This day we passed Weddell's farthest south, and have consequently now only ourselves to beat in our first voyage. Our latitude at noon to-day was 74° 26′, long. 182°.

Friday, 18th.—Whilst sitting at breakfast this morning, the quarter-master of the watch came down to the gun-room to tell me that a lestris was hovering over the mast-head, and I lost not a moment in reaching the deck, gun in hand, and shot it, when luckily it fell into the boat astern—a splendid specimen and a new species or variety, like the one I shot on Possession Island last year; and the only pair I have been able to obtain, so scarce is it; yet a bold, piratical bird, quite capable of defending itself against any feathered foe; hence I christened it the " Rover of the South Pole."   I shot four white petrel, three of which fell on board, and were added to the collection, but the fourth and last fell overboard, and consequently was lost.   A great number of the brown and white petrel were about the ship to-day, also three or four *Diomedia fuliginosa*, a *Procellaria gigantea*, a tern, and a pair of stormy petrel; the latter, so unusual with them, flying higher than the mast-head in their evolutions to leeward, as they are usually seen skimming in the wake of the ship, occasionally sweeping round the sides.   The blue petrel have now left us.   We sounded to-day in 300 fathoms, green mud, a bright ice-blink in the horizon extending from E. by N. to W.   Chiefly making eastern, to-day cleared up fine, with a fresh breeze.   Lat. 76° 52', long. 178°, thermometer 25°.

Saturday, 19*th.*—Sounded in 240 fathoms; white petrel very numerous this evening, and I shot three, two falling on board and one overboard.   On skinning the lestris I found a ball of penguin feathers in its stomach, a convincing proof, were any needed, of its predatory habits, and also its power to destroy such strong, muscular, vigorous, and horny-feathered birds as the penguins.   Lat. 76° 41', long. 173° 48', thermometer 33°.

Monday, 21*st.*—It blew a very heavy gale of wind last night, the ship rolling heavily.   At four p.m. we tacked ship off the pack-edge, and at five p.m. wore ship, crossing the *Terror's* bows rather close in doing so.   Her sides were completely encased in ice, and our own bows and bowsprit presented a dense mass of crystallization. The force of the wind in the height of the gale during the night was ten; the barometer falling nearly as low as 28°.

Tuesday, 22nd.—Passed several bergs and heavy pieces of ice during the day, and at six p.m. sounded in 190 fathoms, green mud and small black stones. A very bright ice-blink along the horizon from east to south; and about midnight we again came in sight of our old friend the Great Barrier to leeward, extending from south to east. It presented a more undulating summit outline than last year, having intervals forming bights between. One remarkable abutment, or promontory, at the entrance to a bay or inlet, bore a striking resemblance to the one where we made the nearest approach to it last season. Several bergs were lying in front of this huge perpendicular wall of ice, one certainly not less than three miles in length. We passed through a number of streams of young ice of a yellowish-brown colour, but of the pancake form, very thick, from a foot to three or four in diameter. The night, though cloudy, was fine and free from mist. On the weather-bow a group of clouds might easily have been mistaken for distant high land. The ship was going at five knots before the wind under studding-sails, with a fresh breeze. Thermometer 27°, lat. 76° 42', long. 165° 50'.

Wednesday, 23rd.—At one a.m. passed a large piece of ice, having two black points of rock projecting from its side. At 3.30 a.m. I shot a white petrel, which fell on deck. The birds are now unusually scarce and shy—a penguin or two, sooty albatross, gigantic petrel, white petrel, and brown and white. Turned in at five a.m. within four or five leagues of the barrier, having remained on deck all through the first and middle watches to see all I could of the barrier. Upon going on deck again at nine a.m. the barrier was astern, the ship having been put

about at 6.30 a.m.   The morning was fine, with a bright
sun and fresh breeze.   Saw a whale blowing.   At
1.40 p.m. we tacked, and stood towards the barrier on an
E.N.E. course, passing through a great deal of young
ice, and close to a large berg.   At seven p.m. we hove-to
off the face of the barrier, with the ship's head to the
north-west, about two miles from a promontory in it,
bearing east, and got soundings in 290 fathoms, fine
green mud and stones.

Whilst the line was running out I seated myself in the
stern-sheets of the boat, on the port-side of the quarter-
deck, to take a sketch of one of the most novel and extraordinary scenes I, or any one else, ever witnessed.   The
day was cloudless, a bright sun in a clear blue sky, the
rays of which, falling on the barrier, gave a beautiful effect
to its steep, indented sides, the various angles and abutments of which stood boldly out in relief, alternately in
light and shade, forming a long, zigzag, perpendicular
wall of ice, upwards of 100 feet in height, extending from
S. 40 W. to N. 21 W.   Along its base numerous fragments of ice, of every form and size, were scattered or
piled together in the wildest confusion, in many places
appearing as if quarried out, leaving recesses in these
stupendous cliffs, hollowed out by the terrific power of
those heavy seas which gales of wind have set in motion
when sweeping over the vast and mighty surface of the
southern ocean, the sea in front of the barrier being covered
with young ice of the pancake pattern, and amongst which
the ship was hove-to.   The extreme of the barrier to
the right had the horizon studded with bergs, both large
and small, resembling, as the sun's light fell upon them,
so many white marble buildings in the distance.   To
the left a huge berg had posted itself in solitary grandeur
in front of the barrier, inside of which we passed at
7.15 p.m., in again making sail, and ran along the latter
at about a league distant, forcing our way through vast

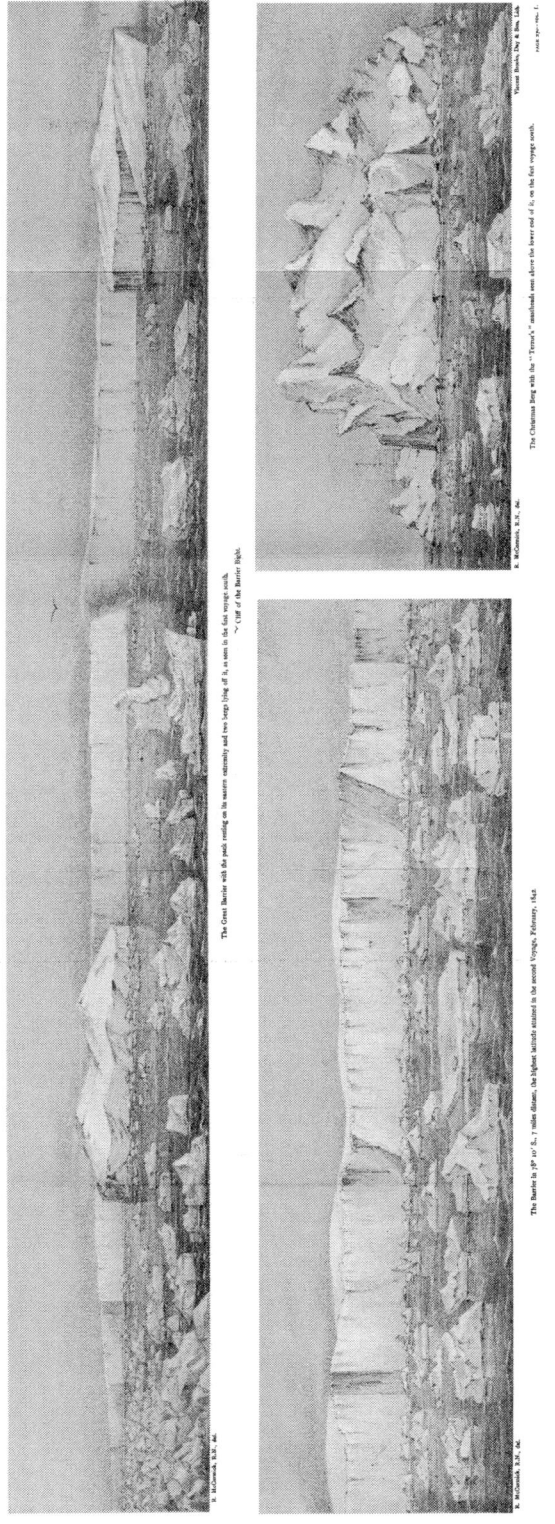

The Great Barrier with the pack resting on its eastern extremity and two bergs lying off it, as seen in the first voyage south.
~ Cliff of the Barrier Right.

R. McCormick, R.N., del.

The Christmas Berg with the "Terror's" masthead seen above the lower end of it, on the first voyage south.
Vincent Brooks, Day & Son, Lith.
TERRA NOVA. I.

R. McCormick, R.N., del.

The Barrie in 78° 10' S., 7 miles distant, the highest latitude attained in the second Voyage, February, 1842.

R. McCormick, R.N., del.

quantities of thick, pancake ice, which became thinner and thinner as we increased our distance. Saw two or three small penguins on it, rising and falling with its waving motion.

Just as we had made sail our consort, the *Terror*, which we had run nearly hull down, came up with us, and went about close under our stern. Our latitude here was 78° 7', the *Terror* making it 78° 9', so that taking the mean of the two observations, would place the face of the barrier in 78° 8', long. 161° 27', and that we have attained about some half a dozen miles higher latitude than last season, our farther progress towards the Pole being checked by the barrier. About 130 miles further to the eastward the summit of the barrier could be seen from the mast-head, looking like a vast plain of ice in one direction, having much the appearance of land in the distance. During the first watch the white and brown and white petrel were very numerous, flying about the ship. I shot three from this most interesting locality, all of which I was so fortunate as to have fall on board, the last one, and the best specimen, at 11.30 p.m. I had also a shot at one of the largest kind of stormy petrel as it was flying over the mast-head, but missed it. Saw a *Procellaria gigantea*, a *Diomedia fuliginosa*, a small penguin, and a seal or two on the ice.

At 10.30 p.m. the sun set in a beautiful red and purple horizon astern of us. The sky altogether presented an extraordinarily beautiful, yet wild aspect, the lovely blue being mottled in many places with fine red clouds. This has been fortunately the finest day of the season. Lat. at noon 77° 49', long. 162° 35', thermometer 29°, wind W. to N. by E., dip and var. $= \frac{85.5}{84.4}$. I turned in a little after midnight. Hove-to eastward of the Barrier Bight.

Thursday, 24th.—This morning, on my going upon deck, I found that we were standing to the northward under a press of canvas, studding-sails low and aloft, and royals. About 10.30 a.m. the weather suddenly changed, becoming overcast, with heavy squalls, which soon reduced us to our top-sails, and a reef in them, running along the edge of the stream ice. I saw a lestris in full chase of a brown and white petrel. Last evening a cask containing a paper with the latitude and longitude and signatures of all the officers was thrown overboard.

Saturday, 26th.—An "ice-blink" all along the horizon to leeward, indicating the situation of the pack. We have been passing through large streams of ochreous-brown-coloured ice, covering the water with a film, thickly studded with small fragments, having the under-surface of an iron-rust colour, the undulations of the water giving it the appearance of oil in motion. To-day I saw an ash-backed petrel for the first time within the Antarctic Circle, and a second one in the evening, with a solitary Cape pigeon.

Monday, 28th.—Fine day. Bergs numerous; we must have passed at the least a hundred large and small, some of them very large ones. A whole cluster of them passed us in the evening, sailing through thick streams of young pancake ice. At six p.m. running along the pack-edge, about a league distant, with many bergs posted along its margin, with others all round the horizon. During the latter part of the first watch and the commencement of the middle one the night was very dark, although but three or four days ago it was broad day-light throughout the twenty-four hours. When off the barrier, lat. 70° 54′, long. 175° 36′.

Tuesday, March 1st.—Passed several large bergs; sketched one, with numerous caverns in it. Snow in the evening. Running along pack-edge.

Friday, 4th.—It blew a fresh gale last night, ship rolling heavily. The blue petrel have made their appearance again in considerable numbers, with a few Cape pigeon. Sea clear of ice.

Monday, 7th.—Crow's-nest got down yesterday, as we have recrossed the Antarctic Circle, and no bergs or ice in sight, so that the white petrel, a true ice-bird, have all left us.

Thursday, 10th.—Last night was beautifully clear and starlight. The Southern Cross appeared high in the zenith, having a very bright appearance. I dined in the cabin with Captain Ross to-day, and we had a fine roast goose for dinner, notwithstanding we have been upwards of 100 days at sea. About midnight, on quitting the cabin and going on deck, I saw a fine, large, perfectly flat-topped barrier berg, upwards of a mile in length. Lat. 68' 18', long. 156° 7'. We have ceased now to alter our latitude materially, our object being to run down our longitude to the " line of intensity," in about 125°, and in the latitude of 60°.

Saturday, 12th.—Weather gloomy, with drizzle during the day; night squally and thick; thermometer 36°, wind northerly, lat. 60° 12', long. 147° 25'.

CHAPTER XXVI.

Collision with the *Terror* off the icebergs—Our perilous situation—
Our ship all but crippled—Captain Ross's self-possession—
Doubling Cape Horn—Loss of a sailor—Land in sight—Hog
Island.

A FEW minutes after I had turned in, and not long after
the first watch had been relieved by the middle one,
about one o'clock in the morning, I heard an unusual stir
on deck over my head. The hurried tread of the watch
at once suggested to my mind that something was going
wrong ; and, whilst listening in doubt as to the cause, the
voice of the officer of the watch calling down the main-
hatchway to "Turn all hands up," became very audible.
It now at once occurred to me that we were about running into an iceberg, as they had been far too numerous
of late in our course to render such a catastrophe at all
unlikely. After hurrying on only a portion of my clothes,
I had but just reached the top of the companion-ladder
to the quarter-deck, when I was all but hurled to the
bottom of it by a shock we received, not from an iceberg, indeed, but from our consort the *Terror* falling
across our bows.

On my gaining the deck what a scene met the eye !
First, the massive hull of the *Terror* surging heavily in
the swell on our starboard-bow, carrying away our bowsprit, and with it our fore-topmast ; whilst, above all,
towered through the mist of a dark, gloomy night, the
stupendous form of an enormous iceberg, whose perpen-

dicular adamantine sides loomed in terrific grandeur high above our mast-heads, and threatened both ships with instant destruction, our own more especially, from the tangled wreck of spars, sails, and rigging hampering us forward, which for the moment rendered our position a most helpless one ; a moment of one's life, indeed, never to be forgotten, as the two ships' sides were furiously grinding against each other in the heavy sea that was running, blowing as it did, at the time, a hard gale of wind.

Our consort, as she rose above the surging swell, showed her bright copper sheathing, and the greater portion of her keel, above our gunwale ; at one moment appearing as if her massive hull itself would come on board of us, and the next moment, in her descent, as we surged upwards, both hulls met together in fearful collision, but most fortunately in the encounter the yards of the ships escaped entanglement. The *Terror*, having received no injury to her spars and rigging, when her sails filled shot ahead, and, by way of a parting farewell, bounced her heavy stern into our starboard quarter-boat, crushing it like a nutshell, as she disappeared in the surrounding darkness and gloom, like a phantom of ill.

No sooner had we got clear of one danger than we were approaching another, of a far more threatening character. With all our head-sails a cumbrous wreck, we were fast closing in with the berg, its lofty, over-hanging crest, frowning destruction, as it loomed over the very trucks of our mast-head, and the ship drifting within the surf so angrily, surging around its hard-washed base. Reduced to such a helpless log in close proximity to such an enemy as we were, our narrow escape seemed indeed truly miraculous. The rebound of the waves, or in nautical language the under-tow, from the granite-like wall of ice alone saved us, humanly speaking, from being dashed to fragments against the smooth, hard, polished blue sides of the berg.

> By a perilous stern-board, through foam and surf dashing,
> Our yard-arms with the berg's frowning sides all but clashing,
> With the under-tow receding alone we are saved,
> To be on memory's tablet for ever engraved.

At such a perilous crisis a captain's position is assuredly not one to be envied, the responsibility is an awful one. We had no room to wear ship, the only alternative was making a stern-board, a hazardous one, in the sea that was running, and justifiable only by the circumstances of our position. However, Captain Ross was quite equal to the emergency, and, folding his arms across his breast, as he stood like a statue on the afterpart of the quarter-deck, calmly gave the order to loose the main-sail. His whole bearing, whilst lacking nothing in firmness, yet betrayed both in the expression of his countenance and attitude, the all-but despair with which he anxiously watched the result of this last and only expedient left to us in the awful position we were placed in. This feeling, I believe, pervaded all; as but for the howling of the winds, and the turmoil of the roaring waters, the falling of a pin might have been heard on the *Erebus's* deck, so silent and awestruck stood our fine crew in groups around, awaiting the result. So sudden was the collision that there was scant time for dressing, and an officer might have been seen clinging to the capstan in his nightshirt only.

I had myself seen much service in our old class of corvettes, very deep-waisted ships, frequently over-masted, with a great spread of canvas, rendering them very ticklish craft to wear in a heavy sea, from their decks holding so much water between their high bulwarks, should they get too much sternway on them. I fully realized all the perils of the manœuvre of a stern-board in such a sea as this with the addition of another element of danger, icebergs, in our drift; so that when the order was given to loose the main-sail, instinctively alive to our dangerous position, I flew to the main-tack, being the first to lay hold of it, followed by a lot of our fine fellows, by whom it was hauled sharp a-back with a will. The good old ship soon gathered sternway, and, as Fenimore Cooper humorously has it in the " Red Rover," began " ploughing the waves with her taffrail," meantime brushing the huge sides of the berg with our lower yard-arms till we reached its western extremity.

Here we encountered another large berg, lying athwart the course of our drift, and on which we were dropping ; and we had to get the ship's head round as well as we could, by means of the after-sails still available. Our situation now became a very critical one, with the loom of a third berg on the other side of us. At this moment, as I was looking aft, I saw our ice-master, Abernethy— one of the most experienced icemen of our day, who had been in the old *Hecla* with me in Parry's attempt to reach the North Pole, ever vigilant and on the watch— extended full length on the ice-plank, with his gaze intently fixed on the berg, when suddenly he reported to Captain Ross an opening between the two bergs, through which we were soon running before the wind, Providence thus directing our course through this narrow, surf-beaten strait, little wider than our own ship's length, none too soon before the meeting of the two bergs, which were rapidly approaching each other. As it was we passed so close to the vertical sides of the bergs that the foam and spray I felt rebound in my face, with such force as nearly to arrest the act of breathing when looking over the quarter.

Having run the gauntlet between these two giant bergs we emerged in the open sea beyond, and through the murky gloom of night, the *Terror's* light, all at once, burst upon us at some distance ahead, hove-to under the lee of a berg ; she having very suddenly disappeared after getting clear of us, must have, after her sails filled, instantly dashed through the same channel which afforded us an exit at least half an hour later, during which, our light having been extinguished from her view by the group of bergs within the circle of which we had been enclosed, they had been under the impression we had gone to pieces under the berg, where she parted from us. And short indeed would have been the work had we been fated to come into collision with its adamantine sides.

At daylight the bergs were just visible, about five in number, forming a chain. I turned in again about three a.m., and upon going on deck after breakfast on the morning of the 13th I saw our anchor sticking in the ship's side, under the fore-chains, and beneath the surface of the sea, where it had doubtless been driven by the *Terror's* hull, sledge-hammer fashion, and now remains, with both its flukes embedded to the depth of our doubling, a thickness of eight-inch planking; all the filling-up work on the starboard bow and rail above the gunwale knocked away.

Monday, 14*th.*—Last night we got a new bowsprit rigged out, and it being dark and misty, the ships at eleven p.m. lay-to for the night. To-day we passed several bergs, some blue petrel, and several sooty and black-backed albatrosses; also a black and white long-pointed-wing bird in the distance, resembling the sheer-water or skimmer. On the following day we got up a new fore-topmast, and a topsail yard, setting the fore-top-sail. Passed several bergs. Our other damages from the collision were, gig destroyed, first cutter stove in, with the loss of a six-pound mortar and the life-buoy.

Friday, 18*th*—At 5.15 p.m. the bower-anchor, which has been sticking in the ship's side for the last six days, and which we have been so carrying for some 700 miles, was observed to work about, and on getting a spar wedged against it, it was cleared from the ship's side, and sunk in deeper water than it had ever been in before.

Saturday, 19th.—At 3.30 a.m. a berg was reported ahead, bearing N.W. by N., and all hands turned up. Upon going on deck I found it broad on the starboard (weather) bow. It was blowing a fresh gale, with a heavy sea running, and misty weather. Having wore ship, we burnt two blue lights to warn the *Terror*, on our weather beam, of her danger. Passed near to another berg during the day, and I saw the first little petrel during the voyage. It was swimming within shot on the port beam, and dived as the ship passed it. At 9.30 p.m. we rounded-to on the starboard tack, the night being dark and thick, blowing a gale of wind, with a long heavy sea running and the chance of running into bergs before they could be seen in time to avoid them. Indeed, so anxious a time was it, that very few of the ship's company turned in for the night; lat. 60° 2′, long. 118° 55′.

Wednesday, 23rd.—On the line of uniform temperature, and Beauchene Island in sight, after our having been out of sight of land for the long space of 136 days.

Saturday, April 2nd.—For the last ten days we have been much favoured by the moonlight nights and fair winds, often going before it under a press of canvas, studding-sails low and aloft. Some time this morning we doubled the much celebrated Cape Horn, but owing to a sudden shift in the wind during the night, we passed at too great a distance from it to get a sight of it. The fresh gale and heavy, breaking sea, reminded us that we were in its traditionary stormy vicinity; and unfortunately, at 1.45 p.m., whilst taking in a reef in the main-sail, we were destined to meet with a painful remembrance of it, in the loss of one of our poor fellows overboard—Angelly, a steady, good old quartermaster of the watch, who, as he was stepping from the yard, slipped and fell, rolling down the main-rigging, and striking against the gunwale in his descent; the sound of which I heard, although in my cabin below at the time.

On my reaching the deck and looking over the stern, I saw him clinging to the life-buoy, which had been let go, and which he had almost instantly reached by swimming. Abernethy, ever the foremost in all emergencies, was already in the port quarter-boat, with a volunteer crew, all ready for lowering. But Captain Ross, deeming it impracticable for a boat to live in the sea that was running, wore the ship round, with the intention of passing close enough to Angelly for a rope to be thrown him; but unhappily the ship fell off before reaching him, though only a cable's length to windward. We tacked at 2.40 p.m., and in ten minutes we fetched the life-buoy, passing to windward of it very close. But the poor fellow was no longer on it. Nothing but the two breakers, with the small red flag waving over them, of which, only a few days previously, the buoy had been constructed, were now to be seen. Doubtless he had sunk exhausted and benumbed with cold in the heavy sea, in the interval taken up by the board we were compelled to make. This is the third of our crew lost by drowning since our departure from England; so unlucky has this ship been through accidents, for a fourth was suffocated in a tank in the hold at Hobart Town. Yet not a single death has occurred from disease or sickness, throughout the voyage. Several Cape petrel both hovered and swam round the buoy, as it drifted by us; as if singing a last sad *requiem* over the sailor's grave. Soon afterwards the sea rapidly went down, ending in an evening of drizzling rain. Lat. 57° 25', long. 67° 36'.

Sunday, 3rd.—Saw a chionis for the first time since our departure from Kerguelen's Land. It was hovering over the mast-heads. The ash-backed and the Cape petrel very numerous. At 6.30 p.m. blowing a fresh gale; passed a brig on our port beam, standing to the S.W. We showed a light, which she answered by a light astern. This is the first sail we have seen for upwards of four months. The night set in dark with drizzling rain. On the following day I saw two more chionis; wounded both, and lost both, which was to be regretted; I believed them to be a different species from that of Kerguelen's Land. To-day the birds about us have been numerous, skimming in the wake, or wheeling round us. Cape pigeon, ash-backed petrel, giant petrel, stormy petrel, black-backed albatross, and several wandering albatrosses (*Diomedia exulens*). Cloudy, but bracing day, with a fiery breeze, and lump of a sea running, squalls, and showers of rain—altogether a wild-looking day. Black clouds chasing each other over the blue sky, with a yellowish-red glare in the horizon, the fiery breeze scattering the foaming crests of the waves with a smoke-like spray.

Tuesday, 5th.—I was aroused from my sleep a little before three a.m. with the report of land in sight. I saw Beauchene Island on the weather (port bow), bearing N.N.E., about three leagues distant. It was one of the most lovely mornings we have experienced during our long voyage. Soon after six the sun rose in a clear blue sky; the water smooth, although there was a fresh breeze, the ship being under her studding-sails, low and aloft, going about seven knots; a number of fish leaping out of the water, and birds numerous. Albatrosses and petrel, with a solitary chionis and lestris, flew past the ship; whilst whole squadrons of cormorants came off from the islands, wheeling around us several times in their habitual prying, curious manner.

About 2.30 p.m. the mainland of East Falkland rose like a faint cloud in the horizon to windward. It was a most charming day, like a midsummer one, to our feelings; thermometer 44°, lat. 52° 36′, long. 58° 42′. At eight p.m. sounded in thirty-eight fathoms. Beauchene Island was the first land we had seen for the last four months and upwards. We have now been between four and five months without seeing land.

Wednesday, April 6th.—Beating up for Berkeley Sound, with a fresh breeze and thick weather, passing many patches of seaweed. After rounding a point in very thick fog, we passed Hog Island at four p.m. The land here has much the aspect of the Shetlands or the Orkneys, in undulating slopes of brownish-green peat, and interspersed with tufts of the tussac grass. We came to an anchor in five fathoms at 5.10 p.m., off a small creek near the settlement, about two miles distant. The small Government-house is indicated by a flagstaff, having an adjacent store-house and long peat-roofed shed. As we came up the sound, saw a troop of the wild horses of the island.

CHAPTER XXVII.

The Falkland Islands—Promotion to all except the medical officers— A shooting excursion—Natural history of the islands—The place described—Bullock-hunting—Black-necked swans.

Wednesday, April 6th.—On sending a boat on shore to the governor's, Lieutenant Moody of the Engineers, a Navy List, with the promotions of some of the officers of the expedition, was brought on board; gratifying to those whose services have met with such prompt acknowledgement for our great discoveries, as it was disappointing to those left out in the cold. For although all the senior executives, eligible for advancement, received a step in rank, the senior medical officers of their department were, as usual, left out, and their services ignored, notwithstanding they had extra duties to perform, natural history, in addition to their professional routine.

Friday, 8th.—Squally and showery. At five p.m. I landed at the Government Creek for the first time, and on the following morning, Saturday, 9th, weighed at 5.40, and warped the ship in shore, opposite the pond, above Government House. The secretary brought me a note from the governor, and I landed with him; had an interview with the governor, and returned on board at four p.m.

Monday, 11th.—At nine a.m. I landed on a shooting excursion, calling upon the governor on my way. I reached St. Salvador Bay at noon. Found rabbits numerous on the low sandbanks skirting the beach. Shot nine, black and white in colour. I saw a number of the island geese, steamer-ducks, hawks, thrushes, and sand-pipers. Started on my return at 3.30 p.m. ; shooting a few birds on my way back—a hawk, four thrushes, and, from the boat, a male upland-goose ; saw only one snipe. St. Salvador Bay is about four miles distant ; I got on board at six p.m. Day gloomy, and night with rain.

Sunday, 17*th*.—Captain Gardener, R.N., the missionary from Patagonia, arrived here in a schooner.

Tuesday, 19*th*.—Gloomy day. At nine a.m. I landed in the cutter at Hog Island, a long and narrow strip of land, boggy in the centre, and belted with the tussac grass, growing in tufts, close together, five or six feet in height, along the cliffs bordering the sea. The rocks on the beach are composed of clay-slate and quartz, strati-fied at a considerable angle. The island is separated from the mainland by a fordable creek. I shot a heron and two chionis, and, as I had anticipated, the latter proved to be a different species to the Kerguelen's Land one. I also shot eight cormorants at one shot, which, with a duck, made up my game-bag. A party of gauchos had been lassoing cattle, seven out of a herd of 135, of all sizes, ages, and sex—from the old bull and cow to the young calf. After throwing the bolas round the animal, they cut the ham-strings and then the throat. I returned on board about four p.m. ; and at six p.m. dined with the governor at Government House; met Cap-tain Gardener, and Captains Ross and Crozier there. I left at 9.30 p.m.

Wednesday, 27*th*.—Dined in the cabin with Captain Ross to meet the governor (Lieutenant Moody), Captain Gardener, and Captain Crozier ; also the governor's secretary.

Monday, May 9*th*.—Left the ship at 9.35 a.m., in the seal-skin punt, and pulled myself across the sound in thirty-five minutes from our little jetty. Fine calm morning. Hauled the boat up on the beach, where I shot three black oyster-catchers, and crossed a creek to the southern ridge of hills. Reached the summit at two p.m. : here I shot a rabbit and two black hawks. Saw a troop of a score of wild horses : they scampered off at a great rate, accompanied by three or four foals, making a curve round the base of the hill in single file. On my return, at 2.30 p.m., I came across them again, when they moved off in the same way.

The summit of the ridge consists of quartz in ruin-like masses, having numerous large fragments scattered about. A deep valley divides the ridge from Mount Vernet, through which descends one of those remarkable streams of stones, perhaps a mile in length, so peculiar to the Falklands, and for the origin of which it is not an easy matter to account in any very satisfactory way. The summit of Mount Vernet was concealed in mist, but I had a fine view of the arms of St. Salvador Bay, extending deeply up. I shot an ash and white-backed hawk, on my return round the head of the sound, as I had to leave my little boat behind, a head-wind proving too strong for her to buffet with. I got on board at 5.10 p.m.

Thursday, 19*th*.—Fine bright sun with light airs all day. At 9.30 a.m. I started from the landing-place on the beach on an excursion round the top of the bay or sound ; on reaching the south beach, I found the skin-boat I left there last week all right ; and at 11.30 a.m. launched her and shoved off, pulling along the beach to the south creek. On passing a small cove at the head of a valley, I saw a number of rabbits sporting about among the low underwood ; and on a remarkable green patch of grass sloping down to the beach, about a score geese were reposing, and amongst them three or four upland ones. Passed numbers of cormorants on the rocky ledges, so stupid or indifferent to danger as to allow the boat to pass within an oar's length of them. I pulled to a small tussac-clad hill, on a small island having a low rocky beach, near the upper end of the creek, only about two miles from the entrance to the sound. Left the top of the creek at 3.15 p.m., and pulled round the point into the sound. At 3.45 p.m., passing whole squadrons of steamer-ducks swimming off the kelp, and through which I had some difficulty in getting the light punt, and more than once ran aground in shoal water on a reef of gravel and shells, extending out from the point for several hundred yards on the falling of the tide.

Numbers of kelp-geese (*Anser antarctica*), the gander pure white, whilst the goose is elegantly barred with black and white, a remarkable contrast, and from their habit of feeding along the shore, the gander is seen a long distance off, whilst the goose is invisible from her plumage, so much the colour of the shingle, concealing her in the distance. After a long pull I reached the ship at 5.20 p.m. The moon shone brightly in a cloudless azure-blue sky—altogether a lovely night; and I was fortunate in having a light air for my light skiff, as I had both wind and the ebb-tide to pull against. My game-bag contained four upland-geese, and besides the value of their skins for the collection, their bodies when roasted supplied a delicious variety of food for our mess; four kelp-geese, handsome birds for their skins, but too fishy in flavour to be of use for the table; a steamer-duck, a brown hawk, two black oyster-catchers, three ducks, three cormorants, and two of the beautiful chionis.

Monday, 23rd.—Fine day; I left the ship again in the skin-boat, at 9.10 a.m. and landed at 9.55 on the opposite side of the sound at Greenbank Cove; and before I reached the land was tossed about in all directions by one of those sudden squalls of such frequent occurrence here. I shot one black and three black and white oyster-catchers, a chionis, and a small gull on the reef. Pulled inside the kelp to the entrance of the creeks, and at noon hauled the punt up near the Greenbank. Shot a male and female upland-goose at one shot, and four rabbits in the valley; which I left at three p.m., with the punt hauled up on the beach, as it blew too hard to venture back in her. I therefore walked all the way round the head of the sound, back to the ship, which I reached at 5.15 p.m. Shot an ash and white hawk and a snipe returning. Met the governor in the bay at the top of the sound. Night moonlight.

Thursday, 26th.—Landed at nine a.m. and walked round the top of the sound to the skin-boat I left on the opposite side, and which I reached at 10.45 a.m. Shot two rabbits and five finches in the Greenbank ravine. At noon I launched the punt, and pulled into the creek, when on rounding the rocky point, a sudden violent squall compelled me to haul her on shore on the opposite side, where I left her at the edge of the green bank above high-water-mark. I shot a female Antarctic goose here, and then started for Urania Bay, so named from the loss of a French frigate of that name here a few years ago, commanded by Captain Freycinet. It is about eight miles from our anchorage; I reached it at 1.30 p.m.; one arm of the creek runs up to within about a hundred yards or two of the bay, boggy and swampy.

Urania Bay forms a white sandy beach, between one and two miles in length, flanked by a narrow strip of low sand-dunes, covered with a small berry-bearing plant, and from fifty to a hundred paces across. I shot a teal on a small lake here, and ten small plover from a small flock on the beach, twice four at each shot; they were not at all shy, running along the sands again after the report of the gun. The weather cleared up fine, but still blowing hard. At 3.30 p.m. I started over very swampy ground, along the opposite side to the creek, skirting the base of the hills. Saw several upland-geese, and shot four, two at one shot, and the other two at single shots immediately afterwards. Had to leave them *en cache*, covered with bushes at the entrance of a rabbits' burrow at the head of the main stream of the creek. About a quarter of a mile above it shot three Caracara hawks, as they were watching my proceedings at a respectful distance, with the intent of making my geese their prize, I well knew, as soon as my back was turned. But I reversed this by making prizes of them for the ornithological collection. Only saw one snipe, which I shot, also a hawk and a brown thrush. I saw three wild horses. Reached the shingle beach bay at the head of the sound at dusk, about five p.m., but it being high spring tides, I could not return by the beach, and had a very rough journey over the top of the cliffs, amongst long grass, tufted hummocks, and swampy bog, alternately.

The night was dark, with a black, threatening sky to windward, from which, every now and then, flashes of sheet lightning issued. I reached the landing-place at six p.m., and had to wait some time for a boat, all hands on board being employed in mooring ship; a tremendous squall with heavy rain came on after I got on board. My excursion exceeded sixteen miles.

Friday, 27th.—This forenoon Captain Ross gave me the whale-boat, with two hands, to bring back my skin-boat with the game I left behind me the other day; and soon after we left the ship a heavy squall, accompanied by snow so thick that we entirely lost sight of both ships and land, shaping our course by the wind, and approached very near to the opposite shore before we saw it. Landed and walked over the hill for about a quarter of a mile to recover the four geese I left behind yesterday; shot two rabbits, a black and a grey one, in the brushwood, and the first snipe I have met with in such a place; also shot a red-breasted starling, and in pulling down the creek I shot five ducks and a kelp-gander. On reaching the spot where I had left the seal-skin boat, she was no longer there, although I had taken the precaution to place her some feet above the water-mark of the highest tides, keel upwards, and being also in a bight of the creek, the wind blew dead on ; indeed, it is inconceivable that even the gale and unusually high spring-tide of last night could have carried her away, yet not a vestige of her oars or the game left with her could be discovered along the shore. I shot two red-breasted starlings at a shot here, and a black and white oyster-catcher. We shoved off at 4.15 p.m., and reached the ship just after dusk.

Tuesday, 31*st.*—At 10.30 a.m. started on an excursion to Johnson's Harbour ; reached it at 12.30 p.m. ; shot a rabbit, and on the beach two black oyster-catchers at a shot, and two brown thrushes in the ravine. At 2.30 p.m. returned across the hills to Fisherman's Creek ; the whole of the low grounds, with the ridge of hills bounding them to the north, presented the most dreary aspect, with the stillness and silence of death, there being nothing to relieve the eye from the white mantle of snow which enveloped the whole, and rendered travelling most laborious work, sinking at every step mid-leg deep in snow, amongst long grass, bushes, and other vegetation, passing on my way a small pond or two frozen over.

Wednesday, *June* 1*st.*—A blowing, stormy day, cold and piercing, with a quick succession of hail-storms and driving mist. I started at eleven a.m. for St. Salvador Bay ; it took me one hour and a quarter, so trackless was the road rendered by the snow. I shot five red-breasted starling, three at one shot, and two at the next. Commenced my return at 3.45 p.m. amid snow-drift and dark passing hail-storms, the whole snow-clad country and sky above presenting the most wild and monotonous aspect ; reached the ship at five p.m.

Monday, 20th.—At five p.m. I dined at the governor's, and returned on board at 8.15 p.m.

Friday, 24th.—The *Carysfort*, of twenty-six guns, arrived from Rio Janeiro with supplies for us, on her way round the Horn ; she was commanded by Lord George Paulet.

Thursday, 30th.—I dined on board the *Carysfort* with the gun-room officers from five p.m. to 12.30.

Thursday, July 7th.—At 7.30 a.m. I left the ship with Abernethy and his foraging party for wild cattle for the use of the ship. At 8.30 a.m. we crossed over the narrow neck of land to St. Salvador Bay, and at 9.40 embarked there in a boat ; blowing fresh with drizzling rain. Passing between two islands, we landed on the one to the left for about twenty minutes, to allow the boat's crew to dine. The birds here, especially the oyster-catchers, were unusually tame and inquisitive. We next proceeded up a creek to the encampment, where the tents suddenly burst upon us as we rounded a point, and landed at 3.40 p.m. The two men left in charge of the dogs were on the look-out for us. The two tents were placed on the beach, backed by a bank, having a creek running up on the right. We had some cold beef for supper, and I distributed two bottles of port wine, which I had brought with me for the purpose, amongst the party, and at eight p.m. turned in. Rainy night.

Friday, 8th.—After a steak breakfast I started at ten a.m. on an excursion to the hills at the back of the tents. Walking round the creek to the eastward saw an old black bull under the hills ; ascended a ridge to the height of about 500 or 600 feet, composed of quartz, more crystalline and veined than at Berkeley Sound. On the hill-side a herd of seven cattle were browsing. Commenced my return by a watercourse and stream of stones at 2.35 p.m., and reached the tents at 4.20 p.m.

Saturday, 9th.—We all started at 9.15 a.m. over the

hills to the westward, cattle-hunting. The day was cloudy and squally, with a strong breeze and snow showers. In about an hour, after going some three miles, three herds of cattle appeared in sight, feeding on the sides of a ridge. When we had got within a quarter of a mile of one of the herds, the dogs were slipped, four fine, powerful hounds, named "Laporte," "Brigand," "York," and "Tom," the latter a cross of the pointer and bull-terrier, nicknamed "Bully," remarkable for its ugliness. Before the dogs reached the herd an old, black bull, stationed on an adjacent hill as sentinel, gave the alarm by running towards the herd and making a loud bellowing. The dogs, however, after a short chase, secured a heifer and a calf. About a mile further on three large cows appeared on a hill above us. I wounded one, and Abernethy shot another of them, soon after which we came upon another herd, near which several old bulls had stationed themselves, and I went alone in pursuit of two standing together on a ridge at some little distance off.

Having fired at and wounded one of them in the mouth, the ball passing through the upper lip and tongue, he instantly charged me, but the ball from my second barrel went through his heart. It was an anxious moment though, as the infuriated beast, tossing his horns, and snorting the blood from his nostrils, tore down upon me, bending his head to the ground for the final charge, for which I had reserved my fire till he had approached very near me. As my own life depended on the accuracy of my aim, this was the moment I seized upon to bring my second barrel to bear upon the region of his heart rather than his head, which had been tossed and lowered too abruptly for a sure aim. He gave a spring upwards as I fired, and came down with such thundering force within a few feet of me as to make the earth tremble beneath my feet. Abernethy, good fellow, who had been a distant observer of the risk I had incurred, and, trusting to his own unerring aim, scarcely allowed time for the report of my gun to reach him before he fired, his ball whizzing rather too close to my ears to be altogether pleasant, striking the bull, however, as he was in the very act of falling with my own ball in his heart. On opening him we found Abernethy's ball (which was a rifle one, and very different to my smoothbore double-barrel one) in the lungs. He was a fine black bull, with a splendid pair of horns, which I preserved as a memento. His companion prudently made off.

Three of the herd were killed by the dogs. In a valley about half a mile off I saw what I took to be a young black bull, and went in pursuit, accompanied by the dogs, and whilst our party were engaged opening the animals already killed. Having crossed a rivulet, the dogs seized the animal, which, proving too powerful for them, dragged them over a rising ground beyond, and I lost sight of them for a time, but was directed to the spot by the baying of the dogs and the bellowing of the beast, and as I came up with them the animal was again in the act of shaking off the dogs, three of whom had attacked it in front, and the fourth had hold of its tail.

I fired both barrels, the first shot brought her on her knees, and the second on her broadside dead, for it turned out to be a fine young cow, I am sorry to say, with calf, but a most fierce and powerful beast. The young dog " Laporte," a fine, strong animal of a light fawn colour, and with a smooth coat, exhibited much spirit in this encounter, and acquitted himself well, this being the first time he has been engaged in an attack on wild cattle, we having brought him with us in the boat. The dogs next started off to attack another herd on an outlying ridge, and protected by two or three old bulls. I had gone within two miles of the Coral near Swan Bay. In returning, two more were killed out of a herd, and at three p.m. we shaped our course for the tents, through snow and wind ; got back at four p.m. with twelve head of cattle killed in all. The early part of the night was one of wind and rain, followed by snow and frost.

Monday, 11*th.*—At 8.15 a.m. I made an excursion to Swan Bay, with fine weather and a fresh breeze. Crossed over the hills to the westward, and at eleven a.m. reached a bay east of the Coral. Saw several upland geese here, and walked along the shore to the bay, above which is the Coral. I observed recent tracks of the wild bulls on the sandy beach, and saw an old bull on the cliff above me. In passing round the points, and along fine sandy and gravelly curved beaches, I saw some Brent geese, but I did not reach Swan Bay. At 1.40 p.m I struck off across the hills, from a remarkable black-looking wall of rock, excavated below, and on the opposite side of a stream, on my return, and reached the tents at 4.30 p.m. Shot a brace of grey ducks here.

Tuesday, 12*th.*—At nine a.m. I again started for Swan Bay, accompanied by Abernethy and two of our dogs. The latter soon left us to rejoin the party in pursuit of the cattle. Eight fine bulls, posted on a rising ground not half a mile from us, catching sight of us, started at once across the intervening space, shaping their course direct for us at a hand-gallop, apparently with the intention of charging us, as they formed themselves into a compact squadron, led on by a fine, grizzly-coloured old bull, with magnificent spreading horns. We stood our ground on the very edge of the ridge to receive them, with our guns ready cocked, having a ball in each barrel. However, after getting within some ten or twelve paces of us, they suddenly thought better of it, altering their course, wore round, and descended the ravine, looking very hard at us as they wheeled about, and directed their course for the herd, then being attacked by the dogs and our party. Had we attempted a retreat, I make not the least doubt but they would have attacked us.

I shot a teal in a rivulet we had to cross on our way to the black wall of rock, where we found an old black bull lying down on the opposite side within rifle shot. Having fired several shots at him without effect, he got up and walked a few paces towards us, when we forded the stream, and he made a stand, until I got within some fifty yards of him, fired, and shot him through the centre of the forehead. He fell, and died with scarcely a struggle. Having a fine pair of horns, I intend preserving them. We now crossed over the grass-clad hills, fording two more rivers before we reached Swan Bay.

Here I found the object of my search, the black-necked swan (*Cygnus nigricollis*) — seven or eight of these most elegant and beautiful of all the swan family swimming on the lake. I had a very long shot at them, but they were so shy and wary they kept out of shot, in the very centre of the lake. I killed a brace of the smaller species of upland geese, not larger than the sheldrake, at one shot. We next walked over to Fox Point, where our people had stated they had seen a fox only a few days previously; but we could not discover any trace of him. I here shot another brace of the small geese at a shot. Started from Fox Point at two p.m., and, it being low water, we waded across the arm of the bay, by the black rock, where the river enters it. We saw a great number of cattle in various directions to-day. One herd to the westward must have numbered upwards of 100 head. The night was moonlight, with occasional snowstorms. We reached the tent at six p.m., it was freezing as we waded across the neck.

Wednesday, 13*th.*—Shot an owl near the tents, and walked to Black Rock River and back, bringing with me the horns of the black bull I shot yesterday.

CHAPTER XXVIII.

Trip to Swan Bay—Description of the camp—Streams of stones—Bird-shooting—Departure.

Thursday, 14*th*.—Fine day. At 8.30 a.m. I started on another excursion to Swan Bay, accompanied by Abernethy and two of the dogs. We saw about a dozen swans, but they were sagacious enough to persistently keep to the centre of the lake. I fired both barrels at them, as chance shots, being very desirous of obtaining a specimen, which would be a great desideratum for our ornithological collection. We walked round Fox Point, saw many teal in the inlets, and I shot one of the smaller kind of upland goose as we passed the cascade where I shot the old bull, and brought away his fine bushy tail. Cloudy, but moonlight evening. Reached the tents at 6.30 p.m.

Friday, 15*th*.—Gloomy day. At 9.15 started for Swan Bay, accompanied by " Laporte." Had a shot at our old acquaintance the large grizzly bull, but he took to his heels, declining any further encounter. Saw six swans in the bay, but they were far too wary to be approached near enough to secure a specimen. I chanced two distant shots as before, with a similar result. I reached the bay at noon, and left it at 1.30 p.m. Shot four teal, three at one shot. Saw several old bulls during my excursion to Swan Bay, which is nine miles distant from the tents, and Fox Point ten miles, over ridges and valleys of grass, and boggy ground, interspersed by streams and rivulets, Black Rock River being six or seven miles distant. On my return at 4.45 p.m., I found a fine calf tethered near the tents, which had been caught by the dogs during my absence.

Saturday, 16th.—Saw a pair of swans flying overhead. I shot two upland geese at a shot near the creek, and one of the smaller kind on the beach. As soon as the tide floated our boat, at 10.30 a.m., we started on our return to the ship, well freighted with fresh beef for the ships' companies, leaving two men in charge of the dogs and tents. Landed upon an island clothed with tussac-grass, where I shot a grey duck. At 1.30 p.m. we landed at St. Salvador Bay, and I shot an upland goose as we beached the boat. At 2.30 p.m. we followed the track to the ships, driving the poor calf before us. Met Captain Ross at the observatory, and got on board again at four p.m, after an absence of ten days in the camp, as the surrounding country is here termed, and the whole aspect of which presented the same kind of moorland region as around Berkeley Sound; the lower land of alternate ridges and valleys of the clay-slate and sand-stone formation abounding in fossil shells, and clothed with the withered-looking brown grass everywhere met with, intermingled here and there with the diddle-dee of the settlers, a small berry-bearing plant. Narrow water-courses intersect most of the valleys, where the soil is generally swampy and boggy.

This tract extends for two or three miles from the shores of the bay, backed by a range of quartzose hills, rising to an elevation of 500 or 600 feet, at the base of which I observed the streams of stones so remarkable in the range of hills south of Berkeley Sound, all the fragments composed of the same kind of angular-shaped quartz. But I did not meet with any traces of organic remains in the clay-slate or sandstone in the vicinity. Birds were more abundant, more especially the smaller kind of geese, rarely found about Berkeley Sound. Upland geese, too, were far more numerous, as well as the teal, and much less shy. The cattle were generally dispersed over the hills and ridges skirting the valleys, which, affording them an extensive field of vision all round, made them safe from any sudden surprise. They roved in herds of from seven to eight to upwards of 100 head in each, the old bulls invariably keeping apart from the herd, stationing themselves on a ridge not far distant, either singly or in pairs, ready to give the alarm on the approach of danger; this they announced by bellowing and running towards the herd, which instantly, taking the hint, made off at full speed, whilst the old bulls, having performed their duty, kept their ground, disdaining to fly, and thus falling an easy prey to the hunter who may be a good shot, and cool enough to put a ball through the heart or brain. Short of this, if otherwise wounded, they are most formidable opponents, giving but brief time to reload for a second attack. The tents were about twelve miles from St. Salvador Bay, and twenty head of cattle were killed during these ten days.

Monday, August 1st.—I received a note from Captain Allen Gardener, asking me to call and see his little daughter, who was unwell, which I did at ten a.m.; returning on board again at noon. It blew too strong afterwards to get on shore again as no boat could leave the ship.

Wednesday, 24th.—Captain Allen Gardener came on board to-day to thank me for my attention to his little girl, during her recent indisposition.

Saturday, 27th.—At two p.m. McMurdo, the first lieutenant of the *Terror*, was invalided. In him I have lost an esteemed friend, and the expedition one of its best officers, whom we could ill spare; but he has been suffering from an internal complaint, which left him no other alternative.   Had a long chat with him on deck afterwards.

Tuesday, September 6th.—At 9.15 a.m. I landed at the head of the sound, crossing over to St. Salvador Bay. Whilst pulling across the sound, I shot two black-headed, pink-breasted gulls. When about three miles north of St. Salvador, I passed a "Fachine" ravine swarming with rabbits.   In crossing over a sloping ridge, clothed with the long yellowish-brown grass which, with the tussac-grass (*Dactylis cæspitosa*), the balsam bog (*Bolax glebaria*), and *Veronica elliptica*, form such prominent features in the vegetation of these moorlands, interspersed in places with a sprinkling of pretty flowers, I shot three brace of snipe and two coral red-legged gulls; returning on board at 5.30 p.m.—my last shooting excursion here. It has been a beautiful summer day, a bright warm sun, yet a few patches of snow still lingered in places.

Wednesday, 7th.—Called this afternoon at Captain Gardener's, to take leave of him and his family, and afterwards on board the *Terror* to see McMurdo, whom I found taking a solitary dinner; and on returning on board I wrote a letter home.

R. McCormick, R.N., del.

Vincent Brooks, Day & Son, Lith.

Cape Horn.
With off-lying rocks,

CHAPTER XXIX.

Hermite Island, Cape Horn—Tierra del Fuego—The natives and their wigwams—My solitary exploration of the island—Ascent of Forster's Peak.

Thursday, September 8th.—At 9.40 a.m. we sailed for Hermite Island, distant about 400 miles, a boisterous passage of twelve days, encountering a succession of south-westerly gales. We made all sail down the sound under topgallant and studding-sails, before a fresh breeze, and a misty, drizzling rain. Got clear of the sound soon after noon, our stock of provisions consisting of 100 rabbits, two pigs, and fresh beef. Night cleared up fine and moonlight, the Southern Cross appearing very bright.

Monday, 19th.—At nine a.m. made the land, the Barnevelts and Cape Deceit on the weather-bow, Cape Horn, being six or seven leagues to leeward, under a mantle of snow. About two p.m. we doubled the celebrated Horn, passing within half a mile of it. It is a bold, bluff-looking headland for its height, barely 700 feet. It presents a precipitous, hard, greyish-looking face of rock to the southward, variegated in places with yellowish-brown vegetation. The S.W. extremity presented a rugged, broken, picturesque aspect. Lying off it are two small rocks, over which the sea was breaking. From four to eight p.m. we were employed beating up the Bay of St. Francis for St. Martin's Cove, Hermite Island, off the entrance to which we came to an anchor for the night, in seventeen fathoms and a half, sand. A thick mist descending from the hills, veiled the land. After dark we tacked close under Hall Island, a high, steep, and bold rock; and soon after we had anchored at eight p.m., a boat was sent up the cove, and reported having seen a light on shore.

In the gloom of the night the high land forming an amphitheatre in front of us appeared very close, like scenes on the stage. The *Terror* anchored about a mile outside of us, near Chanticleer Island.

Tuesday, 20th.—All hands employed warping the ship up the cove, and at three p.m. she was moored about one-third of a mile from its head. The best bower on the north shore, in ten fathoms, and the small bower in twelve and three-quarter fathoms on the south shore, on a bottom of sand, shells, mud, and black specks. The morning was fine; both captains landed before breakfast to select a spot for the observatory. I could see from the ship the smoke curling up from the native hut or wigwam, and five of the Fuegians on the beach, all naked, with the exception of a seal-skin mat thrown over one shoulder, the shoulder to windward. They shouted for us to come on shore, and exhibited no appearance whatever of fear or astonishment. Their canoe was lying on the shingle beach. The weather in the afternoon became overcast with snow squalls.

The only birds seen were a solitary pair of the Antarctic goose on the rocks, and a pair of steamer-ducks, with a cormorant or two swimming on the water, an occasional hawk or vulture hovering over the mountain heights, and the note of some small bird was heard in the woods.

Wednesday, 21st.—I landed for the first time at eight a.m. on the shores of Tierra del Fuego in the boat taking the observatory on shore. Three of the natives met us as we landed, saluting us with their customary expression, whatever it meant; the words were "Yammerskooner." I shook hands with them all round, giving each an anchor-button from my uniform jacket. In their canoe I saw the white pigment with which they smear themselves over. I went to the wigwam, a dome-shaped hut, about the size and form of an ordinary haycock, formed by a number of branches of trees driven into the ground and forming a circle, having the ends brought together at the top, and the interstices filled in very rudely with the smaller branches, leaving a small opening fronting the beach as a doorway. In the centre of the interior a few embers were still smouldering on the bare ground; a bag made of the skin of the steamer-duck, having the feathers inside, and a piece of the skin lay in one corner; from the roof was suspended a calabash-shaped bottle made of kelp, near which stood two spears with barbed bone heads, and on the ground lay four pieces of sticks tied together at their ends by means of rushes, the whole forming an oblong square, the use of which I could form no notion of.

Whilst I was examining the spears, the youngest of the three Fuegians, a youth of about seventeen perhaps, entered the wigwam, and I struck a barter with him for one of them, giving him half a dozen anchor-buttons from my jacket in exchange, with which he appeared to be highly satisfied. He had a goodnatured expression of countenance.

On leaving the wigwam, I took a ramble up the beech woods, some hundred yards up the valley above the cove, amongst bushes and rushes, and came upon a number of stake steps and bridges over the watercourse, indicating the site of some observatory erected here years ago by a former shipmate of mine in the *Hecla*, Lieutenant Foster, when he was in command of H.M.S. *Chanticleer* in those seas. I ascended the steep side of Mount Kater, first through a belt of the evergreen beech and some dead trees; then through a zone of arbutus, holly-leaved barberry, and dwarf deciduous beech, for about two-thirds up the hill, above which was the steep and all but barren peak, strewed over with a few projecting fragments of rock, intermingled with tufts of moss and lichens.   Here and there a half-withered, tortuous branch of the deciduous beech struggled for existence through the stony soil, in which its roots were embedded.

The air on the summit was very keen, the ground frozen hard, coated in places with ice and patches of snow, rendering the ascent and descent very slippery, with scant hold for either hands or feet; the projecting fragments of rock very treacherous to trust to, often giving way under the slightest pressure.   When within a hundred feet of the summit the gusts of wind, here called Williwaws, became so violent, it was somewhat difficult to cling to the sides of the peak against them. However, I gained a ledge of rock on the right, from which I had a fine view of the surroundings.   The sky was of a clear blue, a bright sunshine; a large hawk hovering overhead.   The snow-clad, rugged peaks of Tierra del Fuego bounded the distant horizon to the north, St. Martin's Cove lying immediately beneath me, with four small lakes on the ridge on its opposite side, backed by Mount Forster and the Conical Peak; a bay appearing on the north side of Hermite Island, and another to the westward, which, with the adjacent islands, completed the panorama.

I next descended, and on my way through the woods found the winter's bark (*Drimys Winteri*) and the holly-leaved barberry (*Berberis ilicifolia*), both budding.   On reaching the head of the cove I fired at a tree to observe what effect it would have on a Fuegian who was standing close to me, but he did not even start; quietly walking up to the tree with me, he picked out one of the shots. The birds were very scarce, and difficult to find in the dense thickets; I only saw two kinds of small birds, a creeper and a small ash-coloured warbler (or sylvia) similar to the Falkland Island beach-bird; the latter I shot, and returned on board at five p.m.

Thursday, 22nd.—Fine, mild, sunny day. At nine a.m. a canoe containing six Fuegians, three of them men, two children, and one woman, came round the north head into the cove. They first went alongside the *Terror*, and then paddled towards us. Two of the men came on board, the oldest remaining with the woman and children in the canoe. One of those on board had several necklaces, and a fillet bound round his head, and the hair smeared over with the white pigment, giving him the appearance of being powdered. He had a mat over one shoulder in the usual fashion, and, with the exception of this, naked as he was born; his face was streaked transversely with bars of some brown pigment. They made very significant signs for something to eat, and accompanied by their stereotyped expression of "Yammerskooner," which probably meant "Give me;" so I gave them some biscuit, which they ate readily enough. In the centre of the canoe was a fire placed on some clay.

In the after-part of the canoe lay four or five spears, near them the old man was seated, his face barred with black pigment. In the bows the woman reclined with a paddle in her hand, and a young child in her lap; near her lay an old rush basket or two and a spear-head. I handed her a few anchor buttons, and part of a gilt percussion-cap case for the basket and the spear-head. The cap case seemed especially to take her fancy, for she at once suspended it round the child's neck. In the bottom of the canoe lay a pile of firewood, with some rabbits and upland geese they had got from the *Terror*, and under which a dog had ensconced himself out of sight.

Their coarse, black hair, daubed over with the white pigment, contrasted strangely with their dark, copper-coloured skins, all smeared with paint, which together with their very crooked, spindle-like, ill-shaped lower extremities, gave them no very prepossessing aspect. All these defects, however, were counterbalanced by their good natured, inoffensive manner, and singular talent for mimicry, repeating every word spoken in the most clear and distinct imitation. On leaving us they paddled towards the wigwam containing the cove family of natives, but their meeting seemed one of a very apathetic character; not a word appeared to be exchanged between them. The woman and children did not land at all, never leaving the canoe. About noon they paddled round the north headland, the way they came, and were soon afterwards followed by all the party from the wigwam in their own canoe, leaving the cove wholly to ourselves; I heard the dog barking on shore before they left.

On Saturday, the 24th, the two canoes returned to the cove about eight a.m. with six men, four women, and four children, but left again at one p.m., after calling alongside both ships. Weather overcast, with drizzling rain.

Monday, 26th.—Gloomy and overcast, with drizzling rain. I landed at eleven a.m. on an excursion up the wooded ridge to the right of the observatory. Passing through a dense wood, I came upon a lake about a rifle shot across, from which I kept ascending to a gap or saddle between the cone-shaped peak on the left and the high ridge extending from Mount Forster on the right, and about an hour's walk from the observatory. The ground spongy and boggy, covered with lichens and mosses and patches of yellow grass; wreaths of snow amongst the trees.

When on the summit of the saddle, looking north, a low tract of boggy land extended to a bay on the opposite side of the island, distant about two miles. Saw two small lakes here, and shot a thrush of the same species as the Falkland Island one near the observatory, and after picking up a few shells from the rocks, I returned on board at five p.m.

Tuesday, *27th.*—Very fine, bright, sunny day. At ten a.m. I left the ship very quietly and alone, with the intention of following up my discovery of yesterday, and crossing the island to its opposite shore, for, owing to the somewhat sudden disappearance of the natives after our arrival in the cove, Captain Ross had entertained some apprehensions as to their feelings towards us as intruders upon their domain, and consequently would not give his sanction to excursions out of hail or sight of the ships on the part of either officers or crew, for fear of treachery if they came across the Fuegians. But I felt that in my own case, with the special duties of naturalist to perform, I had good reason to infer that he would not expect this general restriction to apply to or affect my movements in the pursuits of science, provided I took the responsibility upon myself, and, having done so, subsequently made him acquainted with it.

Accordingly I started at ten a.m. of the 27th. Following the native track from the observatory to the right of the watercourse and the lake, up the North Saddle, which I had reached yesterday, I now descended it to the other side through dense woods of the deciduous beech (*Fagus antarctica*) and the evergreen beech (*Fagus Forsteri*), then to the left along a watercourse over moist and boggy ground. Passed over two or three open glades in these woods, swampy, and covered with long rushes, not unlikely spots for moorhen.

At one p.m. I all at once suddenly emerged without any previous sign whatever from the densest thicket, upon a shingly beach of the bay, which I had seen from the summit of the saddle yesterday. On the beach a pair of kelp geese were feeding, as usual with them, amongst the seaweed, and just as I was about firing at them a pair of grey ducks caught my eye, swimming in the

VOL. I.   X

bay about 100 yards to the westward ; and thinking they might be new to me, I skirted the bushes till I came within shot, fired and killed one ; at the very same moment a fine caracara hawk flew overhead before I observed him. and just as he was about alighting with outspread wings upon a dead tree I brought him to the ground with the remaining barrel.

At this very instant, much to my surprise, I now dis-covered that I was in close proximity to a native wigwam, just within the skirts of the bushes behind me, and which had been concealed from view by a shell mound, the soil of which was grown over by a profusion of wild celery, giving it a green appearance—a very " kitchen-midden " in short. Picking up my dead hawk, which had fallen at the foot of the withered tree a few paces to the left, I lost no time in reloading my gun before I ventured farther in exploring the precincts of the native hut, first, however, securing the remaining duck, which had never left its dead companion, by firing one of my barrels I had just loaded. The wind being on the shore, both birds drifted upon the beach, and having transferred them with the hawk to my haversack, I reloaded the barrel, and thus armed I approached the wigwam, the door of which fronted the bay. On entering I found that its owners must have only recently left it, as the remains of a fire, with the ashes still smouldering, occupied the centre, which, with a few limpet shells scattered about, was all it contained.

It is about five miles distant from the ships, the North Saddle being about midway between. I left the wigwam at 1.30 p.m., as I was satisfied that its inmates were lurk-ing about not far off, and I guarded against any sudden surprise by avoiding as much as possible the thickets, keeping to the beach. I was not, however, kept long in suspense, a canoe hove in sight, paddling along shore from the opposite side of the bay, and making for the wigwam, so that probably the owners had been on a fishing excursion when my unexpected visit to their hut had attracted their notice, for they are very keen-sighted, and no one from the ships had before this crossed the island. In returning over the North Saddle is a platform free from bushes, in shape resembling a ship's deck ; here I shot a male teal, out of a pair swimming in the lake below, and got on board at four p.m.

Wednesday, 28*th.*—No one from the ships having yet attempted the ascent of Forster's Peak, at 10.15 a.m. I landed with that intention. The day proved squally, with showers of rain and hail alternately, and sunshine at intervals. I commenced my ascent at the watercourse, and on gaining the first ridge I continued along it till I reached the base of Mount Forster. Passing by a lake, I shot a tree-sparrow and the Falkland Island sylvia. After rising a second ridge I gained the summit soon after noon. It is formed of a very hard, compact, close-grained greenstone, highly magnetic, and forming a rugged pile of projecting masses and detached fragments of rock, the ascent being at about an angle of 45°. The deciduous beech even here struggles for existence on the bleak and barren summit, entwining itself in a tortuous network amongst the rocks in the scanty soil, in company with a few lichens and mosses.

On its south-east aspect is a deep, wild-looking, rugged, perpendicular precipice, like a yawning gulf as it descends to the shores of the cove. Cape Horn stands boldly forth in the distant horizon. On the north side Mount Forster forms a gradual sloping descent to the low, boggy ground, bounded by the north bays, and on the distant horizon by the mainland of Tierra del Fuego. I could see the shell-mounds of celery in front of the wigwam I visited yesterday, and counted six small lakes on the low ground north of the peak; to the north-east deep water and the narrow boat entrance to Maxwell Harbour, when an opening between the hills revealed a glimpse of the bay to the westward, and I counted some twelve islands around.

Whilst on the summit I had an opportunity of witnessing the formation of one or two of those local squalls termed here "Williwaws," as they rose in the form of a light white, vapoury kind of cloud, gradually expanding in size till, reaching the opposite western hills, they burst forth in all their concentrated strength, gusts of wind so violent that I was compelled to cling to the bare rocks to avoid being blown over some precipice.   These gusts are followed by hail and rain in very large drops, the air becoming all at once cold and most pinching to the fingers.   This peak is very nearly as high as Mount Kater.   At 1.40 p.m. I descended towards the North Saddle, which I reached at three p.m., along the wooded side of the hill, under Conical Peak, and west of the lake, by a watercourse, through tall trees, and got on board at five p.m.

Monday, October 3rd.—Gloomy and overcast morning. At 9.15 a.m. I landed at the watercourse, shot a tree-sparrow, and the female of the teal I shot here the other day, and also a yellow-breasted fringilla.   After proceeding for some hundred yards through the dense woods and thickets, I missed my geological hammer from my shot-belt, in which I had fixed it, as I thought, securely, but the struggle through the bushes had dragged it out. I felt much mortified at this discovery, as it had been a favourite hammer of my old friend Dr. Fitton, formerly President of the Geological Society, and who took the liveliest interest in the expedition, and told me, when he presented me with it, after cutting an inscription on its handle, that he had done most of his geological work with it.   I began to retrace my steps at once, as nearly as I could guess, in the intricate maze of wood and shrub, with but the faintest hope of recovering it.

On regaining the margin of the lake, however, I was fortunate enough to catch sight of the handle sticking up just above the level of the brushwood, and now securing it more carefully, I shaped my course up the North Saddle, and through the wood, up the ridge to the Conical Peak, a remarkable-looking mass of greenstone, highly magnetic, and some half-hour's walk from the saddle. I descended near a deep narrow cleft or gully. On the west side several gullies were filled with snow, and a dense mist overspread the hills, accompanied by fine rain, so that I could with difficulty find my way along a precarious footing.

At one p.m. I got a momentary glimpse of the north bay, through the drifting mist. Shot a small, dark-coloured sylvia ; and amongst the underwood and beech-trees, when between the saddle and the lake, I met with a great prize, shooting a brace of woodcock, the first seen here ; the first bird I flushed rose close at my feet. The rain continued till I got on board at 4.30 p.m, the trees showering down upon me their drippings as I struggled through them, drenching my specimens and myself to the skin. Captain Ross this evening pointed out to me the polarity of the many magnetic specimens of the greenstone rocks I had given to him.

Tuesday, 4th.—The morning threatening and gloomy. At 11.30 a.m. I landed on an excursion up the high hill at the back of the observatory, the rival of Mount Kater in altitude. After passing over a ridge of deciduous beech-trees, stunted and straggling, I reached the summit of the peak by a crescentic ridge of snow, at 1.45 p.m. ; the sun shining forth through a break in the clouds, I had a finer prospect than from any of the other hills.

Below me, to the eastward, St. Martin's Cove with the ships at anchor ; on the west side, a deep, broad-bottomed valley, clothed with grass, rushes, and underwood—the latter chiefly of the deciduous beech ; be-

yond which, on this side of Cape Spencer, St. Joachim's
Cove, terminating in a fine white sandy beach, separated
by a narrow neck of low land covered with underwood,
rushes, and grass, and nearly a mile in breadth, from a
deep bay running up on the opposite side of Cape
Spencer, and terminating in a cove at the upper end,
bearing a striking resemblance to St. Martin's Cove,
the hills bounding it being high, and much the same
in character of outline. At 2.30 p.m. I descended along
a ridge which extends in a westerly direction towards
Joachim's Bay, and dividing the valley into two portions,
forming a slightly curved ridge in the centre.

Reached the hill above Joachim's Cove at three p.m.
Saw a pair of remarkable birds here, in size and appearance between the partridge and the quail. I shot one
of them, and in my descent to the valley on my return,
I flushed a woodcock from the rushes and underwood of
dwarf beech, and after shooting it, had it not fallen on
a white patch of snow, amidst the impervious thicket I
might have lost it. I reached the West Saddle at 4.45
p.m., descending along the margin of the snow patch,
to the observatory. Saw an arbutus in bloom here, and
got on board at six p.m., well pleased with my day's
sport, having obtained two rare and beautiful birds, a fine
specimen of the perdix and the woodcock.

Wednesday, 5th.—I gave to Captain Ross one of the
woodcocks and some specimens of magnetic rocks, of
great interest to him. I landed at eleven a.m. Went
up Observatory Valley, found a whole line of trees blown
down and uprooted here, by the late williwaw, sweeping
down the valley with overwhelming force ; and got so
entangled in the intricate matted thickets on the side of
the ridge bounding the western valley in the south,
consisting chiefly of beeches, six feet high, with their
branches interlaced at the tops, that I was obliged to
cross the valley to the opposite ridge, through patches of

rush and underwood. I saw two more of the quail-like birds in the place where I first met with them yesterday. Descended nearly to the sandy cove, but a thick mist, with drizzling rain, accompanied by heavy gusts of wind, arrested all further progress, and at 3.15 p.m. I returned, flushing a woodcock, and hearing the voices of the natives, although I did not get sight of them ; and got on board at six p.m.

Friday, 7th.—One of the Fuegians, the youngest man, came down the valley to the observatory this afternoon, and made a fire in the wigwam, where he remained all night. I skinned eight birds to-day, and found the elytra or wing-cases of a small beetle in the stomach of the woodcock.

Monday, 10th.—Very clear, sunny day. Landed at the rocks under Kater's Peak, and at nine a.m. started up the north face of the mountain, through thickets of evergreen beech, from twenty-five to thirty-five feet and upwards in height; passed some deep gorges and fissures, about two-thirds up, found the first inaccessible ; tried to scale an intervening mass, but found it too steep, with no hold for either hands or feet to trust to. Descended round the point, and effected the ascent by a gorge to the west of it, and reached the summit at 10.45 a.m. This is from twelve to twenty feet higher than the point seen from the ship. It is composed of sienite of a lighter colour, from having a greater proportion of felspar than at the base, which contains more hornblende.

I had a fine view from the summit of the Capes Horn and Spencer. At 12.15 I descended on the south-west side, over rugged ledges of rocks, following the ridge to Joachim's Cove, where it terminates in a steep wood-crested hill, having a perpendicular escarpment next the sea. Passed several large boulders of very fine-grained greenstone. On board at five p.m.

Tuesday, 11th.—Landed at nine a.m. Crossed over the West Saddle. Saw a greater number of small birds amongst the tree-tops than I have observed before. I shot a wren and finch in Sandy Cove, and a black and white oyster-catcher. Had a shot at an owl. I noticed a wigwam on the other side of the neck of land, but a heavy fall of snow, rain, and thick mist compelled me to return at two p.m. On passing by Sandy Cove I discovered a wigwam embosomed in trees above the bank, having a mound in front covered with celery. Met with large boulders of granite in Sandy Cove. The air became very raw and cold in the latter part of the day. Got on board at 7.45 p.m.

R. McCormick, R.N., del.

Vincent Brooks, Day & Son, Lith.

Cape Spencer and Joichim's Cove, Hermite Island, Tierra del Fuego.

CHAPTER XXX.

Excursion up to the summit of Cape Spencer—Bird-shooting—Difficulty with a native as to plunder— My diplomatic manœuvres—We leave the island.

Wednesday, October 12th.—Morning cloudy and threatening rain. I landed at 8.45 a.m. amid a fall of snow. Going up Observatory Valley, saw many small birds, but limited to two or three species of the fringilla. Shot a thrush after I had crossed the West Saddle. The weather cleared up, and I followed the course of the Western Valley to Joachim's Cove, where I shot a polyborus hawk. Crossed Sandy Cove, and at 11.30 a.m. I began the ascent of the ridge on the opposite side of the cove leading to Cape Spencer, to reach the summit of which was the object of to-day's excursion. First passed over a short scrub, mixed with brown grass and moss, and a few patches of dwarf beech. On gaining the rocky top of the ridge I continued along its sharp crest, having an almost perpendicular escarpment on the left, skirting the deep valley beneath, in which was a lake. This ledge had an undulating outline, rising in one part to a considerably elevated point, with corresponding depressions.

The rock was granitic, very distinctly marked by divisional planes in the escarpment to the south, and dipping in that direction at an angle of 45°. Enormous boulders lay scattered about in the wildest confusion the whole of the way to the base of Cape Spencer, where, at

12.50, I descended a saddle, or depression, three miles from Sandy Cove, at the head of Joachim's Bay; many of the granite blocks were intersected by veins of a close-grained greenstone, from three inches to three feet in diameter. Here I could trace the recent impressions of the naked feet of the natives, where the soil was soft and plastic all along the ridge, and in an upward direction towards the summit of the cape. At the before-named depression the actual ascent of the promontory constituting the cape commences, near two small pools in a peaty soil from this point. I reached the summit in twenty minutes, at 1.15 p.m., over an elastic moss, filling up the interstices between the fragments of rock, which are piled one upon another to the summit, approaching which, however, the vegetation entirely disappears, and you step over the bare rock, from one block to another, till the top of the ridge is attained, when a perfectly circular bowl, an extinct crater, in short, presents itself, between 200 and 300 feet in depth, and upwards of a mile in circumference, the bottom of which is filled by a large lake, its greatest length being from north to south, coated at the time of my visit with ice on the northern side. From the margin of the lake to the upper rim of this evidently once submarine crater, the whole was formed of loose stones and fragments of rock, composed chiefly of sienite or greenstone, whilst the base of the mountain is of granite, similar to the ridge from which it arises.

The highest part of the rim is to the westward, along which I proceeded, following its narrow crest to the point overhanging the sea to the south, about five minutes' walk from its highest part. I now found myself standing on—with one exception, the celebrated Horn itself, which stands out a few miles further south—the southernmost extremity of the vast continent of America, or rather, islands lying off it.

As no one from either of the ships had preceded me here, I emptied some shot from my belt on the rock I had seated myself upon, placing an anchor button from my jacket in the centre, as a memento of my visit.

With the great southern ocean spread out before me the scene was one of unrivalled grandeur and sublimity, much enhanced by the cheering rays of a bright sun, which shone forth from a clear azure-blue sky, the day having cleared up charmingly fine. In the distant horizon to the south-west the Diego Ramirez rocks appeared like faint hummocks, and the western extremity of Hermite Island low but distinct, whilst Cape Horn stood boldly out to the south-east, breaking in upon the monotony of the vast expanse of waters.

One of those squalls for which this region has become so celebrated passed before my eyes, presenting a very singular appearance as seen from the spot where I stood. It commenced as a light, white, vapoury cloud, the out-line of which was so well defined that it assumed all the appearance of an iceberg, for which I should most cer-tainly have taken it, had I not witnessed its beginning, and watched it through all its phases of development, and its effects upon the surface of the water beneath it in its transitions, as it very slowly moved in the direction of Cape Horn, which it doubled, not materially altering its form, that of an iceberg, a phantom one, it is true, and of thin, aerial aspect.

Looking landward, too, there was no lack of incidents to render the day's excursion an impressive one. Below me, in the valley beneath, the dusky forms of half a score of the Fuegians caught my eye at the distance of about a mile from me. They were winding along in single file by one of their tracks from the rocky ledges skirting the bay, where doubtless they had been collecting limpets —their favourite food, and, as bread is with us, their staff of life—from the rocks along the shore. Notwithstanding my elevated position, and all but concealed by the rock's fragments grouped around me, I really think their keen, sharp eyes discovered me, for they came to a halt more than once, as if reconnoitring the spot on which I stood. The solitude and deathlike silence reigning around was only broken by the presence of a large polyborus hawk hovering immediately overhead, and a small warbler which flitted by me at the moment.

At two p.m. I began the descent. Reached Sandy Cove at four p.m. Flushed a woodcock amongst some bushes, but missed it. Saw a vulture, and shot a small hawk I had not before seen on the beach. Crossed the West Saddle at 6.15 p.m.. Shot a thrush in the valley, and got on board at 7.15 p.m. Soon afterwards it came on thick with snow.

This evening I volunteered to Captain Ross to be one of a party in a boating expedition to Cape Horn. The distance which I went over to-day must have exceeded twenty miles, as the summit of Cape Spencer cannot be less than nine or ten miles from the ship. On the following day two of the natives returned to their wigwam in the cove, and lighted a fire in it, and left again on Sunday, the 16th. On the following day, Monday, the 17th, I again landed, on an excursion round the shores of the cove, returning on board at six p.m.

Tuesday, 18*th.*—Was the most boisterous day we have as yet experienced here. The barometer fell nearly to 28°, blowing in furious squalls or williwaws from the south-west, lashing the surface of the cove into sheets of spray and mist, accompanied at intervals by heavy falls of snow, which covered the hills to the very margin of the cove, with one mantle of white, so that I could not communicate with the shore. No boat could land; and to-morrow is "term-day."

Wednesday, 26*th.*—We have had a whole week of stormy weather, notwithstanding which I made excursions on shore both yesterday and the day before, but of no great interest. To-day I made one to Maxwell Harbour, starting at 9.45 a.m. I landed at the watercourse and crossed over the North Saddle and the flanks of Mount Forster, where I met the surgeon of the *Terror*, who joined me in my excursion. At 1.45 p.m. we reached the ridge above Maxwell Harbour, passing a lake and over moorland ridges, and across a deep ravine.

Maxwell Harbour is land-locked, formed by three islands—Hermite, Jerdan, and Saddle Islands, affording a safe and commodious anchorage for a number of ships, protected on the north by a long reef, and having the principal entrance from the eastward; but there is also a narrow strait for boats from the southward. The shores are skirted by woods, having a pretty appearance from the ridge above, on which we stood. The furious williwaw, accompanied by a pelting shower of hail, compelled us to seek shelter on the cover of a block of granite, of which the rocks here are composed, intermingled with greenstone. We returned across the moorland, which is studded over with small lakes, at five p.m., in a hail and snow storm. Huge boulders of granite are strewed along the whole line of coast. I shot a fine specimen of the vulture flying over the tops of the cliffs above us, also a tringa, or sandpiper, a finch, a tree-sparrow, and a fringilla. The distance to Maxwell Harbour is about eight miles. Got on board at seven p.m.

Friday, 28th.—Fine day. Employed on board skinning birds, eleven in all. Both captains left in the gig, on an excursion to Maxwell Harbour, and Captain Ross on his return gave me a penguin which I had never seen before, and on the following day another penguin, two dotteril, and the quail-like bird of the island. The gunner of the *Terror* found two woodcock's eggs in the wood to the right of the Observatory Valley to-day.

Tuesday, November 1st.—Being a very fine day, I landed

at 9.30 a.m. by the watercourse for the purpose of paying a visit to the Fuegians at the North Bay, on the other side of the island, whose wigwam I had discovered when there last. My way at first lay along a deep watercourse, flanked by tall trees. Reached the top of the saddle at eleven a.m.; shot three dark-coloured sylvia here. Descended on the opposite side, through the dense wood, to the valley running between Joachim's Cove and the North Bay, the recent impression of the feet of the natives directed towards the bay; following up their footsteps led me through a thick wood, guided by a watercourse, till I emerged upon the rocky beach of the bay not far from the wigwam. I now shaped my course round by the western extreme of the bay, doubling the granitic and greenstone point, fringed with trees and underwood, in search of eggs; the birds I met with were black-backed gulls, grey ducks, steamers, kelp geese, and an oyster-catcher or two.

Repassing the wigwam on my return, as I anticipated, the Fuegians were at home. They at once came out to meet me; one, advancing towards me, proved to be the young man with whom I had exchanged anchor buttons for a spear on the first day of my landing in the cove. I shook hands with them, and gave to each a copper nail or two, with an anchor button from my handy store, the uniform jacket, in reply to their ever-ready expression, " Yammer-skooner." When, soon after taking leave of them, I had got little more than 100 yards to the eastward of the hut, a kelp goose rose from the long brown grass on the bank above the shingly beach, and alighted on a rock just projecting above the sea within gunshot of the beach. The nest of which she had betrayed the site by rising from it, contained seven eggs, cosily laid in the soft down from her own breast, within a tuft of the long grass.

Whilst I was in the act of stowing them away in my

haversack, the elder of the two natives came running towards me, so that my poaching on their domain had not escaped their sharp and vigilant eyes. But I had made up my mind not to easily part with a prize so valuable to me; therefore, on my friend "Copper-skin" beginning his "Yammer-skoonering," and, at the same time very significantly pointing to my haversack, and suiting the action to the word by going through the pantomime of sucking eggs, smacking his lips, and patting his stomach, I was left in no manner of doubt of his meaning and his wishes too ; also that he was fully aware of the nature of my find; therefore I felt that all my powers of diplomacy would be put to the test to avoid a rupture with this unsophisticated savage, and its consequences, should I refuse to give up to him what he naturally, and doubtless justly enough too, considered the property of himself and tribe.

Armed as I was, with a double-barrelled gun, active, and self-possessed, I could have no apprehensions for the result personally to myself, in the event of an attack from these two poor, naked savages, provided there were no others in ambush with their spears ; but at the same time I felt that if I gave them provocation to attack me, and that it happened, even in self-defence, one or both were injured or perhaps killed, I should never after pardon myself. Such were the thoughts that made the rapid circuit of my brain, when the poor goose, still perched on the rock, caught my eye, and suggested the idea of a compromise, and I made signs to my competitor for the eggs to that effect. Frankly opening my haversack and showing him the eggs, and then pointing over the hills in the direction of the ship, I endeavoured to make him understand that they were needed on board, and pointing my gun at the goose, accompanied by significant signs, which these children of nature are very expert in the interpretation of from their keen instincts, left him in no doubt that the goose was to be the substitute for the eggs for his dinner ; and as a proof that I was right in this impression, I had no sooner levelled my gun to fire than he pushed me along by the shoulder nearer to the water's edge till I was ancle-deep in the surf, evidently for the purpose of getting me nearer to it to insure my killing it. It was rather a long shot, but the bird fell winged into the water.

I made signs to him now to launch his canoe, lying at some distance on the beach, to retrieve it, and this suggestion of mine called forth an amount of cunning and observation really amusing in this poor, untutored Indian ; for, pointing at once in the direction of the wind, he then swept his hand round towards a curve in the coast-line on which it blew direct, and picking up a bit of seaweed from the shore, he shut his eyes and made a gesture, in imitation of the death struggle, and moving his jaws, as if in the act of mastication, implying by all this panto-mime that he knew as well as I did that the wind, being on the shore, would drift the bird into the curve of the beach lower down, and that he could then get it without the trouble of launching his canoe, when he would finish its existence and eat it for his dinner.

All this being now amicably settled, he accompanied me as far as my course lay along the sea-shore, all the way goodnaturedly singing and admirably mimicking every word I said, whilst not a look or gesture escaped his prying gaze. We reached the point at right angles with the beach, when I was about to enter the thick woods through which my course lay in the direction of Mount Forster for the ships. Night approaching, and not deem-ing it desirable to have his companionship any further, I resorted to a dodge to shake him off, which succeeded. For, as I stepped from the shingly beach into the thicket, I stooped to cut a good-sized sapling from a clump, well aware that he would most assuredly follow suit, and whilst he was intently occupied stooping down to cut his, I embraced the opportunity of cutting with him by goodnaturedly slapping him on the shoulder with my hand, and bidding him good-night, which he repeated in the same tone of mimicry. I was soon lost to his sight in the labyrinth of the woods, keeping a watchful eye and ear, both ahead and astern of me, to guard against any surprise; getting safely on board at seven p.m.

Wednesday, 2nd.—Very fine day; a bright sun shining in a clear, intensely blue sky. Last evening, as I was returning over the side of Mount Forster, between the lake and the opposite side, I shot a small, black creeper, having a grey pate, in a deciduous beech bush, and which I had never met with before, and one of the gallinaceous or quail-like birds allied to the tinamides of South America, and on the boggy ground saw four couple of upland geese, a vulture, and a swallow.

This morning at 8.30 I landed upon my last excursion in the island. Shot two yellow-breasted fringilla in Observatory Valley; crossed over the saddle into West Valley, through the underwood of beech, rushes, and grass, to the swampy ground, and a deep and rapid watercourse, to the rocky point of Joachim's Bay in search of eggs; unsuccessful however. Shot a kelp goose, saw two or three pairs of black-and-white oyster-catchers, who, from their very clamorous manner, were doubtless breeding there. Searched the southern range of grass-covered hills for the quail-like bird, but in vain; cut some walking-sticks in the valley, and got on board at 7.30 p.m.

Sunday, 6th.—Captain Ross to-day brought on board with him the two stranger natives I saw last evening. They walked about the decks dancing and singing and eating biscuit, seemingly quite happy. The captain gave them two small saws and some fish-hooks, and I gave them a knife and some buttons, of which they are exceedingly fond. One of them in return took a fillet of beads from his head and presented me with it. I measured their heights; one stood five feet one inch, and the other four feet ten inches, and they were the best-looking we have seen. They left the cove this evening.

Monday, 7*th.*—At about seven a.m. we stood out of the cove, with a fine fresh breeze; and as we crossed the Bay of St. Francis I took a sketch of the general outline of Hermite Island. At nine a.m. we doubled the Horn at about a league distant, and I saw the last of it at four p.m., a barque in sight at the time.

Thursday, 10*th.*—I prescribed for Captain Ross, who was very unwell. On the following day we sounded over Burwood's Shoals, the depths varying from twenty-six to eighty fathoms. Next day we passed Beauchene Island, the passage to the Falklands occupying a week.

Along the east coast of Tierra del Fuego, from Cape Horn, we found an undercurrent of very cold water coming from the south; and subsequently crossed the line of equal temperature of the ocean throughout its entire depth in lat. $55°$ $48'$, and long. $54°$ $40'$ W.; the thermometer standing at about $40°$ at the surface, and about $39°$ at the intermediate depths from the surface downwards; also met with an easterly current running at the rate of twenty miles a day.

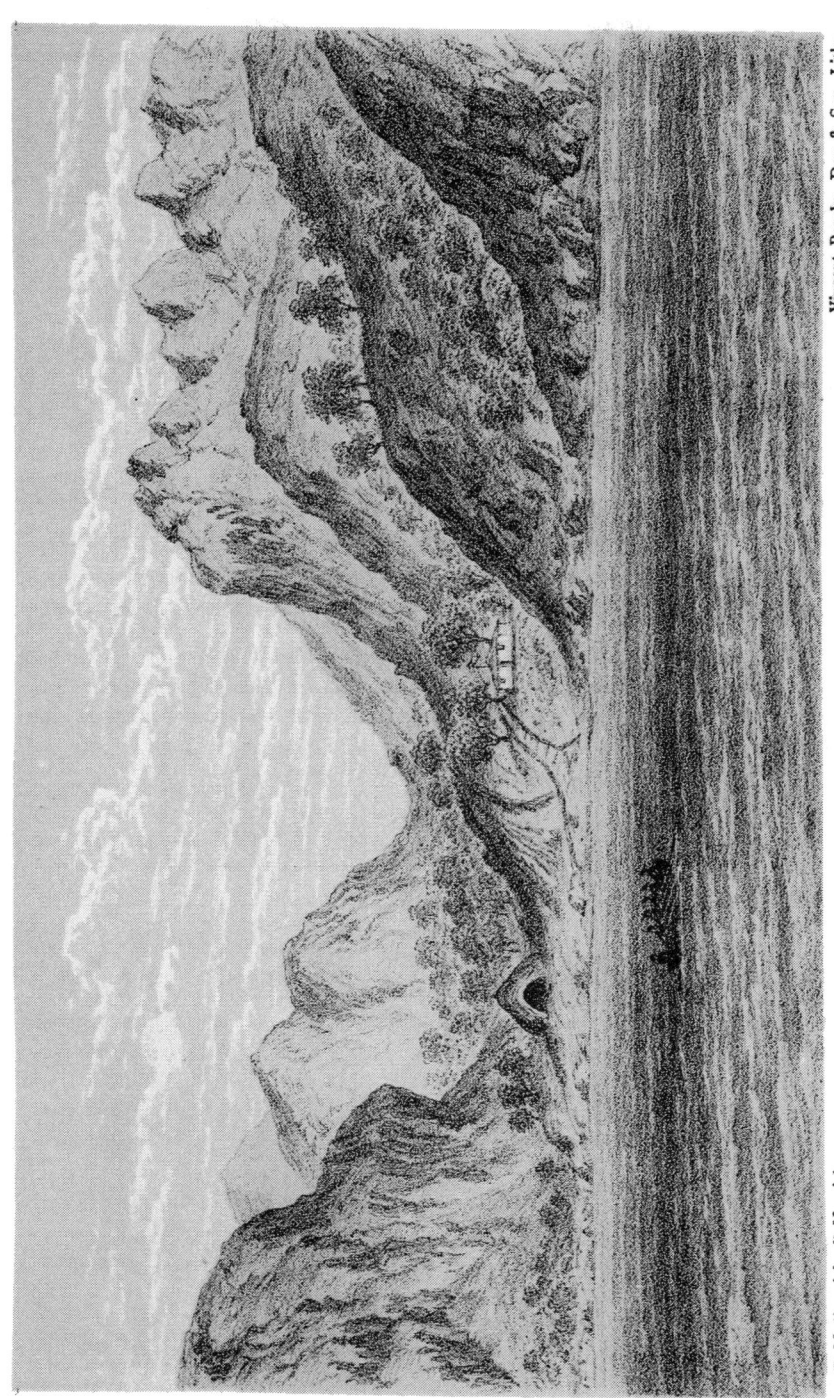

R. McCormick, R.N., del.

Vincent Brooks, Day & Son, Lith.

St. Martin's Cove,
Hermite Island, Tierra del Fuego.

PAGE 322—VOL. I.

323

CHAPTER XXXI.

Summary of Hermite Island—Geology—Ornithology—Botany.

THIS pretty and picturesque island, named by the Dutch, I believe, after one of their early navigators, lies about ten miles north-west of Cape Horn, is of irregular form, deeply indented by coves and bays on either side. Its shores are bold and steep, rising into peaks, such as Mount Kater, to 1742 feet above the sea. Its greatest length from east to west is twelve miles, and is situated in the lat. of 56° S., and long. of 67° W., the variation being 23° E. The rise and fall of tide in St. Martin's Cove is about eight feet ; its southern extremity, Cape Spencer, is next to the Horn, the southernmost point of the Americas.

The geological structure is plutonic, consisting mainly of granites and greenstones, the latter varying in appearance with the relative proportions of hornblende and felspar, passing into syenite. These rocks occur in very irregular order, the granite, however, forms the basis, and breaking out at Joachim's Cove, forms the ridge between it and Cape Spencer, where it is capped by syenite, composing the crater-formed summit of the promontory. The granite again appears forming the low land between Maxwell Harbour and the North Bay, with large boulders of it piled along the beach. All the high peaks, such as Mounts Kater and Forster, are constituted entirely of various modifications of syenite and greenstone in which

Y 2

small particles of pyrites are occasionally found disseminated. Some masses of purely white quartz occur in and near Deep-water Bay. Huge blocks or boulders of granite, traversed in many instances with veins of greenstone from three inches to three feet in diameter, lie scattered in the wildest confusion along the ridge terminated by Cape Spencer. Near to South Head is a close-grained quartzose cliff, traversed by a vertical greenstone dyke two feet in diameter, with a felspathic vein.

The vegetation is really abundant considering the bleak and stormy nature of the climate exposed to the violent squalls called " williwaws," which sweep over the hills from the westward in hurricane-like gusts, though sometimes lasting for only a few seconds. From the 20th of September to the 7th of November, the period we passed in St. Martin's Cove, the weather was for the most part wet, either rain or snow, the prevailing winds being from the N.W. and S.W., the thermometer ranging from 30° to 56°, and the mean temperature taken from our daily observations, maximum and minimum, gave 42° Fahr.; northerly winds generally brought thick weather with drizzling rain. When a fine day occurred it was accompanied by an atmospheric clearness, the sun shining brightly in a sky of the intensest blue, producing an elasticity of feeling which rendered existence on such occasions simply delightful.

The sides of the hills to within about 400 feet of their summits are clothed by dense woods of the two beeches, the evergreen (*Fagus Forsteri*) and the deciduous (*Fagus antarctica*), frequently attaining a height of forty feet, and often having a remarkable parasite, " misidendron," naturally engrafted on its stems, of which I noticed two species. Fungi also sometimes formed globular-shaped excrescences of a hard woody fibre, having a wrinkled, reticulated surface encircling one of the branches.

The ordinary size of the trunks of these trees was about one foot in diameter ; but trees were found upwards of ten feet in circumference. The deciduous beech was leafless at the time of our arrival, appearing amongst the bright-leaved evergreen-like large patches of brown on the sides of the hills and the upper portion of the valleys, giving a pleasing variety and tint to the landscape. It attains quite as large a size as the other kind, the stem is of a dark brown, mottled over like the common hazel. At the time of our departure its bright green leaves had fully unfolded themselves, giving one uniform tint to the woods.

The Winter's bark (*Drimys Winteri*) grows profusely amongst the beeches, like a laurel in aspect but more erect, the ordinary height being from six to seven or even twelve feet ; and now and then may be seen a good-sized tree, rivalling the beeches themselves in magnitude. The shrubs forming the underwood consisted chiefly of the holly-leaved barberry (*Berberis ilicifolia*), which at the time of our visit was covered with clusters of bright orange-yellow flowers, the wood itself also having a deep yellow tint, with the bark encrusted over with lichens. It is a crooked, crawling bush as prickly as the bramble. One or two other species occur bearing red berries (*Berberis parvifolia* and *Arbutus rigida*). The tussac grass was in bloom, growing on the north shore; and the celery (*Apium graveolens*) was found growing on the mounds of shells in abundance in the vicinity of every wigwam, yet the natives did not appear to eat it. Scurvy grass (*Cardamine hirsuta*) was very generally met with along the watercourses, as was also the fascine of the Falklands.

The south-western aspect of the hills, where exposed to the effects of the williwaws, are for the most part barren. The bottoms of the valleys are usually clothed with a tangled underwood of the dwarf beeches, intermingled with long grass and rushes. On the low boggy

tracts the surface is studded over with tufts of a vividly
green plant, forming discs of various magnitudes (*dona-
cea*).   Mosses and lichens flourish in the greatest
profusion in the woods, beautifully embossing over the
fallen trees in a state of decay, and in many spots
forming fanciful and richly embossed arches.   One or
two very pretty species of ferns are met with, and a small
creeping plant bearing a red berry.

Birds are more numerous in species than in individuals ;
there are two kinds of hawks, a polyborus, the turkey
buzzard or carrion vulture, an owl, a thrush, three species
of fringilla or finches, three warblers (*sylvia*), two
creepers (*certhia*), a woodcock, and a remarkable quail-
like bird, in appearance and habits allied to *Tinamides*, two kinds of oyster-catchers, a dotterel, a tringa,
a grey duck, a seal, a steamer-duck, a cormorant, a
bittern, a black-backed gull, a gigantic petrel, a swallow,
a burrowing petrel, a penguin, a few upland geese (*Anser
leucoptera*), and the Antarctic goose, which frequents all
the shores.   In all perhaps about thirty-four species.

Of mammalia an otter and a seal or two frequent the
bays, and with a mouse complete the number met with.
The shell-fish are limited chiefly to limpets and mussels.
Of insects a pretty green beetle occurred amongst the
moss.   The birds, from their extreme shyness, are
evidently much harassed by the natives, and their numbers kept down by them.   We met with about three
families, numbering in all about eighteen natives on the
island.

CHAPTER XXXII.

Second visit to Falkland Islands—Berkeley Sound—Search for specimens—Birds and eggs—Summary.

Sunday, November 13th.—Beating up against a strong breeze for Port Louis in Berkeley Sound, where we anchored at 5.30 p.m. in our old place. On the following day the barque *Governor Halkett*, whaler, arrived, on her way from New South Wales to England, having amongst her passengers an old brother officer of mine, Thomas Gibson, on his return from Hobart Town, where he had taken out the *Somersetshire*, convict ship, a medical superintendent. He came on board to see me in our own boarding-boat, and accompanied me in my excursion to St. Salvador Bay, where I shot three brace and a half of snipe, and a dotterel, and he afterwards dined on board with me. We had not met for many years past, and consequently each had much to relate since we were fellow-passengers in the *Sandwich* packet as supernumeraries for the West India Station, having both entered the service at the same time in the year 1823.

Tuesday, 15th.—Gibson beakfasted with me, and afterwards accompanied me on a shooting excursion to the head of the sound, Green Bank, and the Creek, and, returning on board, dined with me at the Creek Point. A steamer duck rose from a tuft of long brown grass, thus discovering her nest, from which I took seven eggs, and on returning found another nest with four eggs, which had been deserted. I shot an upland goose and gander at the lake, and another goose, a brace of snipe, a smaller goose, and a rabbit, at the Green Bank. An immense squadron of steamer ducks were swimming under the bank. We got on board at 5.30 p.m.

Thursday, 17th.—This evening I tried the effects of hydrocyanic acid on three penguins, to ascertain the speediest and most humane method of ending their existence. One dram of the diluted acid destroyed a bird in one minute and fifty seconds. I sent Captain Ross five brace of snipe.

Monday, 21st—Made a shooting excursion to Uranie Bay. Shot one black-headed gull and four tern, and found some steamer's eggs.

Tuesday, 22nd.—Fine day. Started at ten a.m. for St. Salvador Bay. Found a steamer's nest with eggs, and just above high-water mark I found the first eggs of the black oyster-catcher, two in number, having a whitish ground, and speckled over with blackish-brown dots, lying on the bare shingle. The parent birds evinced great anxiety for them, crouching along the ground with outspread wings, uttering lamentable cries, and would not quit the spot. Following up the south arm, about half way along the beach, I picked up two more eggs, and soon afterwards came upon the nest of the black-and-white oyster-catcher, having also two eggs, but with larger blotches on an olive-coloured ground. They were deposited on some dry seaweed at high-water mark. I also found two tern's eggs laid in the same way in the shingle, a steamer's, with five eggs, and kelper's nests, one having five and the other four eggs. I found a dotterel's nest, with two eggs, prettily blotched with blackish-brown on an olive-green ground. They also were laid in a slight depression of the ground by the side of a gum-plant. The parent bird discovered them by rising from her nest. I shot three brace of snipe, a rabbit, a kelp goose, two grey ducks—the latter at one shot—and one tern.

Gibson and the surgeon of the *Terror*, who had accompanied me in the excursion, dined on board with me, on our return at six p.m. H.M.S. *Philomel* arrived here this morning at nine.

Thursday, December 1*st.*—At 9.30 a.m. the barque sailed with a fair wind, the master having come on board of us for his despatches. I was employed all day in skinning birds, and preserving eggs.

Wednesday, 7*th.*—Cold, raw, gloomy day, with drizzling rain. Landed at nine a.m. Crossed from the head of the sound to the south arm over the hills. Found two black oyster-catcher's nests, with one egg in each, sheltered by a small stone, a black-backed gull s with two eggs in a nest of seaweed under the cliff, just above high-water mark. I also found a kelper's with five eggs, but as they were in an advanced state of incubation, which I found out by placing one in water, I of course left them in their nest. Shot a rabbit and eight brace of snipe. I observed several snipe going through evolutions in the air, at a great height, soaring upwards for a considerable time, and then darting downwards, with a whirring sound, followed in the subsequent ascent by a shrill whistle or cackle, just as I have in my youth seen them do in the Norfolk marshes. I shot one out of a pair whilst thus going through its evolution. Returned on board between seven and eight p.m.

Tuesday, 13*th.*—Landed at ten a.m. on another egg-hunting excursion. Found two dotterels in a depression at the foot of a tuft of diddle-dee. When near the creek at Salvador I shot an upland goose out of a flock of six or seven. In a bay forming a fine, curved, sandy beach, several pairs of oyster-catchers of both kinds, with some tern and gulls, were hovering about. I found two tern's

eggs in one nest, and one in another, also a black oyster-
catcher's, having two eggs, and a black and white one's,
with two eggs also. Here I witnessed a most remarkable instance of affection and devotion on the part of a
pair of plovers Having picked up their young one, only
covered with down, although it could run very fast, and
putting it into my trout-basket I generally carry with me
for putting plants and eggs in, the parent birds manifested
an anxiety and pertinacity I never before met with, one
of them fluttering in front of me, flapping its pinions, and
convulsively throwing itself on its side, as if badly
wounded, utterly reckless of its own safety, and making
use of every device it could think of to attract my attention. I could no longer resist such powerful pleading,
and, to its delight, I permitted the young one to escape
from my basket. I shot a pair of grey ducks at one
shot, and got on board at 7.20 p.m.

The Malouines, as they have been called, are said to
have been first discovered by Amerigo Vespucci in the
year 1502, and in the year 1700, Beauchéne anchored
off them, and Bougainville founded a settlement at Port
Louis in 1764.

The general aspect of the country is dreary, naked,
and unprepossessing in the extreme. A monotonous,
undulating moorland, consisting of peat-bogs, swamps,
and rivulets, or tracts covered with yellowish-brown grass,
relieved only by the central ranges of hills of grey quartz.
Its geology is very simple, clay-slate and greywacke,
passing into sandstone, and the latter again into quartz.
The higher ridges being of quartz, and from which the
so-called streams of stones descend, having the sound
of running water between them, or rather beneath them,
the clay-slate and sandstones containing abundant
organic remains of spirifera, orthes, orthoceratites, and
stems of encrinites, embedded in them, forming the
lower tracts, and extending from the base of the hills,

which attain an elevation of from 600 to 1000 feet to the sea-shore, dipping at various angles, having a mean of 45°, and of very irregular superposition.

The vegetation is for the most part herbaceous. The valleys and ravines are filled with the fascine, a bush of the composite order, bearing a white blossom, attaining a height of only three or four feet, and much resembling a rosemary bush. The other most universally distributed plant is the diddle-dee, a heathlike-looking plant, growing in large, spreading tufts, and bearing a red berry, about the size of the cranberry, on which the upland geese are fond of feeding, indeed almost exclusively in the months of April and May, rendering these birds delicious eating. The gum plant and the tussac grass attain a height of some six feet.

The balsam bog (*Bolax glebaria*), belonging to the order umbellifera, forms hemispherical hillocks of every size, from that of a mushroom to a mass four to six feet in diameter, of a bright green colour, and giving a very peculiar character to the landscape, as it is thickly scattered over the hills.

The birds are very similar to those of Hermite Island, but more numerous and not so shy. The snipe migrates from June to August, lays its eggs in September, and the young birds are strong on the wing in November and December.

CHAPTER XXXIII.

Third attempt to reach the South Pole—Our farmyard of live stock—
Our Christmas dinner-party—Icebergs—Louis Philippe Island—
Taking possession of the Pyramidal Island—Thirty-eight days in
the pack.

Saturday, December 17*th*, 1842.—At 4.30 a.m. we hove
short, and at 6.30 a.m. weighed from East Falkland,
firing three guns, in acknowledgment of the same number
we were saluted with from the shore. At ten a.m. we
had cleared the sound under a press of canvas, all studding-sails set, and going before a fine westerly breeze.
The weather began very auspiciously for this our last
voyage southwards, being the finest day, with a bright
sun, that we have experienced of late.

Our decks formed quite a farmyard. In the boat
amidships five sheep were stowed away, the same
number of wild pigs, with a litter of young ones. In
the port waist were three calves, in the quarter-boat two
turkeys and a goose ; and a sea-stock of tussac grass as
food for them ; each quarter was festooned with dead
rabbits, geese, seal, and snipe, with a quarter of beef
and veal, and dried fish in every direction. Abaft the
quarter-deck a wild colt was tethered. With this colt
is associated an episode I much regret having to refer
to here. Yesterday I was asked how I relished a beef-
steak at breakfast, and my reply having been that I
thought it juicy and good, barring a somewhat saline
impression it left on the palate, but that I had eaten

much worse, when, to my surprise and astonishment, I was told that I had been partaking of horse-flesh, and that a young colt had been thoughtlessly, to use a mild expression, shot, on the previous day, by a party from the midshipmen's berth. It appears that two of those beautiful wild creatures had been shot from the troop I had been so often in the habit of meeting with in my almost daily rambles, and a colt brought on board alive.

Now there was nothing whatever that could in any way justify the taking the lives of these harmless, in-offensive creatures, and it is very sad to reflect that the happy life of freedom led by these noble animals in the wilds of nature should have been closed by so wanton an act of cruelty ; for we had an abundance—as I have only just stated—of both fresh beef and a variety of game on board. I had, with my own gun, alone, contributed no less than four dozen upland geese, forty brace of snipe, two dozen rabbits, besides two dozen and a half of the Antarctic geese, and other edible birds, teal, plover, and grey ducks, without limit to our mess.

Saturday, 24th.—In lat. 61° 23′, long. 52° 19′, we fell in with our first iceberg about nine a.m. I saw one on the weather beam, three or four miles distant ; and at noon another astern ; also passing some heavy pieces of ice. We have had westerly gales and boisterous weather. The birds accompanying us were the wandering albatross, cape pigeon, blue petrel, ash-backed petrel, and a few stormy petrel following, as is their habit, in the wake of the ship. Last night we passed within a few miles of the South Shetlands, but owing to the thick state of the weather did not see them.

Sunday, 25th.—The morning was threatening, but cleared up a fine Christmas Day, the latitude at noon being 62° 14′, long. 52° 5′. We had another heavy westerly gale last night, when the ship rolled so heavily, that I found a little pet of mine, a young oyster-catcher I had brought from the Falklands with me, unable to stand on his legs, and panting and gasping for breath. Up to this time he had appeared lively, and ate readily; but now he only took a very small bit or two of food. He lingered through the day, his eyes gradually becoming dimmer, and on my turning in at night, I found him out of his basket, dead on the deck.

At three p.m. Captain Ross and the gentlemen from the berth dined with us in the gunroom, the president's chair falling to my lot on the occasion. Our Christmas dinner was really a sumptuous one for these regions, consisting of veal, calf's head, teal and snipe, &c., with a liberal allowance of champagne; and after a late supper I turned in at four a.m. Several bergs were seen during the day, and at eight p.m. the extremes of the pack bore from S.W. round to E. Saw the first white petrel to-day, and on the following day we got up the crow's-nest.

Wednesday, 28th.—I counted no less than twenty bergs around the horizon, half a dozen of them very large ones. Several had the appearance of having been capsized. Two large finner whales were blowing astern of us. Latitude at noon 62° 44′, long. 53° 43′. About 6.30 p.m. I saw Louis Philippe Land, bearing S.W. by W. to S. It appeared from the deck and ahead, a bank of misty clouds suspended over it, which rendered its outline, clad as it was in one dense wreath of snow, very indistinct. About a league from its eastern extremity a snow-clad islet appeared, resembling a berg in the distant horizon. We were now really encompassed by bergs, some of them of huge magnitude, and in every direction around the horizon.

As we ran along the land about eight p.m., it had the appearance of one vast continuous bank of snow, perfectly smooth in outline as a snow-wreath, everywhere, save where it showed the action of the waves, at the margin of the sea, where bergs had been separated from

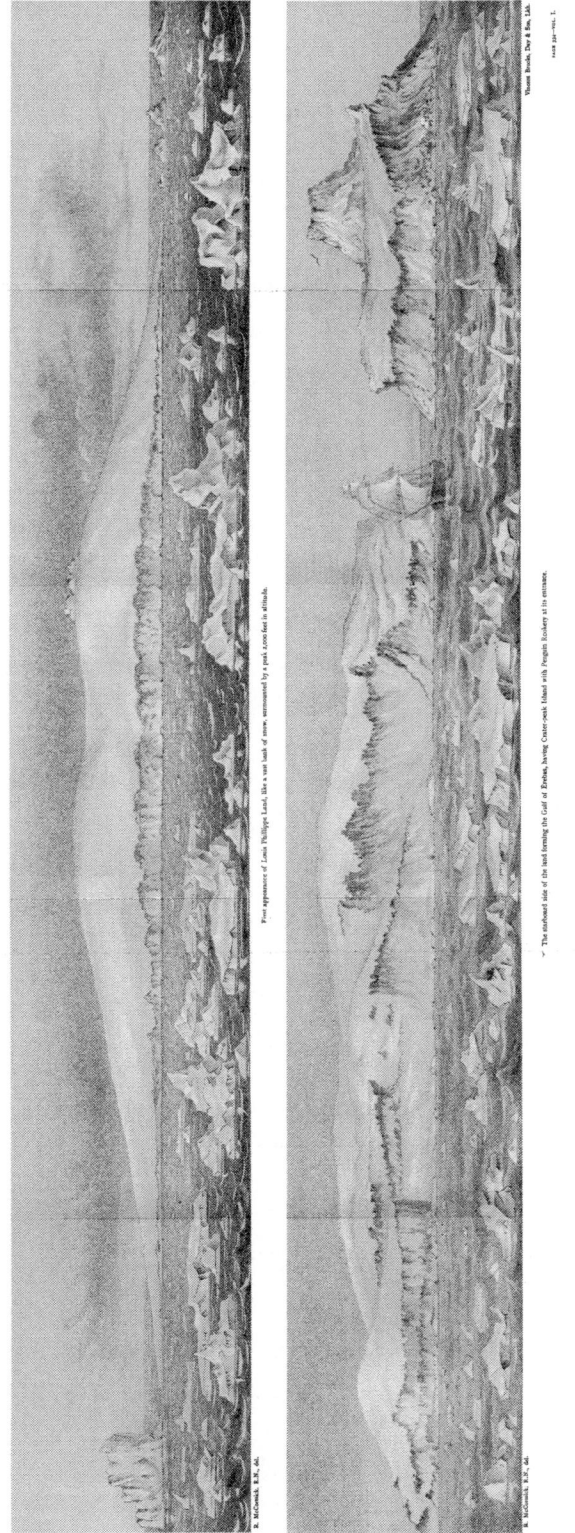

R. McCormick, R.N., del.

Front appearance of Louis Phillippe Land, like a vast bank of snow, surmounted by a peak 2,000 feet in altitude.

R. McCormick, R.N., del.

The starboard side of the land forming the Gulf of Erebus, having Cutter-peak Island with Penguin Rookery at its entrance.

Vincent Brooks, Day & Son, Lith.

PAGE 331—VOL. I.

The material originally positioned here is too large for reproduction in this reissue. A PDF can be downloaded from the web address given on page iv of this book, by clicking on 'Resources Available'.

it. From the centre it gradually sloped down to a point, running out very low and long to either extremity of the island. In one bearing only could a particle of the land itself be seen, and that was at the highest elevation of the ridge, appearing like two very small oval hummocks close together.

As we neared the southern extremity, five black-looking, small, low islands formed a chain at various distances from the low point, and in the midst of a labyrinth of bergs; some of these so darkly shaded as to be with difficulty distinguished from the islands themselves at a distance. We passed the last of them, distant some six or seven miles, about midnight. When I turned in, the highest part of the mainland I estimated at 2000 feet, and we passed within about three leagues of it. Many whales where spouting, and I noticed a seal on the top of a berg, also a chionis and a lestris. Penguins were cawing and quarking in all directions, sometimes jumping out of the water like skip-jacks, and moving along in a line or single file like that of fish. On one berg I noticed upwards of 100 collected on the summit.

Thursday, 29th.—Fogs and southerly winds, and a current setting N.N.W. half a mile an hour. Sounded in 162 fathoms, sand and small black stones. At six p.m. tacked close off the pack-edge, our progress in this quarter obstructed by it. About eight p.m. I saw a very long, large berg, its shadowy outline but just visible through the fog ahead; passed it to port. Our wild colt died to-day, apparently from cold and confinement in a cramped space; poor thing, so different from its life of freedom among its fellows! A hard fate.

Friday, 30th.—Pack close and heavy, extending from east to west; tacking about off its edge observed several large bergs within it. At three p.m. we passed the six islands we saw on the 28th, the largest, bearing N.W., having a remarkable broad belt of snow down it. At

5.30 p.m. I saw the extremes of the land from North Cape to south extreme, and a pyramidal-shaped island bore W. by S. Working to windward through lanes of water towards the land, blowing fresh; the mainland in W.S.W., high, covered with snow, having only a few black promontories of rock peeping out near the coast-line, which was indented with two or three deep inlets filled with ice and snow; lat. 63° 36', long. 54° 33'. My pet dotterel died to-day.

Saturday, 31st.-- Fine clear day. I took a sketch of the land: one portion had all the appearance of black lava streams in terraces singularly waved in outline to the southward, and having a small crater-formed hill. Here the black rock appeared conspicuously through the mantle of snow: a yellow blink indicated the trending of the land a long way, apparently forming a deep bight. Sounded in 207 fathoms, green mud. Fine clear night: I sat up and saw the old year out and the new one in. Many whales about us.

Sunday, January 1st, 1843.—We were not destined to cross the Antarctic Circle this New Year's Day, as on our preceding trips to the southward; but found ourselves, by observations at noon, only in the lat. of 64° 14', long. 55° 54', with Louis Philippe Land on one side of us and closely packed ice on the other. Many of the hills to-day had the appearance of smoke issuing from them, probably caused by the ascent of light vapoury clouds or small particles of fine drift snow. The weather was delightful; a bright sun shone forth in a clear blue sky, with scarcely a breath of wind, rendering the opening of water between the ice smooth as a lake. Large black whales numerous.

Three grey seals and three small penguins were caught. At 1.30 p.m. I went on board the *Terror* with Captain Ross, returning again in about an hour. We had all of us our New Year's Day's dinner in the cabin with Captain Ross at four p.m., and broke up at mid-night. Ice close and heavy; saw an immensely long barrier berg, quite flat topped as a table, and exceeding five miles in length. The customary warm clothing was served out to the officers and ship's company.

Monday, 2nd.—We got beset within the edge of the pack, and I walked over the surrounding ice for about a mile, as far as the western edge of the floe; the surface soft and hummocky, sinking up to the knees in places. I saw a white, or albino, gigantic petrel, and caught a small penguin. At 3.30 p.m. a boat was lowered after we had cast off from the ice, and three of the larger kinds brought on board.

Thursday, 5th.—Quite a summer day, so mild and calm, with a bright sun. Standing in for the land, through lanes of open water, saw a great number of the large whales spouting. What a fine field this would afford for the whale ships. A flock of fifty or sixty white petrel soared to a great height to-day. We were off the entrance to what appeared to be a large strait, running deep up, and having high bold land on the starboard side, and a bay filled with ice near its entrance. On the port side the land covered with snow sloped down to the upper extremity, where it seemed to cross rather than unite with the opposite side; the breadth of the entrance might be from four to five leagues. This strait separates Joinville from Louis Philippe Land, off which lies a bold, black-looking island (Pyramidal Island, 2760 feet high), rising steeply from the sea, and terminating in a fine peak, having a well-marked crater-formed summit almost bare of snow. At its base a large colony of penguins had established their rookery. I took a sketch of the land, and at 6.15 p.m, I went up to the crow's-nest for the first time this season, to have a look at the top of the strait. It appeared to be pretty free from ice, and what few streams there were, fast setting out of it. It was a lovely evening; the sea smooth as a lake, scattered over with loose ice; but all round the horizon the pack appeared close and heavy, the open water being only inshore, a long chain of bergs extending along the coast to the northward.

Friday, 6th.—Fine day, at ten a.m., when about two miles from the Pyramidal Island at the entrance to the strait, both captains landed at the penguin rookery, and shortly afterwards a boat from each ship with a party of officers; and they returned about noon, after hoisting the union jack and taking possession of the island.

I never more keenly felt the effects of the narrow-minded policy existing in this expedition; by which I mean the carrying out to the very letter the ordinary routine orders of the general service; that is, that each ship should not be left without a medical officer on board; although, as in the present instance, if any accident were to occur during the brief temporary absence of the medical officer, a signal could be made from the ships, and his return ensured almost as soon as the recall was made.

I personally endeavoured to urge upon Captain Ross, making use of all the influence I might possess with him after so long a service together as ours had been, that this was an occasion when such an order might well be set aside for the benefit and good of the special service we were employed upon. I pointed out to him that such a fine opportunity might never occur again for examining the remarkable crateriform peak now in sight of us, with its surroundings, more especially the incubation of the rare birds now breeding there, and their eggs, so great a desideratum to the ornithologist; such as the Antarctic or great penguin, the chionis, the white petrel, and the new species of lestris of the south. I said, as it was Dr. Hooker's turn now to land, his time would be more than taken up with the botanical researches; and unless I had an opportunity of landing, all else would be necessarily lost to us. But unhappily my appeal failed in its object, and during the absence of the party I was left to glean all I could through the medium of the telescope from the deck of the ship.

At eight p.m. the captain of the *Terror* came on board, and we all spent Twelfthnight in the cabin with Captain Ross; regaled with wine, grog, and the customary twelfth-cake. Working ship between the chain of bergs and land two miles to windward.

Sunday, 8th.—Foggy, thick weather; had to keep company with the *Terror* by means of fog-signals, firing muskets, with bells and gongs going. Between seven and eight p.m. we passed through a narrow channel between an immense berg and the floe-ice rapidly closing, towed by two of the boats ahead. I saw one of the large penguins on the floe walking away upright as a dart, in the most grave and grotesque manner, looking like an old monk going to mass, with a group of the smaller kind on the same piece of ice with him. We made fast to various large pieces of ice during the day, and one piece, having a quantity of dark mud on it, drifted close alongside of us; and whilst we were watering the ship from it, in searching the layer of mud I discovered a fine specimen of silicious rock, of a bottle-green colour, very hard, ponderous, and close-grained—a large fragment.

Friday, 13th.—Fine day. Beating up for the land-ice, tacking about in open pools of water, surrounded on all sides by closely packed ice, or large bergs, the extreme of the land bearing W. by S., a bleak point running out horizontally from a black, bluff cliff, showing its rugged sides through the mantle of snow and ice. This forenoon our surroundings were so striking and peculiar, that I took a very fanciful sketch of the horizon all round, making the ship the central object of a circle.

Z 2

At 2.30 p.m. an unusually strong current drifted us to the eastward, upon the edge of the pack, where we made fast to a large floe, the most level and smooth piece we have as yet met with; this drifted with us very near the large berg, having two caves in its side. The loose ice drifted down so rapidly with the force of the tide, that we were very soon beset, and had to warp along the edge of the floe till 5.30 p.m. before we got into open water. Saw a seal on the ice and another in the water. We left the *Terror* beset, about half a cable's length from us; made sail and ran out into an open pool between two bergs to the W.N.W. Lost sight of the *Terror*, concealed by the berg, from which she got clear about midnight.

Saturday, 14th.—The captain of the *Terror* came on board. Finding the ice so close and heavy to the southward, with the long chain of bergs, and no prospect of penetrating any further in that quarter, we bore up, and at 1.45 p.m. entered the pack with a fair wind and tide. We made but slow progress, however, warping and boring through the heavier portions, and running through the small pools of water. Saw several of the young of the large penguins swimming near the bows to-day, with a grey seal or two on the ice. At nine p.m. both ships became beset close to each other.

Monday, 16th.—Cloudy, and nearly a calm, closely beset with heavy ice all round. Four seals were killed near the ships. At 3.30 p.m. sounded in sixty fathoms, green mud; two more seals killed. Saw several gigantic petrel on the ice. In the afternoon I went upon the floe with my gun, and shot a seal about a cable's length ahead of the ship: the ball went through his body under the left fore-flipper, and although he bled profusely he managed to wriggle himself into the water not twenty paces off, and not having a second ball in the other barrel I tried to stop his way by firing a charge of small shot under his flipper, which was followed by a torrent of the crimson stream, so that he must have expired on his plunge into the water. The seal, unless shot through the brain or heart, is so tenacious of life, that if the water happens to be near it will generally manage to reach it before it breathes its last. I walked to the opposite side of the floe and along the margin on the land side for about a mile, and on returning struck across it for the ships ; lost my powder-flask, and got on board at six p.m.

After tea I took the same ramble again, and found my powder-flask. Observed one or two cracks in the floe, and along its margin, where fragments of ice were piled one upon another like blocks of stone in a wall, the effect of pressure from former tides. The tide was now setting with great rapidity, the brash, small bits of ice, being carried along like a sluice between the floe and a berg. Soon after my return on board, the floe to which the ships were made fast suddenly parted into three pieces, and the canals formed by the breaking up rapidly widened. We drifted fast in shore, and to the northward. Night thick, with a fall of snow.

Tuesday, 17th.—Still drifting about amongst the floe-pieces off the chain of bergs and the pack. I went upon the floe and caught two small penguins, knocking one down with a stick. At 3.30 p.m. we cast off, and made sail to the north-east. I saw about 100 penguins congregated together in-shore in a knot, and whilst I was watching them they marched off in single file, and disappeared amongst the hummocks. We drifted within two or three miles of the shore, passing an inlet, and, as we approached the point, opened Pyramidal Island. The very long berg also came in sight, and the only change in our position is that we are now outside instead of being inside the chain of bergs and the shore, as we were last week. The whales are very numerous in this locality.

Sunday, 22nd.—Hove-to most of the day in a pool of water. The whales very numerous again to-day; and I witnessed this evening a singular habit of the birds I had never before noticed. A flock of about 100 Cape petrel following the whales, hovered over their wake in the water, and the whale had no sooner risen to the surface to blow than these birds pounced down in a body into the very jet of vapoury spray sent up by him in a broad column, when a momentary scramble commenced for something they obtained from the water, either small creatures discharged in the jet, or parasites attached to his skin ; or probably both, as in one instance I saw two birds alight on the whale's back, but from which they were almost as soon unshipped by his diving below the surface ; although only to come up again a few fathoms farther off for another blow, which the petrels anticipating, followed close in his wake, all ready for another scramble as he rose again above the surface of the deep.

Wednesday, 25th.—Saw Pyramidal Island. A solitary lestris flew over the mast-head, glancing down on the deck with that keen, inquisitive look so peculiar to this most rapacious bird, and then instantly shaping his course for the land direct, was seen no more. I also saw a group of the small piebald whales, having a long, black, scimitar-shaped fin appearing high above the water.

Friday, 27th.—Being a calm all day, the boat was employed towing the ship ahead, to keep her clear of the lee-ice, and I took advantage of this to obtain some specimens which I had hitherto not been able to do since we left England ; and between two and three p.m. I succeeded in shooting no less than six stormy petrel (*Procellaria pelagica*) from the stern of the ship, and also a Cape petrel, all of which were picked up for me by the towing-boat during the first watch. I saw seven penguins soundly sleeping on the ice whilst resting on their breasts, and stretched out at full length, the head and beak both in a horizontal position.

On skinning the six stormy petrel on the following day,

I found all their breasts bare of feathers, a sign that they had been recently incubating. Captain Ross himself, following my example, was employed to-day in shooting three or four from the stern.

Monday, 30th.—Fine day. I gave Captain Ross one of my own specimens of the stormy petrel. A boat was sent on board the *Terror*, with orders that in the event of our separation she should, after the 7th of February, make the rendezvous on Weddell's track.

Saturday, February 4th.—Weather overcast, with snow. About noon a heavy squall from the north-east indicated that we were in the vicinity of open water. Made sail, and bored the ship through the streams of ice, getting clear of the pack in the afternoon after some heavy thumping from the ice. Between five and six p.m. we were once more on the open sea, with a heavy swell setting from the north-east, after a detention of thirty-eight days, beset between the pack, the chain of bergs and the land, encountering baffling winds sometimes in our teeth.

Friday, 10th.—When I went on deck after breakfast this morning I distinctly saw the pack margining the horizon to leeward, and apparently very close and heavy. In the afternoon a strong ice-blink was observable from S. by E. $\frac{1}{2}$ E. to E. $\frac{1}{4}$ N. Weather gloomy. On the following day we passed through some loose ice off the pack, and close to a large berg. Overcast and gloomy weather, with snow.

Sunday, 12th.—Gloomy weather, with snow; passing through much loose ice. We have had several blue petrel about the ship for the last few days.

CHAPTER XXXIV.

Our highest latitude 71° 30′, and longitude 14° 51′ 10″—We complete the circumnavigation of the globe and cross the circle—Deep-sea soundings—By avoiding Weddell's track we fail to reach his latitude—Birds—The comet of 1843.

Tuesday, 14th.—This day we attained the highest latitude during this season's voyage, being at noon in 65° 6′, and crossed Weddell's homeward track (from 74° 15′) in the long. of 41°, having a heavy pack to the southward, with a strong ice blink all round from E. to W.S.W.; sailing among loose ice off the pack, contending with a strong wind; so that our prospects at this advanced period of the season are anything but promising. Weather overcast.

Wednesday, 15th.—Gloomy and overcast weather. Still passing through loose ice streaming off from the pack, with strong breezes and a long swell from the north-east. The ice here presents a very altered appearance, evidently caused by the continued action of the waves on the slowly melting ice, the pieces of ice assuming every possible fantastic form, some in the shape of pedestals rising from a broad expanded blue tongue beneath the surface, and supporting a cupola, or table-topped summit, resembling vases or gigantic mushrooms. In others, flat slabs piled on each other at various angles. I saw two tern fly past the ship to-day; and two large penguins were seen by the watch.

Saturday, 18th.—We this morning completed the circumnavigation of the globe in lat. 62° 39′ and long.

31° 44′, dip 59° 11′, and variation 7° 22′. Sailing through much loose ice off the pack-edge, a berg or two in sight, going three knots with a south-easterly wind, the course being E.N.E.; thermometer 32°.

Monday, 20th.—This day, now twenty years ago, Weddell reached his highest latitude, about seven degrees to the westward of the longitude we are in to-day. Yet we have met with a continuous line of pack since our departure from Louis Philippe Land.

Wednesday, 22nd.—We crossed the line of no variation at noon, being 0° 12′ W., and lat. 61° 37′, long. 21° 51′, dip 57° 40′.

Wednesday, March 1st.—We have at last crossed the Antarctic Circle, and for a third time on the first day of the month, although, unfortunately for us, not the same month as on our two preceding trips to the southward; but two months later, and the season closing in just at the moment we have the prospect before us of a fine open sea, and perhaps, for all we know to the contrary, to the pole itself. Our lat. at noon was 67° 6′, long. 8° 35′, dip 62° 42′, and var. 8° 12′. Weather dry and bracing. Sketched two large bergs.

Thursday, 2nd.—This has been by far the finest day that we have experienced since our departure from the Falklands : a clear blue, cloudless sky, with a bright sun, light winds, and a long swell. It is a curious fact that we have always met with the finest weather within the Antarctic Circle. At 7.25 p.m. we had most brilliant sunset, the parting rays from the luminous sphere, as it appeared for a moment to roll along the waters ere it dipped beneath them, in a clear horizon, were dazzling, and reflected a singular bronzed kind of neutral tint on the light hazy clouds in the opposite horizon.

Saw a tern and a sooty albatross or two, and a number of large whales sporting about, spouting up jets of vapour and spray to the height of from twelve to twenty feet.

Friday, 3rd.—Both cutters were lowered at one p.m., and the two captains left their ships to superintend the deep soundings. After running out 4000 fathoms of line off the reel, which occupied an hour and fifty minutes, no bottom was obtained, and 250 fathoms of one inch and 3750 fathoms of three-quarter inch, with a pig of ballast were expended. The current ran 0° 3′ per hour. Whilst the boats were away sounding I shot a blue petrel from the ship's deck, but it fell into the sea, drifted astern and was lost, which was vexatious, for this I believe is the first bird of the kind shot throughout the whole period of the expedition. They are never met with in the pack-ice. I saw several whales, but birds are very scarce.

Saturday, 4th.—Gloomy day, but wind fair for the south; yet we are shaping a south-westerly course to avoid the track of our enterprising predecessor Weddell. This prejudice on the part of the commander of the expedition is to be deplored, but from the first there has been a disposition not to follow in the track of others, which together with having already frittered away the best part of the season in the vain effort to force a passage between the perilous chain of stranded bergs and the broken land of Louis Philippe, mainly for the sake of a display of some new land of a trifling character on the chart, cost us the season, and ultimately proved fatal to our attaining even so high a latitude as that of Weddell himself: apart from the daily risk to the ships, knocking about for weeks together in narrow channels and pools of water, beset with strong currents, pent up between a chain of grounded bergs and a most dangerous coast. Indeed, I believe there were few on board either of the ships, if they candidly expressed an opinion, ever entertained the shadow of a hope that we could ever make our way further south through the inextricable difficulties of a course so ill-advisedly adopted and so pertinaciously followed up in this our last attempt to reach the South Pole. I saw a solitary white petrel which had frequently flown past the ship during the day, a sure harbinger of ice not being far off.

Sunday, 5th.—Upon going on deck after breakfast I saw the pack very distinctly margining the horizon to leeward on the starboard beam, apparently very heavy ice, having a very large berg lying off it. Weather overcast and hazy, but through which a glance of the sun at noon enabled us to secure an observation, which gave the lat. 71° 10′ S., being about Cook's farthest ; our long. was 15° 47′, dip 65° 21′, var. 2° 23′. We continued on with a fresh breeze from the north-east till four p.m., when we were again arrested by the pack-edge, passing through some heavy streams of ice off it, and by several large bergs.

The season being so far advanced, the ensign was hoisted in each ship, and we bore up on our final departure frrom these regions of ice and snow, in which we have now passed three seasons. We had snow-showers during the day, the maximum temperature 32°, min. 29°; that of the sea 32° and 30°. Saw several white petrel and a large flock of blue ones swimming in the water. I also saw a Cape petrel or two, and a pair of brown and white ones. Our furthest south, when we bore up at four p.m., was 71° 30′, and long. 14° 51′ W. The pack extended from S.W. by W. round by S. to N.E. by E. A cask was thrown overboard containing a paper with the latitude and longitude, signed by the captain and officers. I went on deck just as the ship tacked, to take a last look at the pack, which now appeared astern and on the starboard quarter.

Monday, 6th.—Blowing a gale of wird in our teeth, with the pack in dangerous proximity to leeward, and off which we wore ship at six p.m., distant a quarter of a mile. It extended from N.W. by W. to S.W. At 4.10 p.m. again wore ship.

Tuesday, 7th.—Last night, blowing hard, with a heavy sea running, which with the thick weather rendered our situation by no means a desirable one. Passed close to windward of a large berg, in the middle watch, it being difficult to make them out at any great distance. The headwind fell suddenly to a calm at midnight, and we wore ship.

Friday, 10th.—The crow's-nest was got down to-day. Passed several large bergs ; strong breezes, going six and seven knots. About two p.m. I shot a Cape petrel, hovering over the weather quarter, when it fell dead aft, on the port side of the quarter-deck ; this indeed is the only bird I have shot within the Antarctic Circle this season, the latitude at the time being 68° and long. 15° west. Several blue petrel about. I signed a paper this evening, to be thrown overboard in a cask, on our recrossing the circle. Between nine and ten p.m. I saw for the first time this voyage, the Aurora Australis, consisting of one ray only, of a pale yellow tint, rising from a bank of dark clouds in the west, to the altitude of about 20°, at an angle of 40°. Night starlight. Saw the Southern Cross.

Saturday, 11th.—At 6.30 a.m. crossed the Antarctic Circle for the last time, and the cask was thrown overboard with the signed paper, in long. 13° 16′ W. In the first watch about four bells, I had the first sight of the moon, or rather her rays, for she was concealed behind a dark cloud about 3° above the horizon.

Tuesday, 14th.—Cloudy, with a fine fair wind, going before it at the rate of five knots under a crowd of canvas, royals and studding-sails low and aloft. Bent a new main-sail and fore-sail. Passed an unusual number of bergs to-day, certainly not fewer than half a hundred, large and small. I sketched one fine imposing-looking specimen which was seen twenty miles off; it was, I think, one of the highest we have ever met with, considerably over 200 feet in altitude, having two caverns in it, into which the surf was heavily breaking.   As we passed within about a mile of it, on the port beam, I saw a large albatross (*Diomedia exulens*), for the first time on this voyage.   We have not met with many birds of late, only a few Cape and blue petrel, some sooty albatrosses, and a brown and white and ash-backed petrel or two.

Friday, 17*th.*—Signed a paper with the latitudes and longitudes thrown overboard in a cask at noon.

Monday, 20*th.*—At noon exactly, we crossed the meridian of Greenwich; lat. 54° 7′, long. 0′, dip 55° 8′, var. 17° 50′; consequently our time is at this moment precisely the same with that of England.   We are now steering a course to look out for Bouvette's Island.

Tuesday, 21*st.*—Blowing a gale all day.   Took a sketch of a large berg, when two miles on the starboard bow, as it towered high above the rolling surge.   Between eight and nine p.m. I saw what appeared the other evening to have been a single pale ray of the aurora in the west, shooting upwards at an angle of about 45°, and at an altitude of about 30° on a clear and starlight night. This afterwards proved to be the remarkable comet of 1843, with a tail extending from the horizon to near the zenith, and passed so near the sun as almost to graze its surface, and when at its perihelion, might be seen in broad daylight.   From spectrum observations, comets would appear to be both self-luminous and to reflect the sun's light.

Wednesday, 22*nd.*—Blowing a heavy gale all day, with overcast weather and snow.   Having looked for Bouvette's Island where it is laid down in the charts, without seeing anything of it, we shaped our course for the Cape of Good Hope.

Thursday, 23*rd.*—Still blowing a gale of wind, with a

heavy sea running. A large berg appeared very indistinctly through the mist, which I watched the approach of for upwards of an hour, till we passed it on the port side, or rather, weather bow; a little before noon, a tremendous surf was breaking over it to windward. I sketched three different views of it. Two fine large albatrosses, and the elegant little stormy petrel were flying about the ship, with some Cape, black, and also blue petrel, and a dusky albatross or two. Hove-to again at night.

Friday, 24th.—The gale abated last night; fell in with the long stems of the *Fucus gigantea* seaweed, called after the naturalist of Duperrey's expedition, *Lessoniæ,* or *Macrocysti pyrifera,* and saw several sheerwater. Passed a berg on the starboard quarter, during the first watch.

Saturday, 25th.—Fine, with strong breezes, ship going nine knots. Passed a small berg, about half a mile to windward, in the forenoon: the last seen. At eight p.m. rounded-to for the night.

Tuesday, 28th.—Very fine day. Ship hove-to, to get the temperature of the sea at 1200 fathoms, and at two p.m. a boat having been lowered and sent to the *Terror* I took the opportunity so afforded of going on board. Dined with the gunroom officers, and returned on board again at three p.m. I faintly saw the comet again to-day.

Saturday, April 1st.—The day was so fine and warm that I was induced to change my Antarctic rig for one more in keeping with the climate we are now in. On the following morning saw a barque, the first sail we have seen since we left the Falklands.

CHAPTER XXXV.

Summary of the last voyage south—Astronomical observations—The vast barrier of ice and the continent beyond.

In this brief retrospect of our proceedings during our third and final attempt to reach the Antarctic Pole, I can only repeat my own conviction that our main want of success too evidently rested with the course we had been so unfortunate as to adopt, rather than follow in the wake of Weddell, who had, with the limited means at his disposal, attained the high latitude of 74°, with a fine, open sea, free of ice as far as the eye could reach in the horizon view, and had not the interests of his owners in his small vessels, ostensibly employed in the seal-fishery, hampered him with responsibility, he might probably have made even a much nearer approach to the pole. His meridian was 40°, ours 55° W., the meridian of Cape Horn, on which we made the attempt, along a very intricate navigation of the shores of that group of land called in the charts South Shetlands, barren islands and rocks flanked by tiers of huge icebergs aground, between which and this forbidding, desolate, iron-bound coast, rapid currents ran like a race through the narrow, and often tortuous, channels, with which, together with tempestuous gales, amid snowstorms, sleet, and fogs, both ships had to buffet for so many weeks, in the vain hope of forcing a passage through to a higher latitude, and the temptation offered for thus adding perhaps some few new lands to the chart. But after all our efforts the short navigable season closed upon us, and we were no nearer the pole than on the day on which we first made this ill-omened land. A hopeless attempt was at last made to get upon the track of Weddell, which, from the lateness of the season and unfavourable winds, was now so encumbered with vast drifting packs of ice, amid dense fogs and gales, that it barred our course, and precluded all chance of even attaining the latitude of our predecessor, and with no small difficulty that of Captain Cook in 71°, when we had to bear up and bid a final adieu to Antarctic lands, packs, bergs, and seas.

Happily for us the attempt in our first voyage south was rewarded by the discovery of a mighty southern continent, dwarfing the quarters of the old world, and rivalling the new one in its stupendous magnitude and general aspect, capping the pole by lofty mountain ranges, sustaining altitudes varying from 5000 to upwards of 12,000 feet; two magnificent volcanoes crowning all, arrayed in an armour of everlasting ice; a glaciation as complete as ever occurred to the opposite hemisphere in ages past. The perpetual-snow line descends to the very beach. The constant presence of ice and snow keeps the thermometer at the freezing-point; consequently no kind of vegetation exists, not even a seaweed on its barren shores; and but for the animal life which animates the ocean, whales, seals, penguins, &c., and sea-birds winging their way through the air, or skimming the ice-embossed surface of the deep, all would be as desolate and silent as the tomb.

It is even conceivable in such a state of things that the annual unceasing accumulation of ice and snow at the South Pole might, in the course of time, so affect the earth's centre of gravity, as to produce a change in its axis, with all its consequences. But nature, ever fertile in her resources, in compensation has caused the vast barrier of ice, which everywhere girts the whole line of coast, to separate and float off in the form of giant bergs, often miles in length. One of these barrier bergs we fell in with, of which I made a sketch, measured no less than five miles from one extremity to the other: the bergs frequently attain 200 feet and upwards in height above the sea. The vast body of water set in motion in the Southern Ocean by heavy gales beating on the face of the barrier, fractures and drifts away those islands of ice in countless numbers, to be, after long drifting about, dissolved in the warmer climes of lower latitudes, and added to the wide waste of waters.

The sun being nearer the earth when over the southern hemisphere, the earth's velocity is increased, and the consequence is the sun is a week less time in this hemisphere than it is in the northern; thus making the South Polar winter longer by a week than the North Polar one. A constant depression of the barometer prevails in the Antarctic seas to the amount of nearly an inch.

The currents ran very strong amongst the rocks and islands forming Louis Philippe Land, as they did amongst the drift ice in the narrow channel separating Possession Island from the mainland of Victoria. Now the currents of the ocean are mainly under the influence of the winds; about three-quarters of the globe is covered by the ocean. The equatorial current has a generally westerly course round the globe, subject to deflections as it crosses the Atlantic Ocean towards the South American continent, where it divides, one portion going south, the other into the Gulf of Mexico, coming out as the gulf-stream, the temperature, breadth, and extent of which I have given in another portion of this work. This warm surface-current of the ocean sets from the equatorial regions towards the poles, conveying heat from the tropics to the temperate and frigid zones, laving the shores of both Spitzbergen and Greenland. Counter cold undercurrents set from the poles to the equator, keeping the bottom of the ocean little above the freezing-point.

Whilst on the subject of currents, I may here observe that we crossed the circle of uniform temperature of the ocean at six different points during our voyage, in a mean latitude of 56° 14′ S.; this parallel of latitude forming a belt of a mean temperature throughout its depth of 39°, whether the temperature at the surface be 78° or 30°, all round the world.

A remarkable fact associated with the ocean temperature, is that forms of animal life belonging to the Arctic seas have been dredged up from the Antarctic Ocean at depths of 2000 fathoms, and may have passed from pole to pole through the tropics, without having been subjected to a greater variation of temperature than some five degrees or so.

The greatest depth of soundings we obtained during the voyage was on the 3rd of June, 1843, in lat. 15° 3′ S., and long. 23° 14′ W., getting no bottom with 4600 fathoms of line (27,000 feet) run out.

The trade-winds are caused by the rotation of the earth on its axis, which with the great altitude of the sun keeping up a higher temperature at the belt of the tropics than either north or south of the equator, is the cause also of the permanent north-easterly and south-easterly winds.

The greatest height attained by waves in the Atlantic Ocean, according to Scoresby, would appear to be some forty-three feet; whilst in the North Sea the highest probably fall short of twenty feet. The surface of the ocean at the equator is said to be four feet higher than at the poles—some 12,000 miles apart.

The great comet of 1843 was a remarkable incident in this voyage. Its first appearance was as a long ray of light some 50° degrees in length, which we at first mistook for a ray of the Aurora Australis, the nucleus not being visible above the horizon at the time. It has the smallest perihelion distance of all known comets, having all but grazed the sun, moving with a velocity of 366 miles per second, at its nearest approach ; the centre of the comet being only 80,000 miles distant from the centre of the sun; and from surface to surface only some 32,000 miles separated the two bodies from each other. The tail has been said to extend to 150,000,000 of miles in length, and the breadth 3,300,000 miles. In fact it has been inferred that the tail extended beyond the distance at which the earth revolves round the sun, itself of such vast magnitude that it would admit the earth, moon's orbit, and all within its disc.

Yet these distances, vast as they appear, sink into insignificance when we reflect that Sir John Herschel, with his twenty-foot reflector, enumerates no less a number than 20,374,000 stars visible through his telescope, and of this entire number of stars thus seen, only about 5000 are visible to the naked eye throughout the heavens.

At such inconceivable distances in the depths of space are the stars separated from us, that only a few, very few, indeed, have had their distances measured (such as 61 Cygni and α Centauri—both double stars) with any reliable degree of accuracy. The former, a small star of the sixth magnitude, had been observed to be affected by a proper motion, or progressive regular displacement of 5″ per annum, among the surrounding stars ; from which circumstance astronomers were under the impression that it was nearer to our system, and on being submitted to observation it gave the parallax of 0·348 by Bessel. The distance of the two stars of 61 Cygni subtends an angle, rendering it probable that the orbit is nearly circular, with a period of 500 years ; so that these two stars revolve round each other in an orbit far exceeding the dimensions of Neptune's round our sun. This star, 61 Cygni, was the first measured star in the heavens affording a parallax. But the latter-named star, Alpha Centauri, is the nearest of all the stars in the heavens to our system. It is a very bright star of the first magnitude in the constellation of the Centaur, next to Sirius and Canopus in its brilliancy, having a proper motion of nearly 4″ per annum, and the parallax of 0·913. The light of Sirius is, however, four times that of α Centauri, but its parallax only 0″230. It moves through space at the rate of fourteen miles a second, α Centauri thirteen miles a second, and 61 Cygni no less than forty miles a second. Vega, in the constellation of the Lyre, has the same velocity as α Centauri, thirteen miles a second. The planets, to keep them in their orbits, move through space, Mercury at the rate of twenty-eight miles, and Neptune at only about four miles a second.

In celestial measurements, the diameter of the earth (8000 miles) has been used as a base-line in determining the distance of the moon to be 240,000 miles from the earth, or about sixty radii of our globe, taking its radius as the unit of measurement; the distance of the sun being obtained through the transits of Venus over its disc. But the distances of the stars are so great that another base-line has to be sought for, in the distance of the earth from the sun in fact, to determine how many times the sun's distance made up the distance of any particular star, the displacement of that star giving the parallax; the radius of the earth's orbit taken as the unit. The results obtained place the nearest stars to us, in round numbers, some 200,000 times the sun's distance from the earth, or 200,000 times the radius of 90,000,000 miles.

The spectroscope has of late unfolded to us much interesting information with reference to the physical constitution of the stars, and comets too, as well as their movements in space. In fact, it has opened out a wide field of research in the future. Spectrum analysis brings out those rainbow-tinted streaks, crossed by dark lines, where tints are missing from absorptive action in the vapourous atmosphere of the star of bright lines, sodium, calcium, and so forth. The dark absorptive lines of hydrogen vary in breadth and darkness, by some supposed to be characteristic of a high temperature in the star or sun in proportion to their depth and blackness.

Vega and Sirius are types of the bluish-white stars; Capella and Aldebaran of the yellowish-white, like our sun; and Arcturus of the orange-yellow, or third type. Inferences have been drawn from these types as to the relative ages of each : the first type with Sirius, the younger ; the third with Arcturus, the oldest.

Whilst on the subject of the heavens I cannot resist the temptation to allude to the wonderful discovery of the planet Neptune, by Leverrier and Adams, in the year 1846, through the perturbations of Uranus, the most distant planet of our system, having a mean distance from the sun of no less than 2,862,000,000 miles, being 165 years in accomplishing its revolution. It is invisible to the naked eye, appearing through the telescope as a star of the eighth magnitude, having a velocity of above 12,000 miles an hour ; of far greater magnitude than the earth, being the third in size of the planets ; and what is most remarkable, its satellite, and those of Uranus, it has been stated, have a retrograde movement, and not from west to east as other satellites and planets. The newly-discovered moons of Mars, Deimas and Phobos, move around their planet from west to east, as our own moon does round the earth. The Uranian year is eighty-four of ours. Olbers supposed that the numerous asteroids which have been discovered were fragments of a large planet.

In my remarks upon the magnificent comet of 1843 I omitted stating that its orbit would appear to be identified with that of several others observed in the

past.   If so its periods of return would be rendered un-
certain, and considered shorter than at first.   A period
of twenty-one years, and even much less than that, has
been assigned to it.   Only one other comet in the present century exceeded this in brilliancy and beauty,
that was the great comet of the year 1811, the head
of which was 112,000 miles in diameter, and that of
the bright nucleus some 400 miles.   The tail extended
over a space of 112,000,000 miles.   Its period of revolution occupied no less than 3000 years.   This comet was
never less than 100,000,000 miles from us or the sun.
I can well remember its brilliant aspect lighting up the
northern heavens, being at the time eleven years of age,
and that year more deeply impressed on my memory as
an eventful one, by the loss of my father on the Christmas Eve, in the sad shipwreck of H.M.S. *Defence*, off
the coast of Jutland.

Halley's comet, occupying seventy-six years in its
revolution, extends beyond the orbit of Neptune, which
planet is thirty times the distance of the earth from the
sun, and the earth's orbit has a diameter of 24,000
diameters of the earth itself ; and although the sun has a
diameter of 882,000 miles, yet were it removed to the
distance of the nearest fixed star, our sun would appear
only as a star of the second magnitude, shining like
the Pole Star, or the stars in the constellation of the
Great Bear.   It is situated in the angle formed where
the two branches of the Milky Way separate, moving
through space at the rate of four miles a second, whilst
that highly coloured star Arcturus rushes through space
at the wonderful speed of fifty-four miles in the same
space of time.   The Pole Star moves at the rate of
one and a half miles in a second only, the earth's velocity
being nineteen miles a second.   The companion of Sirius,
a minute star, is forty-seven times the distance of the
sun from the earth.   The small star, Alcor, in the tail of

the Great Bear, and so named by the Arabs, is increasing in brightness, and the circular contour of the constellation of Hercules is said to be enlarging from year to year. In the constellation of Taurus the six stars are double ; the finest, Halcyone, is only of the third magnitude. The seventh star of the Pleiades is said to have disappeared at the capture of Troy. The Milky Way is a conglomeration of stars, too faint to be separately discerned. In the language of the North-American Indians it is called the "road of the souls ;" and by the Chinese, the "celestial river." Aristotle's notion of the Milky Way was, that it resembled a large comet constantly reproducing itself.

Our earth having a surface of some 200,000,000 square miles, is indeed the merest point in a universe unfolding to our view 1,000,000,000 stars or suns, as seen through the largest telescopes, in every part of the celestial sphere. If suddenly forced from its orbit, it would in little more than two months fall into the body of the sun. Our sun is supposed to be in the first or glowing vaporous stage of planet life, Jupiter in the second stage, and the earth in the third or life stage, whilst the moon has reached the last or final stage of decrepitude and death. Those meteor systems, about 100 of which are encountered by our earth, have been associated with the tails of comets : the November display called the Leonides, and the August ones, the Persides.

The great comet of 1843 has been said to have been associated with a train of meteors, and that its near approach to the sun's corona has caused a retardation of its velocity to such an extent as to hasten its return long before the period it should become due ; and it may, consequently, be looked for again before the conclusion of the present century, when in passing round the sun its perihelion distance will be lessened.

CHAPTER XXXVI.

Simon's Bay—Excursion to Cape Point—Stellenbosch—St. Helena—
Ascend Ladder Hill—Fairyland.

I NOW resume the narrative of my travels.

Tuesday, April 4th.—Fine day ; upon going on deck after breakfast, I saw Table Mountain and Cape Point. Volumes of smoke were ascending above Simon's Town, from the burning of the bush on the hills beyond. We had to work up False Bay, with strong breeze and squally weather.

At 3.30 p.m., exchanged numbers with the flag-ship. H.M.S. *Winchester*, and about four p.m. a pilot-boat came alongside. A considerable surf was breaking on the Bellows Rock, about two miles from Cape Point. At 6.30 p.m. a blue-light was burnt by the flag-ship, and at 7.15 another, which we answered ; and soon afterwards a lieutenant from the *Winchester* boarded us. At 7.30 p.m. we came to an anchor in ten fathoms, outside of the flag-ship ; and both captains waited on the admiral, the Hon. Josceline Percy, this evening.

Wednesday, 5th.—Fine, warm, enjoyable, sunny day. Moored ship ; only a whaler in the bay. At three p.m. I landed at the eastern point, beyond the fort. After walking through the town, and to some stores, I returned on board at 5.15 p.m.

Saturday, 8th.—The admiral and his family paid a visit to both ships ; they remained on board of us about an hour.

Wednesday, 12th.—I started at 7.15 a.m. from the fort, where I landed, on an excursion to Cape Point : the day very warm, notwithstanding a strong breeze. I first followed a track over the hills, overhanging the sea, and through a gap in the mountains, to a Dutch farmer's house in the valley beyond. On ascending a ridge having a rocking-stone poised on it, I had a fine view of Cape Point to the left, and Cape of Good Hope to the right. Here I put up a covey of partridges, which alighted again in a valley beyond, amongst underwood and fragments of rock. I flushed them again and shot one, which I bagged, after having had some difficulty in finding it, as it flew above 100 yards before it fell. Reached a spot midway between Cape Point and the Cape at 12.45, returning by a different route to Simon's Town at 5.30 p.m., walking thirty miles there and back. Hills, granite base, sandstone summits.

Monday, 17th.—I landed at six a.m., and mounting a horse from Kettley's stables, started from Anderson's Store by the turnpike-gate. The morning cloudy but fine. I had no little difficulty in getting my steed to go ahead, so accustomed was he, according to the hostler's account, to go over the red hill with shooting parties ; however this might be, I certainly never rode a more stubborn, intractable beast. I rode round Eloges Bay, Vish-hook Bay, and Kalk Bay, thence round Muizenberg, to Farmer Peck's, about seven miles from Simon's Town. I alighted at 8.30 a.m., and just afterwards the marine officer of the *Winchester* rode up, and we breakfasted together. Farmer Peck has let his inn for 6*l.* per month, and now resides in the house between it and the toll-bar. We called on him whilst our breakfast was getting ready, and after this repast of eggs and toast, tea and coffee, we remounted our horses at 9.30 a.m. and rode together as far as Maskill's half-way house to Cape Town, whither my fellow-traveller was bound. Here we baited our horses, and had a glass of ale; this house, which we reached in a hour, is fifteen miles from Simon's Town.

At 11.15 a.m. I started by the cart-track to the right, just above the inn, and over the sandy flats, in the direction of Stellenbosch. This sandy tract winds irregularly round to the left, but branches off, and is crossed by many others, rendering the true course somewhat embarrassing to follow. About seven miles from the turnpike-road the sandhills commence, which I reached at one p.m. The road here became heavy from the vast quantities of loose white sand, forming rugged hillocks, studded with bushes and rushes. I passed two or three huts inhabited by the Dutch Africanders, who could not speak a word of English. I had a cooling draught of water from one of them, and from a stone bottle suspended from the side of a wood cart, which I met on the way further on, so thirsty did the heat of these sands make me. Altogether I only passed a half-caste Africander or two, with a waggon, and the Cape Town omnibus, with four horses, moving along at a good pace over the sands.

About three p.m. Mr. Edwards, the Wesleyan missionary of Stellenbosch, overtook my old and stubborn horse, I could hardly get beyond a drawling, walking pace, and we rode on together into the town. On passing over a ridge, both False Bay and Simon's Town came in view. I flushed a covey of partridges near the roadside. Several white farmhouses, surrounded by their vineyards, were scattered on either side of the road, forming a pretty approach to the head of the valley in which Stellenbosch is situated. A few miles to the right a long, narrow strip of wood, studded with white farms, indicated the position of the Eerste Rivier. Passed through a small wooded dell by a rivulet just before Stellenbosch made its appearance.

The entrance is strikingly pretty and picturesque. The group of houses nestled at the head of the valley, bounded on either side by lofty mountains. The immediate approach is by a sort of lane, and then up the main street, which runs straight through an avenue of magnificent oaks, having a stream of water coursing down the centre. At six p.m. I alighted at Kinniberg's boarding and lodging-house, where my companion took leave of me, giving me his name and address, with an invitation to call on him. The day had cleared up very fine and warm. I must have travelled over a distance of not less than forty miles since my departure from Simon's Town. The landlord of the inn I found a hearty old veteran, who had formerly been a sergeant-major in the Horse Artillery, which corps it appears he entered in the year 1793, and has been a resident in this colony since 1806 ; and the old man still prides himself on his horsemanship. I had a cold duck, a bottle of ale, and some Cape wine for dinner. The landlord introduced me to a Captain Cormack, of the Indian Army, who was staying in the house. When I turned in at eleven p.m. I heard the colonial girls in the house singing psalms.

Tuesday, 18*th.*—Breakfast in company with Captain Cormack at eight a.m. Afterwards strolled through the village, the streets of which form fine roads through wide, shady avenues of lofty oaks, streams running through them, crossing each other at right angles. The houses are well built, and uniform in appearance, the doors and shutters of a green or mahogany colour, with the roofs thatched, a raised pavement some feet above the road in front of them.

At 2.30 p.m. I called on Mr. Edwards. A young lady about sixteen or seventeen met me at the door, and ushered me into a room on the left, where she had been employed drawing flowers. And here Mr. Edwards himself soon joined us, and followed by his wife, to whom he introduced me. Some willows in front of his residence give it a pretty, rural aspect. On my return to the inn at 3.15 p.m., Captain Cormack, with two friends of his, and myself, mounted horses, and accompanied by two dogs, Quail and Grouse, rode up the valley above the village to the foot of the hills, partridge shooting; but returned unsuccessful to our inn, where we dined on duck and green peas.

Stellenbosch has the air of a retreat of peaceful seclusion, nestled in the lap of the mountains, amid venerable and picturesque oaks. I noticed several pretty girls tastefully dressed, stealing hasty glances at the passing strangers, from corners of doors and windows. The unexpected sight of strangers evidently occasioned a break in their monotonous surroundings. It is situated about twenty-five miles from Cape Town, and takes its name from a former governor, Simon Van der Stell, in 1681. It contains about 250 houses, and 2162 inhabitants.

Wednesday, 19th.—Rose at six a.m. After a cup of coffee I started on my return at seven a.m. Captain Cormack and his friends accompanied me as far as the Erste Rivier, a distance of seven or eight miles, where I took my leave of them, near some white farms, and struck off along a track over a ridge and across the flats. Saw two herons rise from a paddock, and flushed a covey of partridges. Alighted from my horse, and after tying him, went in search of them, without success, losing an hour or two. On remounting, I had an opportunity of seeing the celebrated secretary-bird of the Cape, striding along, on his long, slender legs, amongst the bushes, about half a mile ahead of me, raising his wings as if in preparation for flight; and fly he did before I could get within shot of him, but alighted again on a rising ground. I fired at him as he again rose; it was a very long shot, but I have no doubt that I should ultimately have secured him had I had a more tractable horse, or been entirely on foot. About sunset I gained the turnpike-road, and passed several of the bullock-waggons drawn by ten or twelve pairs of oxen. In passing round Muizenberg the surf rolled on the rocky beach in brilliant, luminous waves. The night starlight. I reached Simon's Town at 10.30 p.m. Slept at Green's " British Hotel."

Thursday, 20th.—Having breakfasted at the inn, and called at the hospital to see how the gunner of the *Terror* was getting on there, under Shea, the surgeon, whom I saw, I called at Anderson's Store, and purchased an ostrich's egg, returning on board in the afternoon. At six p.m. I dined on board H.M.S. *Winchester* with the wardroom officers, and left at nine p.m.

Wednesday, 26th.—Went on board the *Samarang*, arrived yesterday; and on board the *Acorn*, of sixteen guns, to see her commander, John Adams, who had formerly been an old messmate of mine. Called afterwards on the surgeon of the flagship and his wife, at their residence on shore.

Sunday, 30th.—Fine day. We sailed at nine a.m. With the ship's head now towards England, we may fairly consider ourselves on the passage home, albeit, by a somewhat circuitous route.

Saturday, May 13th.—At eight a.m. we anchored off James Town, St. Helena, after a fortnight's passage from the Cape, during which nothing worth recording occurred, and want of space, as my journal draws near its close, will not permit me giving further details.

Monday, 15th.—Landed for the first time this forenoon; met my genial old messmate, Gulliver, on horseback. I walked out to his pretty cottage, Brookhill, about four miles from James Town, and divided from the Tomb Valley by a ridge, embosomed in an amphitheatre of hills. Gulliver returned home about an hour after my arrival, and I dined and slept there. On the following day, Tuesday, 16th, showery weather prevented our excursion to Fairyland, and at 4.20 p.m. I left for the ship.

Wednesday, 17*th.*—I paid a visit to my friends the Gideons at Fairyland, and dined with the family. In my journey out at noon I ascended Ladder Hill in nine minutes without once stopping to rest.

Saturday, May 20*th.*—At 1.45 p.m. we got under weigh, with the wind at east; a rainbow arched St. James's Valley. At noon to-day I received some specimens of silk, of the national colours of France, taken from the flag waved over Napoleon's tomb, at the time of the removal of his remains from this island to Paris, and sent me as a memento from my fair friends at Fairyland, by whom the flag was made.

CHAPTER XXXVII.

Ascension Island—Up the Green Mountain—Rio de Janeiro—Corvo—
Scilly Islands—Woolwich—The expedition paid off.

Thursday, 25th.—After breakfast, upon going on deck,
I saw the island of Ascension for the first time; it was
right ahead, and distant about a degree. Appearing
high and capped with clouds; I took a sketch of it on
nearing the highest part of the land. A flock of about
a score of frigate pelicans flew off to the ship, and when
hovering over the main-truck, I shot two, but had the
vexation to lose both by their falling overboard, which
was the more to be regretted as I have not hitherto
been able to obtain a single specimen for the Government ornithological collection, of the *P. Fregata*, or
man-of-war bird.

We ran round a low point to North-west Bay, at the
base of the Cross Hill, on which is a signal-station, and
anchored at three p.m. No ships there. The health-
officer, a surgeon in the navy, boarded us.

Saturday, 27th.—Fine weather. Took a sketch of
the island, which is of sub-aerial formation; and at 11.30
a.m. I landed. Visited the hospital with the surgeon,
and also the turtle ponds, two in number, containing
300 to 400 turtle—the largest weighing about 400
pounds. Saw several of them floating with their backs
above the surface; returned on board between one and
two p.m., and sent a supply of medicines on shore for
the use of the hospital.

Sunday, 28th.—After divisions and divine service, I went on shore; called on the surgeon of the hospital, and at eleven a.m. started on an excursion up the Green Mountain: weather fine. Struck off across the scoria strewed flat, round to the right of Cross Hill, 870 feet in height, by a cart-road which led across a plain between it and the base of Green Mountain. Passed the two-mile stone on the right, and afterwards the tank, over rugged ledges of scoria to the base of the mountain, some four miles. The rocks consisting of scoriaceous lavas, studded here and there with straggling bushes of the castor-oil plant, presenting an excessively wild and barren aspect on this cindery island. Saw a goat, and perhaps 100 head of cattle. The only species of bird I met with was the wide-awake or egg-bird, a black-backed tern (*Sterna fuliginosa*), flying overhead, uttering a mewing sort of kitten-like cry. These birds breed in such numbers on one part of the island, as to gain for it the name of "Wide-awake Fair." I saw two mountain butterflies, having a white spot on a dark-coloured wing, numbers of crickets, and a few cock-roaches. I ascended direct for the mountain-house up a zigzag road. The first higher form of vegetation I met with, was the aloe, now in seed. I passed the detachment barracks, and along a very pretty evergreen fence to the officers' quarters, a neat building. Here I partook of some biscuit and cheese and a glass of ale with the officers.

This spot, after ascending from the scorched-up, cindery soil below, appeared to break upon one all at once, like an oasis in the desert. Here is an excellent garden, ornamented with flowers of various hues, the house itself shaded in trees and shrubs. Amongst them the banana grew vigorously, a few pine-apples struggled for existence, and I noticed beds of carrots and leeks. Merely the water falling as it does, drop by drop, from.the rocks into a reservoir, I was told, proved sufficient to supply the wants of the mountain-party. At the extremity of the garden is a tunnel cut through the hill, several hundred yards in extent. The view from this mountain-house is very fine, the smaller hills rising like cones from the surface of the plain-like expanse below: some of a reddish-brown colour, others grey; and the mountain itself appears of a light-green tint: hence its name. The highest peak, at the mountain-house, is 2818 feet above the level of the sea.

There are said to be guinea-fowl, but I did not see any. I arrived at the house at 12.45, and left it at 1.30 p.m. by a somewhat different route for change. The peak soon afterwards became enveloped in mist, and a light shower or two fell. Reached George Town at 3.30 p.m., and dined with the surgeon, Harvey Morris, my brother-officer, in charge of the medical establishment here. We had turtle soup, and fin, pork, both boiled and roast (cold); kumeras and callaboas, with damson tart, ale, port, sherry, and claret. Walked round the hospital, and after tea I returned on board at eight p.m. Mrs. Morris is a daughter of the celebrated Dr. Clutterbuck, of London.

Monday, 29th.—At 8.50 a.m. we sailed from Ascension.

Saturday, June 3rd.—Nearly a calm. Made deep-sea soundings, and it took somewhat more than two hours to run 4600 fathoms off the reel without reaching bottom, expending 600 fathoms one-inch and 4010 fathoms three-quarter-inch rope, with three cwt. of iron ballast. Current W. $\frac{1}{2}$ S.

Wednesday, 7th.—Upon going on deck at nine a.m. I saw the island of Trinidad very distinctly ahead, bearing S.S.W. Lat. at noon 20° 7′, long. 29° 12′, dip 11° 22′, var. 9° 20′. Sketched the land, and at 4.30 p.m. we bore up W.S.W. abreast of the Nine-pin Rock, about a league distant. Saw the beach where we landed three and a half years ago. Our remaining turtle was killed this evening, and I preserved the shell and head, as it was a very large specimen, weighing at least four cwt.

Friday, 16*th.*—This morning saw the revolving light on Cape Frio, on the lee-beam. On the following evening we tacked within three or four leagues of the Sugar-loaf, and stood off the land for the night. Saw the peak of the Corcovada, with black squalls passing over the land, and a dense nimbus suspended over the mountain-tops. After dark saw Raza light.

Sunday, 18*th.*—Beating up all day for Rio de Janeiro Harbour, and anchored at 4.40 p.m. in five fathoms.

This fine capacious harbour derived its name from the discoverer, De Sousa, having discovered it on the first day of January, 1531. The aborigines called the harbour Netherohy, or " Hidden Water," and the Sugar-loaf Mountain, Pao d'Azucer, which attains the height of 1000 feet; the Corcovada Mountain, 2000 feet, and the Organ Mountains, 3200 feet.

Monday, 19*th.*—At eleven a.m. I landed for a stroll round the town. Called at Tross's, and lunched at " Pharoux's Hotel." Visited the fair in the campa, and the cathedral, returning on board at six p.m.

Tuesday, 20*th.*—Crossed over to the Braganza side in the steamer. Walked as far as the old fort and back again, and missed the packet twice. Saw numbers of oranges growing in the gardens. Got back to Rio at 3.30 p.m.

Wednesday, 21*st.*—Landed again at eleven a.m. Fine day. Walked over the hill to the *passeo* for about a mile, returning through the new market-place. Had a lunch of prawns and stout at Pharoux's, a capital large new hotel, at the corner of the Largo de Paro. Went to Madame Finot's, in the Rua d'Ovidor, and purchased a box of feather-flowers and one of insects. She had about thirty young creole girls seated at the farther end of the shop, making flower-wreaths with feathers; and an interesting-looking young damsel, whom she called Catherine, attended as interpreter, apparently of English extraction. I bought also a grey and a green parrot at different shops adjacent.

Sunday, 25th.—We sailed at 8.15 a.m. Rio is much improved in appearance since I was last there in 1832. A new landing-place, hotel, and market-place. The Botanical Garden at Boto-Fogo is seven or eight miles from Rio. We had a remarkable fine week of weather during our stay, and being St. John's week, the city was very lively with ringing of bells and fireworks and rockets at night.

Monday, July 10th.—We have had the trade-winds for the last few days, and crossed the line this evening in about the twenty-sixth degree of longitude.

Friday, 14th.—This afternoon entered the variables with the wind from S.W.; lat. 6° 39', long. 24° 23'.

Tuesday, 18th.—We entered the north-east trades, and Saturday, 22nd, got the temperature of the deep sea at 1850 fathoms, the greatest depth yet, for temperature.

Monday, 31st.—Fine day. Crossed the tropic this forenoon.

Monday, August 7th.—Employed in arranging and packing the Government collection of birds on deck; and sent into the cabin, in compliance with the customary orders, all my diaries and sketches made during the voyage.

Wednesday, 9th.—Two dolphins caught. Had some for dinner, and found it very good eating. Employed arranging and packing the Government geological collections.

Wednesday, 16th.—The calms and light, variable airs for some days past have prevented us making the Azores, and being now in about the latitude of them, there is no chance of our touching at any of them as was intended. This evening for the first time saw the chart of Victoria

Land, with the names of the officers of the expedition attached to it.

Saturday, 19th.—Flores and Corvo in sight. At eleven a.m. passed between them. They are seven or eight leagues apart. We hove-to off Corvo, about a league from the land, for the island boats to bring us supplies;—two goats, thirty-two fowls, eggs, milk, cheese, potatoes, water-melons, onions, and heads of Indian corn. I purchased four fancy baskets, made of red and white cane, for the small sum of two shillings and ninepence; fowls one shilling each, eggs one shilling per dozen.

Corvo rises to about from 1200 to 1400 feet above the sea, presenting a mottled, dull, brownish-green aspect, the surface reticulated with the enclosures of maize and other produce. The cliffs appeared much rent in places by subterranean movements. The village consisted of some 200 sombre-looking, lava-built huts, having tiled roofs and thickly grouped together; having a white-coloured church, and belfry in the centre. At 1.30 p.m., on the *Terror* coming up, we made sail with a fresh breeze for England, nearly 1300 miles off.

Wednesday, 30th.—Between seven and eight a.m. we were boarded by a Scilly boat. St. Agnes lighthouse was seen from the deck this forenoon. At three p.m. we sounded in sixty-seven fathoms. In clear weather this light may be seen eighteen miles off.

Saturday, September 2nd.—Very fine day, but with light, contrary airs. At nine a.m., upon going on deck, I saw the land of old England again, after some four years' absence, extending along the weather, or port-beam : we were off the Bolt-tail coast of Devon, distant some four or five leagues. At the least a score of vessels in sight. After dinner a Cowes pilot-boat came alongside. Saw the Start light on the port-quarter.

Monday, 4th.—At nine a.m. off Bexhill, with a fine view of Beechey Head, St. Leonard's, Hastings, &c. A lovely day; the sea smooth as a lake, and studded over with countless vessels; whilst the line of coast, displaying the rich, golden-yellow fields of corn, some in sheaf, some still standing, altogether gave animation to the scene. Saw the Dungeness, and both the Foreland lights, in the evening.

At 11.15 p.m. we anchored in fifteen and a half fathoms off Hythe. Weighed again on the following morning, Tuesday, the 5th, and stood in for Folkestone, where Captain Ross landed at seven a.m., and proceeded by train to town. At 1.30 p.m. anchored off Walmer Castle, for the tide, and at 6.40 p.m. weighed again, and dropped anchor again in the first watch. But a breeze springing up, again weighed and came to an anchor once more off the flat of the North Foreland. At 4.40 worked through the Queen's Channel, and at 11.30 a.m. anchored in Pan-sand Hole. Saw the Reculvers, and my old station-house, Epple Bay, through the haze over the land.

To-day my South American green parrot, an intelligent, lively creature, followed me into the main-top, and down again by one of the ropes.

Thursday, 7th, 6.20 a.m.—We sailed up the Queen's Channel, and at one p.m. anchored off the Great Nore, and received orders to pay off at Woolwich. At 2.20 p.m. we got once more under weigh, and finally let go our anchor there.

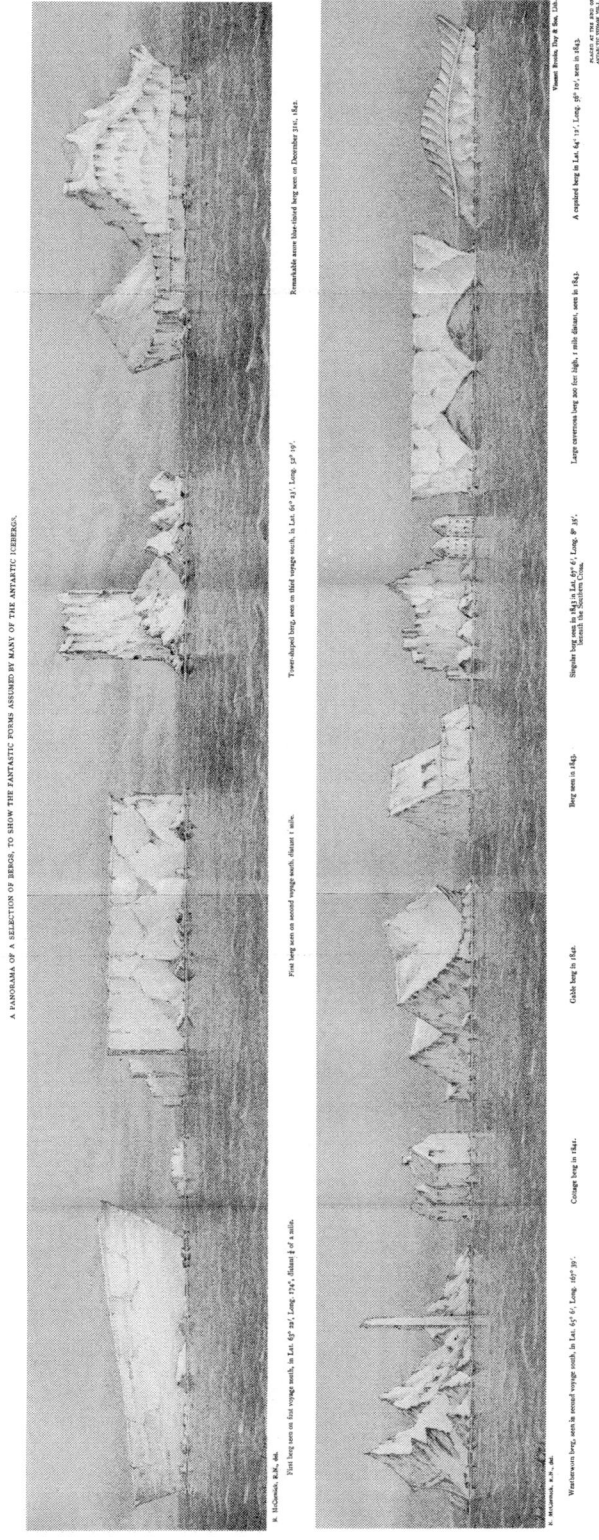

A PANORAMA OF A SELECTION OF BERGS, TO SHOW THE FANTASTIC FORMS ASSUMED BY MANY OF THE ANTARTIC ICEBERGS.

The material originally positioned here is too large for reproduction in this reissue. A PDF can be downloaded from the web address given on page iv of this book, by clicking on 'Resources Available'.

PART II.

NARRATIVE

OF THE ATTEMPT TO REACH

THE

NORTH POLE

IN THE YEAR 1827

BY

ADMIRAL SIR WILLIAM EDWARD PARRY, R.N., F.R.S.

Little Table Island, Spitzbergen, the northernmost land. (*See page* 396.)

PERSONAL NARRATIVE

OF THE

ATTEMPT TO REACH THE NORTH POLE

IN THE YEAR 1827.

———•———

CHAPTER I.

Departure from the Nore— Arrival at Hammerfest—Visit of the " Belle
of Lapland " and her party to the *Hecla*—Ptarmigan-shooting—
Magpies held sacred—Departure—Crow's-nest got up to fore-
topgallant mast-head—First ice seen—Fall in with whalers—Sight
Spitzbergen—Cloven Cliff—Enter the pack in a gale—Red
Beach—The attempt to start in the boats from the pack—Verlegen
Hook—Highest latitude reached—Seven Islands.

On Wednesday, the 4th of April, 1827, at four a.m., we
weighed anchor from the Little Nore, the *Comet* steamer
taking us in tow.   At six a.m. sent her to Sheerness with
the ship's company's wives and families, and stood on our
course, returning the steamer's cheers.   At noon, when
off Harwich, the steamer came up with us again, and took
us in tow.   At six p.m. she finally parted company with

us off Orfordness, with a hearty cheer, having discharged our pilot on board of her. At seven p.m. saw Lowestoft revolving light. Wind S.W., with cloudy weather and light breezes.

Sunday, April 8th.—This forenoon Captain Parry performed divine service on the lower deck, taking his text from St. Paul's Epistle to the Romans, second chapter, and thirteenth verse. He afterwards read the articles of war on the quarter-deck. Wind W.S.W., and fresh breezes, with a moonlight night.

Monday, 9th.—Struck topgallant masts, and close-reefed the top-sails, going before a strong gale of wind from the southward, with a heavy sea running, and dark, cloudy weather.

Tuesday, 10th.—Still blowing strong, with cloudy weather, accompanied by sleet and snow at intervals. The wind at noon N.W. Saw several fulmar petrel (*Procellaria glacialis*) for the first time; and on the following day, Wednesday, the 11th, two solan-geese (*Sula alba*) and a great number of fulmar petrel in the wake of the ship.

Thursday, 12th.—Last night we crossed the Arctic Circle; and I saw a solitary duck. We killed our two oxen this morning.

Friday, 13th.—Last evening all the officers assisted in working the ship. To-day we lighted the hot-air (Sylvestre's) stove for the first time since leaving England; kept it lighted for twenty-four hours to air the ship, when the thermometer in the cabin rose to 50° Fahrenheit. Weather cloudy and squally, with a fresh breeze from the E.S.E.: a few fulmar petrel still following in the wake of the ship, but not in such numbers as before we entered within the Arctic Circle, 66° 32′ N. latitude. Between ten and eleven p.m. saw the Aurora Borealis for the first time, shooting upwards from east to west, a brilliant display of the most beautiful coruscations, changing appearance and position with the greatest rapidity; at times darting upwards in a beautiful arch to the zenith, and finally dying away in faint coruscation. The night was fine and clear; the heavens brightly studded with stars, only slightly obscured now and then by a few passing light, fleecy clouds.

Saturday, 14th.—Weather fine and clear, with light south-easterly winds. The officers to form the travelling-party have been living on their allowance of pemmican for the last four days, to test its effects on them, as it is to form their only diet of animal food whilst away in the boats. The pennant and union jack, made of blue, red, and white silk by Mrs. Parry, to be hoisted at the pole should the expedition be so far successful, were unfolded to-day on the quarter-deck. I had, when first nominated for the *Hecla*, volunteered myself to form one of the Polar party, but, unfortunately for me, Captain Parry had already promised to Mr. Charles Beverley, who had been a medical officer with him in a former expedition, that he should accompany him, in order to give him a stronger claim for reinstatement on the Navy List, from which his name had been removed on his refusal to take up an appointment to a ship when called upon to serve.

Sunday, 15th.—Captain Parry performed the morning service. The weather extremely fine and clear, the sky being of the most intense azure blue, and unobscured by the smallest cloud; the sea quite smooth, and of the deepest, darkest blue; with light airs from E.S.E. Only three or four petrel following the ship to-day.

Monday, 16th.—The travelling-party were mustered on the quarter-deck to-day in their travelling-dresses, and the nature of the undertaking and allowance of provisions explained to them by Captain Parry. At one p.m. made the land; and at six p.m. I saw it from the deck bearing S.E. on the starboard bow. Weather extremely fine and clear, with light breezes from E.S.E.

Tuesday, 17*th.*—Running along the land, with a fresh southerly breeze. The mountains appeared covered with snow. In the evening we were becalmed, when about ten miles from the land. Hoisted a jack at the fore-top-gallant mast-head, and fired two guns for a pilot. Saw several ducks and gulls.

Wednesday, 18*th.*—Beating up the fiord for the town of Hammerfest, with a moderate breeze from S.E. and cloudy weather. In the afternoon a boat came alongside with a pilot. Had the hot-air stove lighted again to-day to air the ship.

Thursday, 19*th.*—At three p.m. anchored off Hammerfest ; and a boat left the ship with Lieutenant Crozier and one of the mates to bring some reindeer from Alten, intended for dragging the Polar boats over the ice. Letters were sent to Alten, to be forwarded from thence to England. Cloudy weather, with snow and fresh south-easterly breezes.

Monday, 23*rd.*—The ship was dressed in flags, and a royal salute fired, in honour of the anniversary of the birthday of his Majesty. The boat returned from Alten ; and the travelling-party were sent on shore to try their snow-shoes.

Friday, 27*th.*—Received eight reindeer on board.

Sunday, 29*th.*—At seven a.m., we got under weigh, but light winds and calms prevented us from getting clear of the land during the day. During our stay here we had visits on board from some of the principal inhabitants of this little out-of-the-way place, who were but just emerging from their winter's hibernation. They are often so blocked up with ice and snow, that during their long inclement winter season they have to dig themselves out. Their floors are strewed with rushes. I met most of the little community on board of a large Russian ship, at a party from the shore to which I had received an invitation. One of the young ladies present, of about twenty years of age, was very pretty and pleasing in her manners, and well deserved the title of the " Belle of Lapland." She lived with her mother, the widow, I believe, of a Russian merchant; and when she came on board the *Hecla* with her friends to see the ship, she placed herself by the binnacle, and was very anxious to have our course to the Pole pointed out to her on the compass; but owing to our not understanding each other's language, this was no easy matter. She was much amused with the chair constructed for lowering the ladies down the ship's side into the boat alongside, and it fell to my lot to have the honour of placing her in it when she left us. On the hills flanking the town there are ptarmigan called " Reepers;" but I did not meet with any. The ascent of the steep hillside of the mountain range, covered with snow and ice, in pursuit of them, was of so very perilous a nature that one of our party, a young merchant of the place, who accompanied us as our guide, when about half-way up the mountain slipped, and after sliding down to within a few feet of a steep precipice, was only saved from destruction by the accumulation of the snow at his feet, and some rugged projection of the rock arresting his downward course. I shot a magpie near the huts before I was aware that the bird was held sacred here. I presented it to Captain Parry for his collection.

Monday, 30*th*.—Land still in sight : the ship's company have been employed for the last three or four days picking the moss brought on board for the reindeer's food, for they will only eat one kind, a white species (*Cenomyce rangiferina*), and of this selected sort each reindeer will consume about four pounds daily. The coast here is of granite formation; and we passed out of the fiord by the island of Soroe, the gneiss rocks, of which it and the adjacent islands are composed, had a cleavage inclining to S.W. The latitude of Hammerfest is in about 70° 40′ N. and longitude of 23° 45′ E.; the dip of the magnetic needle 77° 12′ N.; the variation being 10° 14′ westerly.

Tuesday, May 1st.—We are at last out of sight of land, with fresh northerly winds, cloudy weather, and snow at intervals. Lemon-juice was served out to the ship's company to-day for the first time since leaving England. Several fulmar petrel following in the wake of the ship.

Friday, 4th.—Still northerly winds, with snow, and squally weather. Pea-jackets were served out to-day both to officers and men. For the last few days we have been passing through water of a comparatively warm temperature, the sea-water reaching 40° when the temperature of the air was only 24° Fahr., doubtless an offset of the Gulf-stream, which, laving the shore of Spitzbergen, tends much to moderate the degree of cold of its climate.

Saturday 5th.—When in the latitude of 73° 30′ N., and longitude of 7° 30′ E., we fell in with the first ice in detached fragments. The opaque white appearance it presents gives to Arctic ice a striking resemblance to loaf-sugar. Saw some kittiwakes (*Larus tridactylus*), little auk (*Alca alle*), and ducks, with a number of fulmar petrel; the latter bird, being so voracious, is very easily caught with a hook and line towed astern. As soon as one of these birds notices the bait floating on the water it swims cautiously towards it, looking at it apparently with something like suspicion; but so soon as it observes its companions approaching, rather than they should secure it, it pounces at once upon the much-coveted morsel, swallowing hook and all. I caught two of them in this way for the Government collection, but it is a cruel way of obtaining specimens, and not to be resorted to when they can be secured by shooting them.

Sunday, 6th.—Captain Parry performed divine service, and read the articles of war on the quarter-deck.

To-day we had some Donkin's preserved vegetable soup served out for dinner, and found it excellent. Passed several pieces of floating ice, and an eider-duck alighted close alongside of the ship.

Monday, 7th.—Fell in with large continuous streams of ice to-day of every variety of fantastic form and shape, tinted with the most beautiful azure blue near the surface of the water, and crested by the purest opaque white, like the top of a twelfth-cake. The helm required the greatest care and attention in steering the ship through the narrow lanes between the ice, pieces of which not infrequently came in collision with the bows, occasioning a smart shock. The crow's-nest was got up to the fore-top-gallant mast-head to-day. This is a circular box resembling a cask without the head, covered outside with canvas, having a seat inside, with a telescope for the use of the officer looking out for the lane of water presenting the best chance of getting the ship through.

Tuesday, 8th.—Snow at intervals. Fell in with the ice in much greater quantities to-day, extending ahead as far as the eye could reach from the fore-top. In addition to the fulmar petrel there were some ivory gulls (*Larus eburneus*), and several seals on the ice. Had some of Donkin's preserved meat and vegetables served out to-day, and both good.

Wednesday, 9th.—Fell in with two whalers this fore-noon, passing close to one of them, and sent a boat on board of her; the other ship made fast to a piece of ice at some distance from us, and sent one of her boats alongside of us. They were both from Peterhead, and neither of them had been able to get any further north than where we found them. The ship made fast to the piece of ice was the *Active*. The year before last, she having got aground in Davis's Strait, her crew deserted her and returned home. But on going out again in the following summer they found her in the same state in which they had left her, and brought her home. The ice was so closely packed to-day that we were compelled to make fast to a floe piece—a large flat piece of ice—by means of an ice-anchor. Saw the first flock of snow-buntings ; several little auk and ivory gulls were swimming amongst the ice. Weather thick and cloudy, snowing at intervals—as it has been since we left the coast of Finmark—with northerly winds. The evening cleared up fine, the sun shining brightly at ten p.m., when three more whalers hove in sight. We had the hot-air stove lighted this afternoon.

Thursday, 10th.—Cast off from the piece of ice last night. Had Sylvestre's stove lighted again to-day. A drift of small, fine snow continuously blowing over the surface of the ice. Eleven whalers in sight to-day, and in the forenoon four ships and a brig passed close to us in pursuit of a whale, whose appearance we had made known to them by signal. Captain Parry hailed them as they passed, and acquainted them with the direction in which it had been seen. However, after tacking about in the numerous lanes of water amongst the ice for some time, and the search proving fruitless, they returned. Each ship had a garland suspended from the main-top-gallant stay, which it is a custom amongst the whalers to hoist on a certain day upon entering the Arctic Circle. Each ship had also a crow's-nest at the main-topgallant mast-head. Passed two or three narwhals, or sea-unicorns (*Monodon monoceros*), sporting on the surface of the water, and saw several small flocks of dovekies (*Uria grylle*), and looms, or foolish guillemots (*Uria brunnichii*), flying about the ice this evening. During the first watch we "sallied" ship ; this consists in both officers and men running quickly from side to side of the ship, all starting at the same time, so as to make the ship roll, and so assist her in getting through the bay-ice. This forenoon a boat came alongside from one of the whale-ships.

Friday, 11th.—The land of Spitzbergen was seen during the morning watch, but the weather has been so thick and hazy throughout the day that it has not been seen since. The thermometer having risen 12° since yesterday, the thaw has melted the snow on the decks. Several whalers have been seen during the day. Large flocks of little auk, dovekies, and looms, swimming in the lanes of water between the ice, and sometimes they would dive on the approach of the ship, at others take wing, and alight again a short distance off, uttering shrill and confused murmuring, chattering notes, a few fulmar petrel following in the wake of the ship, as usual with them.

Saturday, 12th.—Very thick and hazy, with sleet and fine snow at intervals. Wind southerly, much obstructed in our progress by the ice. Two of the whalers in com-pany. Towards evening the breeze died away to a calm, accompanied by a thick fog. Saw innumerable flocks of little auk during the first watch. The ship having become nearly becalmed in a hole of water, Captain Parry allowed me to take the dingy, and pull about amongst the ice, to shoot some of them for the Govern-ment collection. But the fog increasing spoilt the sport ; and as I could scarcely see the ship a few yards off, with the risk of a breeze suddenly springing up and losing sight of her altogether, a musket was fired from the ship as a signal for the boat's recall, just before midnight. I shot two little auk at the first fire, two looms at the second, and at the third shot a fulmar petrel.

Sunday, 13th.—This morning I saw the land of that most interesting island, Spitzbergen, for the first time, distant some seven or eight leagues, towering upwards in high and sharply peaked mountains, as its name implies. The angular symmetrical forms of these lofty snow-clad peaks gives a very picturesque aspect to the whole coast-line, in lat. 79° N. The two boats which we have hitherto had on the quarter-deck were hoisted out to tow the ship, and afterwards hoisted up again before the gangways. Captain Parry performed divine service, and we went to divisions. This forenoon I went up to the crow's-nest to have a look at the land, we having to-day got up a second one above the other, at the fore-top-gallant mast-head. Saw three whale-ships fast to a piece of ice. We were beset with ice during the forenoon. About noon we cast off from the piece of ice to which the ship had been made fast, and towards evening the sea became much more clear of ice. Saw a glaucous gull (*Larus glaucus*) for the first time—a noble, dignified-looking bird, the largest of the gull tribe—with vast flocks of the little auk and looms.

Monday, 14*th*.—Having early this morning passed Magdalena Bay, we this forenoon rounded Hakluyt's Headland to look for an anchorage in Smeerenburg Harbour, but which we found filled with the winter's ice, the floes being still unbroken, and we had consequently to make fast to the floe-edge, mid a southerly gale, and heavy gusts of wind coming down the ravines and high land above the harbour. Captain Parry himself went with the boat to lay out an anchor with hawsers, on the land ice, which was accomplished with great difficulty. But the gale increased so much in force, with the swell setting with so much fury on the floe-edge, where we had made our anchor fast, that this part of the floe gave way, and at ten p.m. the ship was adrift.

Several reindeer were seen on the snow at the foot of the mountain, which at first sight the boat's crew took for bears, whilst they were passing a hawser round the hummock of a floe. I saw a number of eider ducks, dovekies and gulls in the harbour; also some walruses (*Trichecus rosmarus*).

R. McCormick, R.N., del.

Vincent Brooks, Day & Son, Lith·

H. M. S. "Hecla"

Driven out to sea from the Bay-floe, Smeerenberg Harbour, Spitzbergen, in a gale of wind.

Being compelled to stand out to sea—blowing so tremendously, in violent gusts of wind, we had to furl all sail, save and except the main-topsail, close-reefed, and the storm-staysail—passing through some stream ice. Saw Cloven Cliff, off which a small iceberg was aground. A faint yellowish tint appeared in the horizon, known as the ice-blink, and indicating a considerable extent of ice in that direction. Passed several sea-horses, or walruses, and after some time spent in endeavours to weather the pack, at the imminent risk of dashing the ship to pieces against some of the larger and heavier masses, by being driven on them in the terrific squalls, which at times nearly threw her on her beam ends—drifting to leeward, and having to wear several times, between Vogel Sang and Cloven Cliff—we eventually had to seek shelter in the pack by entering the loosest part of the margin of it, under a press of sail about three a.m., and within an hour afterwards, found ourselves quietly reposing half a mile or more within its confines. Whilst the gale was raging all round this sheltering haven of ice, but had no effect on us, fixed and stationary as we became by the pressure of the surrounding ice on all sides. Every officer and man had been on deck throughout this perilous night, all the officers assisting in the working of the ship with the men.

We had now time to make ourselves more comfortable by having Sylvestre's stove lighted. The mainbrace spliced, this means an extra allowance of grog and rations of preserved meat served out, which the ship's company partook of on deck. As I did not turn into my cot till eight a.m., I for the first time in my life had an opportunity afforded me of witnessing twenty-four hours of daylight, the sun being now constantly above the horizon : wind, S.W.

Tuesday, 15*th.*—We are now completely beset on all sides by the pack, with fine clear weather and light

c c 2

southerly winds, the very contrast to the stormy, tempestuous scenes we have gone through of late.   Fulmar
petrel flew about the ship in great numbers and I went
upon the ice and shot eight of them and an ivory gull.
In the evening I shot a glaucous gull and another ivory
gull from the ship.   Saw two walruses lying upon the
ice at some distance ahead of the ship.   On the following day a young bear came alongside the ship and was
shot.

Friday, 18*th*.—Being clear and fine weather, a party
consisting of Lieutenant Ross and Mr. Beverley and
three men were sent over the ice to attempt a landing,
the shore being three or four leagues distant.   But they
returned after a few hours' absence, having met with a
lane of water arresting any further progress.   They had
shot a few birds.   Red Beach, on the northern coast of
Spitzbergen, off which we are, is in latitude 80° N. and
longitude 13° E.

Saturday, 19*th*.—The weather became thick and hazy
with falls of snow, the ship's side undergoing heavy
pressure from the surrounding ice, forced up in hummocks
around her, and getting her into a "nip."   The whole pack
was thrown up into rude shapeless masses by the strong
westerly gale blowing, and, covered with a white mantle
from the falling snow, gave to the whole scene more the
appearance, as the crew expressed it, of a stone-mason's
yard, only on an exaggerated scale, than anything else
it could be compared to.   Sylvestre's stove was lighted
to-day.

Sunday, 20*th*.—Divine service was as usual performed
by Captain Parry, the text taken from the first chapter
and fifth and sixth verses of the Epistle of St. James.
Hot-air stove lighted.   Light westerly winds and snow.

Monday, 21*st*.—Hot-air stove again lighted.

Tuesday, 22*nd*.—Lieutenant Ross with two other
officers and three men were sent from the ship early this

morning, and effected a landing over the ice, returning on board again soon after noon, a signal having been hoisted and several guns fired for their recall on a fog coming on. For the last few days the pressure of the ice has been so great as to force up immense hummocks all round the ship.

Wednesday, 23rd.—To-day the ice seemed in motion, separating in several places; one hole of water made its appearance near the ship; so to assist in this movement of the ice, we got sail on the ship, and " sallied " her to loosen her from the ice-cradle in which she rested, but without effect.

Thursday, 24th.--A bear was seen near the ship last night, in the middle watch. The weather extremely fine to-day, and although the temperature was not above 32°, seemed quite mild to our feelings. Sails were loosed to-day, and the wheels for the polar boats got upon deck, from the hold. Having seen a walrus blowing in the hole of water near the ship, I took my gun with me on the ice, and approaching the edge of the hole just as he rose to blow, I fired a ball at and wounded him, but he immediately dived under the ice, and I saw no more of him.

Friday, 25th.—Cloudy, with light south-easterly winds. In removing one of the reindeer out of the launch on deck this morning, its leg was broken, and it had to be killed, poor thing. Captain Parry and Lieutenant Ross, have been busily employed stowing the stores in their boats, preparatory to their departure. Captain Parry finding the season already so far advancing, has resolved upon losing no more time, and consequently, intends leaving the ship where she is in the pack, and commence his journey over the ice in the boats, on Sunday morning next, the 27th. To-day, I shot a fulmar petrel from the ship.

Saturday, 26th.—Captain Parry told me to-day, that,

during his absence from the ship, he should give me the charge of a watch on board the ship. The boats and eight sledges, laden with stores and provisions, are getting ready for his departure to-morrow.

Sunday, 27th.—At four a.m. the two travelling-boats were lowered from the ship upon the ice, and by seven a.m. had all their stores stowed in them, and the sledges loaded. Then the reindeer were harnessed to them, Captain Parry's boat and the dingy were hauled to a floe-piece of ice about a hundred yards from the ship, after a great deal of labour in getting them over the rough, uneven, hummocky ice. On the wheels being attached to one of the boats, they sank up to the axle-tree in the soft snow covering the ice, and resisted the attempts of all hands to move them. And after many fruitless efforts, all unavailing, the deep snow drifted between the rugged hummocks, rendering the boat quite immovable. Captain Parry was most reluctantly compelled to relinquish the attempt as hopeless and impracticable. Never, perhaps before, was witnessed so novel and exciting a scene as the motley group of officers and men, together with the boats on the wheels, the sledges, and the reindeer, presented on the ice, alongside of the *Hecla* this morning. The Polar party themselves were picturesque in the extreme, their travelling-dresses, consisting of a racoon-skin cap, box-cloth jacket with a hood for hauling over the head, breeches of the same material, having their seats covered with a patch of waterproof canvas, of a white colour, in striking contrast to the blue colour of the cloth; these were buckled at the knee with a strap, from which gaiters of the same material encased the legs. After their arduous and laborious morning's work, all hands were piped to breakfast, after which they were assembled on the quarter-deck, and Captain Parry in a few words explained to them the impossibility of proceeding over the ice in its present condition. The whole forenoon was occupied in removing the stores from the boats and sledges, and getting them with the reindeer on board again. A party of five men with the senior mate, under the command of Lieutenant Ross, were sent away with the dingy, to report upon the state of the ice at a distance from the ship. They returned in the evening, having found it better at some distance, but on proceeding still farther on, found it again became worse than even that in the vicinity of the ship. Captain Parry, when the returning party were within a mile of the ship, sent a mate with a party of four men to assist in getting the dingy on board. One of the reindeer this morning, on being hoisted out of the ship, was taken ill on the ice and killed. I went up to the crow's-nest this forenoon, and saw a walrus lying on the ice, about half a mile from the ship. The weather during the day has been very fine and clear. The sun quite warm, with light airs from the north-west to east. Evening cloudy.

Monday, 28th.—Lieutenant Crozier was this morning sent away with a party to the eastward to ascertain the condition of the ice, but a fog suddenly coming on, a gun was fired for their recall. A boat was prepared to-day for the polar party, the captain now contemplating taking only one boat with him. Weather hazy, with some sleet and snow falling, with light airs, ending in a rapid thaw ; the ice separating in a number of places, and forming holes of water, one very large one, on our port quarter, in which several ivory gulls were swimming. Wind N.E.

Tuesday, 29th.—I left the ship at nine a.m., with Lieutenant Foster and a party, to haul a boat over the ice and land her on Red Beach, with stores and provisions, to be deposited there for the use of the polar party on their return, should they need them. After we had hauled the boat over the very rugged hummocks, with great exertion and labour, to a considerable distance from the ship, about eleven a.m. a gun was fired for our recall; but our guns, having got wet in the snow, would not go off, and we consequently answered the signal with three cheers. Met Lieutenant Crozier with the dingy and a party sent to assist us in getting on board. Having left our boat on the spot to which we had hauled her, we had no means of ferrying ourselves over the holes of water between us and the ship, several very large ones having opened; we got on board soon after noon. Lieutenant Foster, on being sent away again, got the boat within about two miles of the shore, and returned on board at eleven p.m.; the forenoon being foggy, and the afternoon clearing up occasioned these changes.

Wednesday, 30th.—The weather being fine and clear, with a light breeze from the eastward, Lieutenant Foster with a party, at ten a.m. again left the ship, to complete the landing of the boat and stores, which they this time accomplished, turning her bottom-up, with the stores underneath upon the shore, returning on board at eleven p.m., several guns being fired from the ship as signals to the party. Just before they reached the beach the ice opened and closed again at intervals near the ship. Saw several eider-ducks and a very bright ice-blink to the northward.

Thursday, 31st.—Employed fitting the polar boats and stowing the sledges. In the evening saw three polar bears prowling amongst the hummocks of ice at some distance from the ship in an easterly direction; a fire was lighted on the ice, and some meat burnt to attract them by the smell, but they kept on their course to the eastward. During the middle watch, however, it appeared that a bear came alongside the ship.

Friday, June 1st.—As we were drifting in shore a party was employed to endeavour to extricate the ship from the ice in which she was imbedded, by cutting through it in the direction of a lane of water on the port side.

Saturday, 2nd.—Still employed cutting away the ice from the hull; and in the afternoon a fresh breeze of wind coming off the land, we took advantage of it to unfurl all the sails and " sally " ship, so that in the evening we found ourselves drifting from the shore.

Sunday, 3rd.—Wind gone down, and weather fine and clear; drifting to the eastward and in shore, very near Grey's Hook. No divisions or divine service; all hands employed on the ice and in " sallying ship." Several lanes of water opened not far from the ship. More glaucous, ivory, and kittiwake gulls have been seen to-day than for some time past.

Monday, 4th. During the middle watch last night the ship entirely cleared herself from the ice that had surrounded her, and passed through the channel cut for her into a lane of water on the port quarter, leaving behind her a kind of dock or ice cradle she had formed for herself in the ice that pressed around her, some seventy or eighty yards on the starboard bow. During the day some ivory gulls collected together about it, to pick up the scraps that had been thrown overboard from the ship. We were now drifting, at the rate of two miles an hour, to the eastward, round Grey's Hook, and much closer in shore. Across the entrance to Weyde Bay, which is not less than two leagues in breadth, made fast head and stern to a floe-piece; water seen all around the pack, with a dark water sky. Killed two of the reindeer to-day.

Tuesday, 5th.—Fine and clear weather, with light, southerly winds; made all sail, but went very little ahead, and in the evening the ice had closed up all the lanes of water, and the ship was stationary off the entrance to Weyde's Bay; made fast to some hummocks on a floe-piece; a number of gulls of various kinds were flying about the ship to-day.

Wednesday, 6th.—Drifting to the eastward and in shore; saw a seal basking in the sun, and rolling over from side to side, on a piece of ice near a hole of water at some distance from the ship. About five p.m. Captain Parry, with a party of officers and men, left the ship, and having hauled their boat over the ice into open water, pulled on shore, and landed near Mussel Bay, two miles in depth, and ten miles S.W. of Verlegen Hook. They returned on board about one a.m., having shot a reindeer and seen several others; lighted a fire on the beach with driftwood, which they found in abundance, and made a *cache* of provisions there, for any future emergencies that might occur; but found no suitable bay for securing the ship in.

Friday, 8th.—Lieutenant Ross and a party went on shore to collect some moss for the remaining reindeer on board, and they shot a deer and brought it on board. In the evening, a strong breeze of wind springing up from the southward cleared the ship of the pack, and left a lane of water in shore of us, after having been closely beset for above three weeks, but with unusually fine and clear weather, for these regions, during the whole period, with fresh water on the floes.

Saturday, 9th.—At three a.m. we rounded the low point of land extending out from Verlegen Hook, in latitude 80° 16′ N.; the sea around us was now clear of ice, the edge of the pack just appearing bounding the western horizon. The land was in sight throughout the day, but in the evening it became too hazy to see it.

Sunday, 10th.—Divine service. Stood in towards the land in search of a harbour for the ship, but found them all blocked up with the winter's ice. Then stood to the southward; and Captain Parry went on shore in a boat, to look for a harbour; and in the evening we shaped a northerly course again, with clear open water in that direction, but a bright ice-blink to the westward.

Land still in sight astern; thus far our search for a secure harbour for the ship has proved fruitless.

Tuesday, 12th.—Shaping a course for the Seven Islands; passed through much stream ice, and came in contact with some of the larger masses, the ship surging considerably as her bows struck against them. Saw six walruses reposing on a piece of ice, at some distance on the port side, and passed two others, having a glaucous gull sitting near them on another piece, between 100 and 200 yards from the ship. Another was seen at the same time swimming on our starboard side. We had a fine view to-day of the Seven Islands, with a part of the mainland of Spitzbergen and the Waygatz Strait. Many glaucous, ivory gulls, fulmar petrel, dovekies, and looms, about the ship. Had Sylvestre's stove lighted to-day, and one of the crow's-nests got down this evening.

Thursday, June 14th.—Standing to the northward through more stream ice than we passed yesterday, with a southerly breeze and cloudy weather. Small flocks of little auk, dovekies, looms, kittiwake, and fulmar petrel flying about. At midnight I went up to the crow's-nest at the fore-topgallant mast-head, and whilst I was there the ship was put about, in lat. 81° 6′ 34″ N., and long. 19° 34′ E., the highest latitude yet reached by any well-authenticated record, perhaps, with the exception of Mr. Scoresby's observation, in 81° 12′ 42″. I saw all around me open water clear of ice, as far as the eye could reach, from my elevated position, encircled on all sides by a panorama at once novel, striking, and deeply interesting, a scene never to be forgotten,

> From memory's tablet ne'er to be erased,
> Though all things else from it shall be effaced.

With the aid of the telescope even nothing but loose sailing ice could be seen in the northern horizon. And oh! how I then wished we could have made a dash for the Pole in the ship herself!

Friday, 15th.—Last night when we tacked, in our farthest northern latitude, we took in the topgallant sails, furled the main-sail, and shaped a course for the Seven Isles. To-day they were in sight; clear water, hazy weather, with small snow at intervals, and westerly winds. Towards evening it fell calm, and Captain Parry gave me the dingy, during the first watch, to go after some birds, and after pulling about until ten p.m., I returned on board, having shot three little auk, a kittiwake, and loom, probably in a higher latitude than ever birds fell to a gun before, and were consequently skinned and preserved at once with the greatest care. One of the little auk I still have in my own collection of Arctic birds, and a second is in the Edinburgh Museum of Natural History, which my late lamented friend, Professor Jameson, exhibited in one of his lectures at the University in the year 1830, when I was attending his course of lectures on Natural History. A light breeze springing up, we stood in for the land. The thermometer on deck was 30°, and below in the cabin 54° Fahrenheit.

Saturday, 16th.—Captain Parry and Lieutenant Foster left the ship in two boats in search of a harbour amongst the Seven Islands, but returned on board in the evening unsuccessful. They landed on Walden Island, which, it appears, is of granitic formation, rising to about 500 feet above the sea in lat. 80° 35′ N., and long. 19° 51′ E. They here made a *cache* of provisions for the polar boats on their return. I went up into the crow's-nest and took a sketch of the islands from this altitude; they are still united together by the winter's ice or land floe. Walden Island is a very barren-looking rock, having scarcely a vestige of even moss or lichen on its cold, grey looking, high and steep, almost perpendicular sides.

Sunday, 17th.—This forenoon Lieutenant Ross and

R. McCormick, R.N., del.

The Seven Islands.

The northernmost land off Spitzbergen. Taken from the Crow's Nest of H.M.S. "Hecla."

Vincent Brooks, Day & Son, Lith.

Mr. Beverley landed on the northernmost of the seven islands, called Little Table Island, and made some *caches* of provisions there. I had a fine view this evening of the whole of the islands from the crow's-nest. When we shaped our course for the mainland of Spitzbergen I saw a flock of eider duck and several dovekies and looms.

Monday, 18*th*.—We were off the entrance to the Way-gatz Strait. Passed a bear on a floe-piece of ice, so close to the ship that she came in contact with one end of it. Weather cloudy, with strong westerly winds. This forenoon Lieutenant Foster was sent away in a boat sounding.

Walden Island, Spitzbergen. (*See page* 396)

Cloven Cliff, Spitzbergen. (*See page* 387.)

CHAPTER II.

Hecla secured in Treurenberg Bay—I land in Hecla Cove, and shoot a
  reindeer— A walrus family circle— Cutting a canal through the ice.
  — Departure of polar boats—Commence my duties in charge of a
  watch —Deer-shooting excursion to opposite shore-—Shoot a polar
  bear—Ship driven ashore—Flora of the islands.

Tuesday, 19*th.*—Two boats left the ship with Captain
Parry, Lieutenant Ross, and Mr. Beverley, to look for a
harbour round a low point of land extending out to the
eastward of Verlegen Hook, which they had the good
fortune very soon to find within a fine bay named by the
Dutch in the chart Treurenberg Bay, or Bay of Sorrow,
and after sounding the entrance to it returned on board.
All hands were now turned up, and Captain Parry,
addressing them, said that as a fine harbour had at last
been found, he expected to have the *Hecla* secured in it
before they slept. I saw the Waygatz from the crow's-
nest to-day. In working in for the harbour I witnessed
a most gratifying scene for the naturalist. As the ship
neared a piece of ice, on which a pair of walruses, with
their young one, were reposing, with a pair of glaucous
gulls standing quietly near them, they became alarmed
for the safety of the cub, probably the only son and heir
to their icy domain. After many and fruitless attempts

to induce the somewhat wilful youngster to take to the water, one of the parents rolled his huge carcase off the ice into the sea and swam along the ice-edge, whilst the other persistently, but carefully and gradually, rolled its little treasure before its own massive form till it reached the margin of the piece of ice, when, with the aid of its flippers, it gently lowered the little creature into the sea, where the other parent was in readiness to receive it; when these not very elegant, but certainly most affectionate, creatures swam away with their pet offspring carefully guarded between them. As we approached the shore I saw a large bear prowling along the beach, and several reindeer grazing on the moss-clad land. We found a great deal of ice in the bay, through which we warped the ship, and made fast to the land ice with ice-anchors, the winter's floe not having yet broken up in this deep bay. We got the ship secured by midnight. A solitary reindeer was feeding at the base of the high mountain above the cove whilst we were warping the ship up to the ice-edge.

Wednesday, 20th.—About half-past two o'clock this morning I walked over the floe between the ship and the shore, and landed for the first time in my life on an island in which I have always felt so much interested—desolate Spitzbergen. All was silence and solitude. As I strolled alone—my only companion my gun—in the direction of the spot where I had first seen the solitary reindeer, and as I rounded the spur of the mountain I sighted it lying down near a low ridge of rocks not far from the spot where it first attracted my attention. On seeing me it suddenly rose and made an attempt to escape the fate that awaited it, but I fired as it was moving off and wounded it so severely that I was enabled by running to come up with it before it could climb, wounded as it was, over the ridge. It turned round upon me, and stood at bay. On its butting at me,

I struck it on the forehead between the antlers with the butt-end of my double-barrelled gun, unfortunately breaking the stock off at the narrow checkered part. I had to seize the deer, which was a fine, powerful male animal, by his handsome branched antlers, still covered with their velvety down, and after a struggle I severed the large artery in the neck with my penknife. I saw two more deer and a pair of brent-geese as I returned on board about five p.m. The ship was at anchor in thirteen fathoms, on a bottom of blue clay, and secured to the land ice by hawsers. The curve in the bay in which she lies on the port side was named after the ship, Hecla Cove. After a few hours' rest the ship's company, both officers and men, were all busily employed in cutting a canal about a quarter of a mile in length, to which the ship was warped to a position nearer in shore, and made fast to the rocks with a cable, having besides this an anchor and chain cable out in the bay. A novel and animating scene the ice presented, the fine old ship herself forming the centre piece of the busy group, as they worked with the long ice saws attached to two triangles in sawing diagonally through the ice across the canal, or with hand-saws, handspikes, boat-hooks, and boarding-pikes shoved the detached fragments out of the canal after they had been separated by the saws. The day proved extremely fine, with a bright sun shining, and all except myself were wearing shades of green crape over the eyes, that had been served out for the purpose; but my own omission of this, in most cases, useful precaution had not escaped the keen, observant eye of Parry, who came up to me as I was towing a large fragment of ice out of the canal, and, goodnaturedly taking off my cap, attached to its peak a small square piece of the green crape which he took out of his pocket, accompanying the action with a friendly rebuke on my reckless exposure of my sight to the intense white glare of the snow under

R. McCormick, R.N., del.

Vincent Brooks, Day & Son, Lith.

H. M. S. " Hecla "
At Anchor in Hecla Cove, Treurenberg Bay, Spitzbergen.

the dazzling rays of the sun. In the evening I went on shore, taking one of the crew with me to bring the deer on board I shot last night. We found a fox enjoying his supper upon it; he had already changed his white winter coat for the more sombre brown summer one, and doubtless little dreaming of having his solitary meal so unceremoniously disturbed by a shot. I fired at him as he bounded off scatheless. Having omitted to grallock the deer when I shot it, the lapse of twenty-four hours even in this climate had rendered it wholly unfit for food. I preserved the fine antlers with their beautifully soft surface, with the feet, and have them still in my collection. I saw some brent-geese and another deer, but could not get within shot of them. Mr. Beverley found an eider-duck's nest with three eggs. Mr. Fiddes, our worthy and ingenious old ship's carpenter, soon repaired my fractured gun-stock with a piece of brass from an old sextant.

Thursday, 21*st*.—All hands employed this forenoon in getting the launch on shore and well hauled up on the beach, hoisting out the polar boats on the ice, and stowing them with seventy-one days' provisions and stores, which, including the two boats and other things, gave a weight of about 260 lbs. per man to draw; each boat had two officers, ten seamen, and two marines. Captain Parry's boat was named the *Enterprise*, and Lieutenant Ross's the *Endeavour*. Between five and six p.m. they took their departure, accompanied by Lieutenant Crozier in one of the ship's cutters, with an extra allowance of provisions, as far as Walden and Low Islands, for their use on their return.

We gave them three hearty cheers from the ship, which was as heartily returned by the boats. I went on shore, and from Flagstaff Point, at the outer corner of the cove, saw them clear of the bay, and on returning on board I went up to the "crow's-nest" at the mast-head, and took a last view of them as they threaded their way through the narrow leads of water amongst the ice.

How gladly then would I have exchanged places with any one of the boat's crew, and cheerfully worked at the paddle or drag-belt, with such a glorious prospect in view as that of unfurling our time-honoured flag on the Pole of our planet!

Friday, 22nd.—The ship's company commenced landing twelve months' provisions and stores to be left on shore as a depôt with the launch, as a reserve for the polar party in the event of the ship being unavoidably driven out to sea from the bay, and by any unforeseen accident prevented from reaching it again, which would necessitate the boats' crews wintering there on their return.

Sunday, 24th.—Early in the morning watch a bear and her cub came alongside the ship, and after having been wounded by the officer of the watch, who had fired two shots at her, made off across the bay ice for the shore. Roused from my morning's slumber by the unusual noise on deck, I hastily dressed, ran on deck, and was soon over the ship's side in pursuit. After tracking them over the floe to one of the hills by Hecla Cove, and getting two long shots, both mother and cub returned to the floe again and made for the top of the bay. In this manœuvre they were intercepted by one of the ship's company, concealed behind a hummock of ice, giving her the *coup-de-grâce* by a ball from a ship's musket; but the cub escaped on shore. Lieutenant Foster this forenoon performed divine service; and in the afternoon pemmican was served out to the crew, with leave to go on shore on a shooting excursion. They saw two bears during their rambles. The weather continues to be very fine and clear.

Monday, 25th.—The *Hecla*, being now left with a diminished number of her executive officers for keeping the watches, I entered upon my new executive duties as had been arranged by Captain Parry before his departure in the boats, by my keeping this morning's watch. These duties will be simple enough, so long as we remain at anchor in the harbour. But should we, by any chance, be driven out to sea, my innate, indeed, I may say, hereditary predilection, for navigation and seamanship, which prompted me, on my first entrance into the service, to commence taking observations for the latitude and longitude, and keeping the vessel's reckoning, will prove valuable to me now. I saw several white whales (*Delphinus albicans*) and walruses swimming about at the top of the canal. About noon Lieutenant Crozier and his party returned from Little Table Island, having deposited a *cache* of provisions on Low Islet, lying just off it. They left the polar boats and party all well at Walden Island at four p.m. on Saturday, the 23rd.

Tuesday, 26th.—I had the charge of the middle watch, and went on shore in the evening with my gun, accompanied by one of my messmates, and each of us about midnight succeeded in shooting a deer amongst the ridges of rocks skirting the bay. We first rolled the carcases of the deer down the precipitous snowy slopes to the floe beneath, and then followed ourselves, by sliding down the slippery, hard surface of the snow to the bottom. Afterwards dragged the deer across the bay-floe to the ship. 1 made mine fast to my shot-belt with my handkerchief. On the following day I again went on shore in the afternoon, and killed another deer on the plain, over by the Waygatz beach.

Thursday, 28th.—Light, northerly winds, with snow and sleet. Shot an ivory gull from the ship. Went on shore in the evening, and shot three sandpipers (*Tringa maritima*) on the rocky point at the entrance to the bay, and saw a brown fox, but could not get within shot of him. I returned on board for the dingy, and pulled myself about amongst the ice in the bay, which is fast breaking up all round the canal cut for the ship, and in motion everywhere. There is open water now to the rocky point, so that we can land in the boats. I shot an eider-duck (*Anas mollissima*), a dovekie, and an arctic gull (*Lestris parasiticus*), and kept the middle watch afterwards.

Friday, 29th.—The boats landed the last of the provisions for the launch to-day. A great deal of hummocky ice set into the bay, with cloudy, squally weather, and snow at intervals in the night. I kept the first watch.

Saturday, 30th.—A shooting party left the ship in one of the cutters, landing on the opposite side of the bay, which appears to afford the best shooting-ground here, the long tract of moss and grass-covered land extending between the base of the ridge of hills and the shores of the bay, where there is excellent pasturage for the deer, of which we met with several herds. I had the good fortune myself to shoot five of them, one being a fawn; five more were killed by the rest of the party. We made a fire with driftwood on the beach near the boat, over which we cooked some preserved meats, with biscuit, and grog mixed with lemon-juice or lime-juice—that invaluable anti-scorbutic, with which all polar explorers should be supplied with daily rations, to be served out with the same regularity as the grog, whether on board ship, or on the floe with boat or sledge. Then they need have no fear or apprehensions of being attacked by scurvy, that sad scourge of all high latitudes.

After the exercise and keen air in the chase of the deer we partook of our excellent dinner with keen appetites, enjoying the novel sort of picnic very much. I also shot an arctic gull, eider-duck, and brent-goose (*Anas bernicla*), for the collection. When returning on board in the evening, we fired at and wounded a large seal (*Phoca groenlandica*) lying on a piece of ice; but before we could reach him he struggled off into the water, and sank immediately. We also fired at and wounded a walrus, swimming near the boat, when he dived and brought up another with him. But they soon disappeared again below the surface, and showed no disposition to attack the boat, although it has been said that when one of these animals happens to be wounded among a herd, the whole of the herd will make a combined attack on a boat with their tusks, which would very soon stave a boat. Weather fine, with light northerly winds.

Sunday, July 1st.—Lieutenant Foster performed divine service, and read the articles of war on the quarter-deck.

Tuesday, 3rd.—I went on shore in the afternoon, and shot an eider-drake ; and again, on the following evening, after two deer, but could not get within shot of either of them. Shot a sandpiper, and brought it on board ; also a glaucous and arctic gull on the beach round the rocky point of the cove, but, owing to their falling into the sea, and having no boat with me, I lost them. I therefore now took the dingy, and pulled myself about the bay nearly all night, but only shot an eider-drake. Immense quantities of the bay floe have been drifting out of the bay. One of our Lapland deer died to-day.

Thursday, 5th.—This evening pulled about the bay in the dingy till midnight. Shot a dovekie, an ivory and a glaucous gull, with two terns (*Sterna arctica*). Fine, clear weather, with light northerly winds.

Friday, 6th.—The strong breezes blowing to-day have nearly cleared the whole bay of ice. I took the dingy in the evening, and pulled about the bay till midnight, returning with an ivory gull, fulmar petrel, kittiwake, and six dovekies, three of the latter at each shot.

Saturday, 7th.—Cloudy weather, with strong, southerly winds ; brought an immense body of ice down from the uppermost part of the bay, which came upon us so suddenly that the ship was driven before it on shore, but

fortunately she grounded on a muddy bottom, not far
from the rocks on the beach.   We struck the topgallant
masts, and hove overboard most of the stone-ballast
we had only taken in a few days previously, and at
high water endeavoured to get her afloat, but failed in
the attempt.

Sunday, 8th.—By carrying an anchor out we warped
the ship off this afternoon, after much laborious exertion
at the capstan and windlass, at which every officer and
man in the ship worked alike with unflagging energy.
Pumps worked, but no water in the hold.

Tuesday, 10th.—A great deal of ice, which had come
into the bay during the night, has been setting out again
throughout the day.   Ship's company employed watering
ship.   I kept the four to six p.m. watch.   A few days
ago I saw a finner whale in the bay, blowing and spouting
up the water, as he rose above the surface.

Wednesday, 11th.—Blowing a southerly gale, drifting
the ice out of the bay.   All hands employed getting the
ballast on board again.   I went on shore and shot a
deer, and two young snow buntings (*Emberiza nivalis*),
and collected a few plants.

Friday, 13th.—Cloudy weather, with fresh breezes,
which, in the evening, increased to a gale of wind.   All
hands employed watering ship, and getting ballast on
board.   I went on shore in the afternoon, and collected some plants, and in the evening pulled about the
bay in the dingy, and shot a dovekie.

Saturday, 14th.—I left the ship about ten a.m., and
walked to the top of the bay, returning on board about
four p.m., having collected some plants, and shot a grouse,
or, rather, ptarmigan (*Tetrao lagopus*).   I saw several
young ones, and also shot a snow bunting, two sandpipers, a kittiwake, and a ringed plover (*Charadrius
hiaticula*), the latter bird was the only one seen during
our voyage.   All hands employed in watering ship and

getting ballast on board. In the evening the ice was again setting into the bay.

Monday, 16*th.*—Warped the ship nearer the shore to a soft mud bottom, so as to secure her from the enormous pressure of the ice drifting in and out of the bay, and prevent her from being forced upon the rocks by it.

Thursday, 19*th.*—We formed another deer-shooting party for the opposite side of the bay, which we have found to be the chief resort of the deer, there being a great deal of vegetation, a rich luxuriant turf, with several lakes of water, much frequented by the wild fowl. Eleven deer altogether were killed by the party, four of them shot by myself; I also shot four looms, an eider-duck, and arctic gull. Some ducks' and gulls' eggs were also found on the beach by some of our party Having taken a tent with us to-day, we pitched it on the beach to dine in, and returned on board about eleven p.m. To-day being the anniversary of his Majesty's coronation, a royal salute was fired in honour of the event at one o'clock, by the small guns at the observatory on shore. The hot-air stove was lighted on board to-day. Fine weather, with light northerly winds.

Sunday, July 22*nd.*—Every officer and man in the ship employed at the capstan or windlass in getting out two hawsers to the shore, and mooring ship afresh— she having during the night shifted the cable from the rocks to which it had been made fast. I had the afternoon watch, and in the evening I went on shore to collect some of the beautiful little primrose-coloured poppies which I had noticed forming a bright spot like a little oasis on a projecting crag, at some height up the almost perpendicular acclivity of the mountain-side, where a rich soil had lodged, producing a luxuriant, moss-clad turf. I found the ascent so steep as I advanced, that it was with no small difficulty I succeeded in clambering up to it, but I was rewarded by the unusual size and fine colour of the flowers, more especially of the poppy (*Papaver nudicaule*), which, from their sheltered position in so rich a bit of turf, aided by the radiation of heat from the side of the mountain exposed to the full glare of the mid-day sun, had attained an unusual size and beauty. It is one of the most characteristic flowers of the Arctic regions together with the little orange-coloured saxifrage (*S. flagelaris*), with its bell-shaped corolla in the centre of its strawberry-like tendrils or runners, and the purple saxifrage (*Saxifraga oppositifolia*), the *Ranunculus nivalis*, the *Cerastium alpinum*, and *Andromeda tetragona*, are also very generally distributed; and amongst the Crucifera, *Draba alpina* and *Parrya arctica*. The most striking flower amongst the Rosaceæ is the *Dryas octopetala*, and of the lichens, the reindeer moss (*Cenomyce rangiferina*) and *Lecanora elegans*. The dwarf willow (*Salix arctica*) is the only representative of arborescent vegetation creeping amongst the moss-clad turf, with its tortuous, root-.ike brown stems, not above an inch or two in height, but rendered conspicuous by its silvery-white, small, flocculent catkins.

Hills west side of Treurenberg Bay. (*See page* 404.)

Remarkable round hill, Treurenberg Bay, Spitzbergen.

CHAPTER III.

Return of the polar boats—Seal-shooting—Ptarmigan-shooting—Under weigh from Hecla Cove—Land at Red Beach—Homeward-bound.

Monday, 23*rd.*—Lieutenant Crozier, with the junior mate and a boat's crew, left the ship to-day for the rendezvous of the polar party, Walden Island, Little Table, and Low Islands. Flags were placed at various points around the bay, preparatory for the survey of it. I shot a seal through the head as he was swimming, but he sank before I could reach him in the boat. I also shot two eider-ducks, two tern, two sandpipers, an arctic gull, dovekie, and little auk. Found a tern's nest with two eggs on the beach, a simple excavation in the shingle, without any kind of lining whatever. The parent birds made a great clamour on approaching their nest, darting in a furious manner at my head, repeatedly renewing the attack, each time nearer, till within a few inches of my hat, until I was some distance away from the nest ; I also found two eider-ducks' nests, with four eggs in each (their usual number), deposited in a very different manner to the tern s—having a warm lining of soft down, which the birds pluck from their own breasts for the purpose. I returned on board about eight p.m., and in crossing over the bay, saw a great number of little auk near the entrance ; the first flock I have seen of them during our stay here, and no king-ducks whatever. Light easterly winds with snow and sleet at intervals.

Tuesday, 24th.—Lieutenant Foster away with a party surveying the bay. The last of our Lapland reindeer died to-day.

Saturday, 28th.—About six p.m. our boat returned from the Seven Islands, and having reported that they had seen a bear walking along the beach, outside of the harbour, about three miles from the ship a party was soon got up to go in pursuit of him, consisting of three officers and three men, starting from the ship in the midst of a snow-storm, sleet, and small snow, almost blinding us. When we had got within about a quarter of a mile of the spot where he had been seen, we discovered him feeding on the carcase of a seal, which had been washed up on the beach, and doubtless was the one I had shot in the bay a few days previously. No sooner had bruin fixed his eyes on us than he began sniffing the air with his snout turned upwards, watching us suspiciously for some seconds, and then resumed his feast. We halted at the base of a rising ground, to examine the condition of our percussion-caps and the priming of the ship's muskets, and to arrange our plan of attack. We separated into three divisions, one to proceed along the beach, the second to make a circuit of the ridges inland of him, so as to cut off all retreat, whilst the third party, consisting of myself and the third mate, with double-barrelled fowling pieces, formed the centre division, advancing in a direct line upon him. When we had arrived within thirty-five yards of him he turned from the seal and advanced a few paces towards us, then halted, opening his jaws and showing his formidable upper and lower tier of teeth, eyeing us with a fierce and savage scowl, evincing some disposition to be the attacking party, or at all events to defend his larder. But we gave him scant time for action, which I anticipated by firing the first shot, grazing his forefoot, which he shook violently, at the same instant, turning his head round, made a bite at his own side, apparently actuated to do so by the pain from the wound. The report had scarcely died away from my own gun, when my companion fired, and his ball missing, I discharged my second barrel just as bruin was plunging into the sea, but with no better effect. By this time the two other divisions had come up and opened a volley upon the retreating fugitive, before which he swam with the greatest rapidity, apparently unscathed. Although we had one or two good shots in our party, our benumbed fingers from the extreme severity of the weather, and the blinding sleet and snow-drift in our eyes, with the somewhat hurried firing, explain our failure. Whilst my companions were discharging their guns, I was reloading mine, both barrels, with some difficulty, in the almost insensible state of the fingers' ends, when the hand was ungloved. Bruin by this time had placed about 100 yards between us, almost beyond range. However, taking as steady an aim as I could when he was broadside on to me, I was fortunate enough to plant a ball behind his shoulder, wounding him so badly, that the blood from the wound dyed his cream-white coat with crimson, compelling him to bear up for the nearest piece of ice aground in the wash of the surf, upon which he managed with no little difficulty to clamber, colouring it and the water around red with his life's blood. The rest of the party, having reloaded, opened a whole volley upon him, and he rolled his huge carcase over into the deep, without a moan or groan escaping him. The sea washed him high enough up the beach to enable us by wading up to our waists in the water to drag his enormous, unwieldy form through the surf, and above high-water mark. This, after no slight exertion, we managed to do, but with some difficulty, notwithstanding there were six of us, with all the health and strength of youth, and there we left his remains for the present.

Sunday, 29th.—Snow, sleet, and fresh northerly winds. Ship having struck the ground last night, we hove her into deep water to-day.

Monday, 30th.—The snow recently fallen on the summits of the mountains has given them quite a wintry aspect. We had Sylvestre's stove lighted to-day.

Tuesday, 31st.—Cloudy, with small rain in the afternoon and light, southerly winds. At ten a.m. a party of us left the ship and crossed over the bay to the opposite side on a deer-shooting excursion, and pitched our tent there, in which we dined. Six deer in all were killed, two of that number shot by myself. I also shot an arctic gull, and had a shot at a seal, returning on board at ten p.m.

Wednesday, August 1st.—I paid a visit to the dead bear, whose carcase had proved a great attraction to the gull tribe, and where I shot a fine glaucous and three ivory gulls, whilst they were feasting on it, and others on the next day.

Monday, 6th.—We left the ship on a reindeer-shooting excursion for two days. Pitched a tent to sleep in. Sixteen deer were killed in all, six of them by myself, and a share in the seventh, a large stag with splendid antlers; but as it appeared to be a debatable point as to who should claim the head, the hunter's perquisite, the officer who had a hand in killing it and myself agreed to present the magnificent antlers to a messmate, who was one of the shooting party. I also shot an arctic gull, two kitti-wakes, and two sandpipers. Saw two foxes, fired at one, but missed him.

Wednesday, 15th.—To-day a man was stationed at the rocks by the signal-post on the point of Hecla Cove, to look out for the return of the polar boats; three of the ship's company were placed in three watches for this particular duty. Surveying parties are out almost every day.

Tuesday, August 21*st.*—Left the ship this morning with a deer-shooting party for the opposite side of the bay. I only saw five deer, and they were so shy and wary, I could only get within a very long shot of one of them, which I fired at, but missed. The deer have now nearly all deserted the bay, either from having been so much disturbed by our shooting parties or the approach of winter, perhaps both causes may have contributed their influence. I only shot an arctic gull and picked up a few plants. When returning from the pursuit of a deer to the boat at three p.m., having only just seated myself on the beach to take some dinner, one of our party discovered the ensign flying on the flagstaff at the entrance to the cove, this being the signal for the polar boats being in sight. We at once got into the boat and shoved off for the ship, without a single deer having been killed by any one of our party. As soon as we had got on board I hastened up to the "crow's-nest," from which I plainly saw the returning boats pulling along in shore, about two miles off. There were three boats, the party having brought back with them the boat from Walden Island. A boat was sent from the ship to meet them, and as they came round the point into the bay, with ensigns flying, we saluted them with twenty-one guns. It appeared that after great exertions they had attained the latitude of 82° 45′ N., longitude 19° 25′ E., 172 miles from the ship, but the whole distance gone over was not less than 569 geographical miles ; the highest latitude was reached on the 23rd of July ; absence from the ship sixty-one days.

Wednesday, 22*nd.*—I shot a seal through the head as he was swimming across the entrance of the bay, and although I lost not a moment in pulling alongside of him, he sank just as I was in the very act of seizing hold of him by the flipper. He was quite dead, and went down with a bubbling sound. The ship's company employed in getting stores on board from the launch. Mr. Beverley and I fell in with a covey of ptarmigan or grouse (*Tetrao lagopus*) in their full, brown summer plumage amongst the rocks at the base of the mountain above Hecla Cove, and out of the covey of nine birds I shot five and wounded a sixth, which escaped. Mr. Beverley killed the other three.

Friday, 24th.—The launch with all the remaining stores were got on board to-day from the beach.

Sunday, 26th.—This forenoon the ship's company were employed in getting the ship ready for weighing anchor early to-morrow morning, but a heavy fall of snow, with a contrary wind, prevented our sailing. This evening a boat was sent on shore to complete the supply of water for the ship.

Tuesday, 28th.—Sylvestre's stove lighted to-day. I went on shore this forenoon, and shot a kittiwake and snow bunting. The snow in places was very deep, presenting quite a wintry aspect.

Wednesday, 29th.—Last evening we got under weigh from Hecla Cove, and made sail on our homeward passage. The sun but just appeared above the horizon, half obscured by a bank of dark clouds, the remainder of the sky above very clear and of a perfect azure blue. In the afternoon we were off Wyde Bay, and at 8.30 p.m. Red Beach. Here we hove-to, and sent two boats on shore to bring off the travelling-boat with her stores, left there on our outward passage. I landed in one of the boats, and found that Red Beach evidently took its name from the colour of the old red-sandstone formation occurring there. I brought off from the beach some of the rolled, water-worn pebbles, banded red and white, having a very remarkable and pretty appearance. Also some mosses and other plants, and shot a kittiwake returning on board. Wind easterly, with very clear, fine weather.

Red Beach, Spitzbergen. (*See p.* 414.)

CHAPTER IV.

Résumé of the geological and botanical features of the island—An
arctic cemetery—Why the expedition failed.

AT Red Beach I for the last time set my foot on the
shores of Spitzbergen, and took a final farewell of that,
to me, remarkable and interesting island, after above two
months' sojourn, very pleasantly spent, with unusually
fine weather for such a latitude. My time had been
fully occupied in collecting specimens of birds, rocks,
and plants for the natural history collection, making
sketches of the land, shooting reindeer, and keeping the
watches, in addition to my ordinary professional duties.

Hecla Cove is a recess or curve on the eastern side of
Treurenberg Bay, between which and the finely-formed
symmetrical mountain, rising some 2000 feet above it,
extends a considerable level surface or plain of rich
alluvium, formed in a great measure from the decomposition of the friable hornblende rock, which enters so
largely into the formation of the mountain, huge masses
of which have rolled down from its summit to a considerable distance from its base : ridges of massive quartz
of a fine pink colour, capable of taking a beautiful
polish when cut, and from which I had a desk seal engraved
with my own initials, which made as good an impression
on the wax as from any gem. This isolated, low ridge

of quartz crops out between the base of the mountain and the cove. The rich plain of alluvium is covered with a carpet of vegetation consisting of mosses, lichens, grasses, and dwarf willows, interspersed with the various arctic flowering plants in miniature, amongst which the saxifrages and the poppy abound. Some fragments of limestone and mica slate occur amongst the *débris*.

Treurenberg Bay in the Dutch language, I believe, means, Bay of Sorrow, or Grief, and must formerly have been much resorted to by Dutch whalers, who suffered severely from that scourge of high latitudes, scurvy. This is evident from the number of graves found here. On one of my shooting excursions, on the west or opposite side of the bay, I came all at once upon a perfect cemetery, no less a number than thirty piles of stones, each having human remains beneath, enclosed in fragments of wood rudely put together. When this spot first arrested my eye in the distance, I was quite at a loss to conjecture what all this pile of broken oars, projecting above the surface on the north point at the entrance to the bay meant. I exhumed a skull from this primitive arctic mausoleum. It was bleached perfectly white, through the many winters' frosts and summers' thaws to which it had been exposed for the greater part of a century. The head-board to this rude grave was the blade of an oar, on which was cut, with a sailor's knife, the name of the ship, and that of the young seaman whose fate it commemorates. The date was 1738. I brought this oar-headboard on board with me, and subsequently gave it to Captain Parry. The sun at midnight at our departure, just dipped his lower limb beneath the horizon for the first time for the last four months, rising again at once.

The climate of Spitzbergen (discovered by Barentz in the year 1596), under the influence of the Gulf Stream, a branch of which laves its shores, is greatly ameliorated, as the heating-power of the warm oceanic currents is very great. The Gulf Stream mainly originates in the Gulf of Mexico, as it passes through the Florida Channel. It has a temperature of some 65°, its mean temperature being about 40°, and it gives to the south-westerly winds their warmth and genial character. This stream is estimated at about fifty miles in width, with a depth of 2000 feet, and a rate of some four miles an hour, of course much influenced by the prevailing winds, which act powerfully on all oceanic currents. Cold counter-currents pass from the poles to the equator, along the bottom of the ocean, and beneath the upper warmer currents. The temperature of the deep-sea bottom is always little above the freezing-point, dark and cold, from the poles to the Equator.

There are said to be interior glaciers in Spitzbergen from 2000 to 3000 feet in thickness, and that the land is rising at the rate of six feet in a century. The fossil remains of trees and plants found would indicate a temperate, even, subtropical climate having once existed here as in other parts of the Arctic regions.

That our expedition failed in its main object of reaching the North Pole, owing to the wrong season of the year having been selected for boating and sledge operations, is clearly evident. The very early months of spring afford the only chance of success. Yet there certainly was, when we left Spitzbergen, so much open water to the northward, that had we been prepared to winter there, the ship herself would have had a fair chance, if not of reaching the pole itself, of attaining a much higher latitude.

As the compass at the pole will cease to act, should plants exist there (which there is no reason whatever to doubt if land exists), their sleeping leaves at the midnight hour, when the sun is in the north, would point the way, and therefore a true course could be steered by the direction indicated by their leaves.

Thursday, 30*th*.—We fell in with a considerable quantity of sailing ice. Passed Cloven Cliff and Smeerenberg Harbour, off which we had the hard gale of wind on our first making the Spitzbergen coast, compelling us to take the pack. We rounded Hakluyt headland, and entered upon the open sea, with a heavy swell. The land along this part of the coast is bold and high, rising to 4000 feet, the summits of the mountains sharply peaked and covered with snow.

Friday, 31*st*.—Land still in sight ; a few loose streams of ice passed the ship this morning; several fulmar petrel following in the wake of the ship, after the manner of the Mother Cary's chickens or stormy petrel. Easterly winds with moderate breezes, and cloudy weather. Yesterday saw several gulls and dovekies amongst the ice.

Saturday, September 1*st*.—No land in sight, but an open sea clear of ice. Strong breezes and cloudy weather, with rain at intervals. Wind changing from fair to foul, S.W.

Sunday, 2*nd*.—Mustered at divisions, and the articles of war read by Captain Parry. To-day we had roast beef for dinner, fresh and excellent, not the least injured by the time it has been kept under the maintop, although we brought it from England with us—a striking proof of the antiseptic powers of the Arctic climate. Having had the sun constantly above the horizon for some months past, it was quite a novelty to-night, being dark, to have to light a candle to go to bed.

Monday, 3*rd*.—Blowing a westerly gale, with a heavy sea running. At night the moon appeared bright through some very dark clouds, half screening it from view, just above the horizon. Saw a star, the first we have seen for a long, long time past.

Tuesday, 4*th*.—The gale abated and nearly fell to a calm wind, drawing round ahead, fulmars following the ship.

Wednesday, 5th.—Arranged the Government collection of birds. Hot-air stove lighted.

Thursday, 6th.—Mustered at divisions; and warm clothing served out to every officer and man in the ship, consisting of a box-cloth jacket and trousers, pair of large waterproof boots, one pair of laced shoes, two pairs of large hose, comforter, mittens, Welsh wig, Scotch cap, and red guernsey frock. Hot-air stove still kept lighted. Lat. 68° N.

Wednesday, 12th.—Crossed the Arctic Circle to-day. Weather, as it has been for the last day or two, cloudy, with fine rain at times, fresh south-westerly winds, and yesterday there was a heavy sea running. A young tern was caught on board this morning. Packed up the Government collection of birds.

Saturday, 15th.—A flock of young kittiwakes, several terns, and a boatswain bird, or arctic gull (*Lestris parasiticus*), were flying round the ship; and on the following day a solan goose.

Monday, 17th.—Land in sight; I went up to the fore-topmast-head to have a look at it. Lambroness, bearing S. 50 W. Saw several solan geese, and at six p.m. a French galliot. Fresh westerly winds.

Tuesday, 18th, ten a.m.—Fired a gun and hoisted a signal for a pilot. At 1.30 p.m. let go the anchor in Balta Sound, Isle of Unst, the northernmost of the Shetland Isles. We grounded for a few minutes in entering the harbour. A number of boats with fowls, eggs, and milk soon came off to us from the shore, which exhibited little signs of cultivation, being for the most part pasture-land, in which a number of cattle were feeding. The hills rise in gently swelling slopes, over which a few houses are sparsely scattered from the harbour, which forms a fine, spacious, land-locked sound, but the water shallow. Westerly winds, with squally weather and rain at intervals.

Wednesday, 19*th*.—Weather much the same, but wind northerly. Some ladies from the shore came on board to see the ship this forenoon. At five p.m. weighed anchor and made all sail from Unst.

Thursday, 20*th*.—Fine, clear weather, with a fresh northerly breeze. Passed Fair Island. Last night an order was issued by Captain Parry for all logs, journals, and sketches kept during the voyage by the officers to be taken into the cabin this evening on or before eight p.m.

Saturday, 22*nd*.—The same contrary southerly wind which compelled us to put into the Shetlands, now obliged us to bear up for the Orkneys. In the evening we had a faint display of the aurora borealis, shooting upwards towards the zenith in faint rays. The night was fine, with a luminous sea, displaying the most brilliant aggregation of bright spots on the surface, both along the sides and in the wake of the ship as she ploughed her way through it. The Milky Way appeared very distinct, and a meteor or two were seen, also a few feeble flashes of lightning.

Sunday, 23*rd*.—Fired several guns, and hoisted the signal for a pilot, running into Long Hope, where we anchored at five p.m. There we found the *Chichester* cutter, a whaler, and several small craft.

Monday, 24*th*.—We got under weigh, and stood out for the Pentland Firth, with light northerly winds, but calms following, and the wind again coming from the southward, we had to bear up for the anchorage in Long Hope again, and at 7.30 p.m. came to an anchor. Captain Parry decided to take passage in the *Chichester* revenue cutter to Inverness, and from thence to proceed overland to London, accompanied by Mr. Beverley. They left the ship at 8.30 p.m. A boat was sent to Kirkwall for letters, and the boats employed watering ship. The laird of the place and his lady came on board to see the ship. The *Diligence* revenue cutter arrived. I went on shore here in the forenoon.

Saturday, 29*th.*—At about six a.m. the wind suddenly changing to the northward, we at once availed ourselves of the favourable breeze, and making a signal for the pilot, got under weigh at 9.30 a.m. This evening poor old George Crawford, our ice-master, died. He had been on the sick-list at intervals throughout most of the voyage, with a long-standing pulmonary affection, asthma, and all its complications ; but having been in five successive voyages to the Arctic regions with Captain Parry, he, out of a kindly feeling to serve him, and gratify the good old seaman's great desire to accompany him in this his last voyage, acceded to his wishes, and in any case, whether he had been at home or abroad, his days were numbered. He left behind him a young widow, and the painful feeling of passing away from this world when so near his home and her, sadly disturbed the resignation of the poor fellow's last moments.

Monday, October 1*st.*—A variety of birds have been flying round the ship all day—hawks, owls, larks, chaffinches, &c. Three hawks, an owl, and a starling were shot ; and a sandpiper, which alighted on the ship's side, was caught by those on board ; we were at the time off the Firth of Forth, but not within sight of the land.

Friday, 5*th.*—Saw an owl and several smaller birds, chaffinches, a thrush, and a robin redbreast, about the ship.

Wednesday, 3*rd.*—At eleven a.m. Flamborough Head made its appearance through the haze, and we hoisted a signal for a pilot, firing several guns at the same time.

Thursday, 4*th.*—At 6.30 p.m. anchored in Yarmouth Roads. About 7.30 p.m. I went on shore, and set off immediately for Southtown, to see my mother and sisters, at the time residing there, and having passed a brief space of time with them, returned to Yarmouth about midnight, and from thence on board.

Friday, *5th.*—Weighed anchor, and made sail at 5.30 a.m. Worked through the New Gat with a moderate breeze and fine, clear weather, running along the land. At nine p.m. we anchored in the Swin. Saw the Harwich and Sunk lights in passing.

Saturday, *6th.*—At five a.m. weighed; at eleven a.m. passed Sheerness, and made our number to H.M.S. *Gloucester*. Arrived at Northfleet at three p.m.

Wednesday, *10th.*—Light southerly winds, with rain. Received orders to proceed to Deptford to be paid off; got under weigh, and as we passed Woolwich returned the cheers of a man-of-war fitting out there. At four p.m. made fast to the *Heroine* hulk at Deptford. Whilst paying off we had the honour of a visit from the Lord High Admiral, his Royal Highness the Duke of Clarence, attended by Sir George Cockburn and Colonel Fitz-Clarence. All the officers were individually introduced to the Lord High Admiral by Captain Parry.

Thursday, *November 1st.*—The *Hecla* was paid off at Deptford.

APPENDIX.

LIST OF REINDEER

Shot in Treurenberg Bay, Spitzbergen, by the officers and crew of H.M.S. Hecla in the year 1827.

	Number.
Lieutenant Crozier	4
Purser Hulse	2
Assistant-Surgeon McCormick	22
Mate Foote	20
Messrs. McCormick and Foote	1
Gunner Brothers	2
Boatswain Smith	2
Carpenter Fiddes	3
	56
Ship's Company	13
Total . .	69

PLANS FOR REACHING

THE

NORTH AND SOUTH POLES,

BY THE AUTHOR.

AFTER a long life devoted to Polar Discovery, both north and south, as mine has been, it may naturally enough be expected from me, that I should refer to the present renewed interest in Arctic research, occasioned by the recent spirited, though unhappily disastrous private expeditions to Franz Josef's Land, sent out by Austria and America, and so lately by our own countrymen. I therefore offer, as a sequel to the foregoing pages, my own opinion of the probability of reaching either pole, and the plans from past experience I would adopt for the attainment of so desirable a result.

PLAN FOR REACHING THE NORTH POLE.

The North Pole being undoubtedly the most accessible of the two poles, I shall begin with it, and briefly review the avenues of approach to it, with the attempts made both by our own country and others, and the cause of their failures originating in the meridian selected for each attempt.

Anterior to my employment in the Franklin search, I had entertained more favourable views of the approach to the pole through Smith's Sound than subsequent experience justified ; and before the American expeditions of search under Kane and Hayes, I had in my plans laid before the Lords Commissioners of the Admiralty, called

the attention of their lordships to the fact of the strong
currents and heavy swell setting out of Smith's Sound,
from which I drew the inference that the so-called
sound was really an opening into the Polar Ocean ; and
as I had always been impressed with the belief that
Franklin would attempt the Wellington Channel, he
might possibly, if beset in the polar pack, be compelled
to make a retrograde movement. Consequently I held
that Smith's Sound should not be omitted in the search,
and long before even the American expedition was con-
templated, I had offered to explore it together with
Jones Sound and the Wellington Channel into the Polar
Ocean.

Subsequent events have not only confirmed the sound-
ness of the views I then held, but, what is of far more
importance, the discovery of the opening of Smith's
Sound into the Polar Ocean, now claimed for the star-
spangled flag, might well have been our own, had my
offer at the time been accepted by the Admiralty.

Be this as it may, the very intricate nature of the
tortuous, long, ice-encumbered channel, so beset with
insuperable difficulties, as the Americans found it—in
Kane's case terminating in a disastrous retreat from their
ship in boats, and in Hall's a still narrower escape on
the drifting floe—convinced me that I could no longer
entertain the faintest hope that the pole would ever be
reached by the Smith's Channel route. This was before
the attempt made by Sir George Nares, who found, as
I had anticipated, after the *Alert* had struggled through
the drifting packs which are disgorged into Baffin's Bay
from the Polar Ocean, through the intricate navigation
of this narrow, irregular, rocky strait, that no secure harbour was available at its *embouchure* as a *point-d'appui*
for the security of the ships, but only a barren, exposed
shore, on which the enormous, hummocky, heavy, polar
pack rested, barring all progress north either of ship,
boat, or sledge.

Now, this vast pack of ice, resting as it does on the whole line of coast—eastward from the north coast of Greenland continuously, to the *embouchure* of the Wellington Channel, and from thence to Behring's Strait—oscillates backwards and forwards from the shore, under the influence of winds and currents, especially during the summer season of these regions, when offsets separate and make their way through the openings of Smith's Channel, Wellington Channel, and Behring's Strait, thus preventing any undue accumulation of ice within the polar area, and by this compensation maintaining an equilibrium.   The position of this polar pack, skirting as it does the whole sea-board, negatives any attempt to reach the pole either by way of Smith's or Wellington Channel, or Behring's Strait ; and I am further disposed to think that the route at present mooted by way of Franz Josef's Land deserves no more consideration, judging from the unsuccessful results of the recent Austrian, American, and English private expeditions to that quarter, in which each of the ships of the respective expeditions were crushed in the heavy packs of ice concentrated in the vicinity of Franz Josef's Land, owing to the peculiar relative position of Spitzbergen and the latter land, resulting in the accumulation of pent-up heavy floes clinging to its shores or drifting about the seas adjacent.

The abundance of animal life met with on the shores of Franz Josef's Land, and assigned as one reason for making that land a base for operations, is, I believe, in no way owing to the land being more productive or better suited for the support of the animal creation than is Spitzbergen, but simply because it has not been, until recently, trodden by the foot of man, whose destructive influence would soon thin the numbers of the animals, as doubtless it has done in Spitzbergen since my sojourn on its shores, when engaged in Parry's attempt to reach the pole, more than half a century ago, when reindeer and wild fowl abounded there amply sufficient for our wants.

Spitzbergen should be the base of operations in any future attempt to reach the North Pole. Parry's failure was wholly and entirely owing to the wrong season of the year having been selected for boating and sledging over polar ice. The plan which I have had under consideration for years past, I now lay before my readers.

PLAN FOR REACHING THE NORTH POLE BY THE SPITZBERGEN SEA.

On the return of Captain Sir George Nares's Expedition, after his attempt to reach the pole by Smith's Channel, I had occasion to address a letter on service to the late Mr. Ward Hunt, dated in 1876, at the time he was First Lord of the Admiralty, in which I emphatically called attention to the Spitzbergen Sea as *the* route to the pole.

Whilst on this subject I cannot refrain from expressing the gratification I felt in having my name associated with my worthy old chief's, the late Admiral Sir Edward Parry, both by Sir George Nares, who has named a valley after me, on the north-west coast of Greenland, and my old friend, Vice-Admiral Sir Leopold McClintock, who in like manner has attached my name to an inlet in Melville Island, and to both those distinguished officers I here record my best thanks for the compliment paid me.

That the North Pole will be reached, though possibly not in my time, yet before the close of the present century, I feel as sure as I do that the unfurling a flag at the Arctic or the Antarctic Pole will stand second to none as a brilliant geographical problem solved; not the discovery even by Columbus of a New World; although its subsequent utility may be of no moment when compared with that of a fourth quarter of the globe opened to our race.

The Spitzbergen Sea, as far as our experience has gone, offers not only by far the best, but, as I believe, the only practicable approach to the pole The Polar Ocean between Spitzbergen and the pole is undoubtedly more open to navigation at certain seasons than on any other meridian of the entire polar area. The current setting to the southwards, as we found it, transferring immense floes and fields of ice from the vicinity of the pole, to be broken up into the huge hummocky packs by the winds and currents as they approach the land, must necessarily leave in their wake corresponding open spaces of water in the north. The inference to be drawn from this is that there is no land sufficient to interfere with the navigation either for ships or boats ; at most, probably, small groups or archipelagos of islets similar to the Seven Sisters off Spitzbergen. The warmer temperature of the Gulf Stream, a branch of which flows northwards, doubtless has the effect of breaking up the ice here, and, as a result, leaving much open water at intervals, and when combined with a continuance of strong northerly gales acting with powerful currents, seasons may occur in which I can readily conceive a small handy steam-vessel, starting from the shores of Spitzbergen, at a favourable crisis, might reach the pole itself, or at all events make a very near approach to it, and be back in one season.

Should the steamer be unsuccessful the first season, I would make Treurenberg Bay my winter quarters, and moor the steamer in Hecla Cove, Spitzbergen, a small curve on the port side as you enter the bay, at the base of a picturesque mountain, and early in the ensuing spring, say the last week in the month of February, before the temperature began to rise so as to affect the ice-floes, the pole might be reached by sledging over the vast ice-fields, which at a distance from the land would in all probability present a smoother surface for dragging the sledge over than the broken-up, hummocky condition of the coast-line floes. If the season proved favourable, with moderately calm weather and a respite from heavy northerly gales, the enterprise might be accomplished of reaching the pole and returning to the ship in the space of three months, that is to say by the end of April. A boat, of course, should accompany the sledging party, and Esquimaux dogs be employed for drawing the sledge.

The outfit for such an expedition should be compact and complete—a small vessel drawing little water, well strengthened by doubling, especially the bows, of from 150 to 200 tons, propelled by a powerful engine ; a brigantine being the most handy rig for manœuvring amongst ice. Such a vessel is more buoyant and less likely to get nipped between contending floes, from the tendency of the hull, having a well-proportioned keel, to rise above the pressure to the upper surface of the ice, her small draught of water enabling her to navigate with greater security the shallow bays and straits, and in any damage to her hull can be hove down with greater facility by a small crew, which should consist of twenty able seamen, and three mates for the charge of the watches ; and, from their experience amongst ice, volunteers from the whale fishery would be most desirable in every way for a service like this.

The vessel should be provisioned for two years, with a supply of coals for that period, and as a safeguard against unforeseen accidents, it would be very desirable to take out the framework of a house, capable of housing the whole party throughout the winter, in the event of any disaster happening to the steamer, which could be put up on the spot selected for the base of operations in Spitzbergen.

All kinds of salted provisions should be rigidly excluded from the dietary composing the outfit, which should mainly consist of preserved meats, poultry, and vegetables, with those excellent farinaceous articles of diet, macaroni, rice, and arrowroot ; and that best of all antiscorbutics, the cranberry, with an unlimited supply of lime juice—our sheet-anchor in cases of scurvy ; malt liquors of sound and good quality will be found valuable resources to " splice the mainbrace " with.   After any extraordinary exertions and fatigue under cold and privation, good old Cognac brandy is the best and only stimulant needed : in cases of extreme exhaustion and prostration, arising from exposure to intense cold and privation, we have nothing in the whole " Materia Medica " to equal it as a prompt restorative of lowered nerve-power, and where the state of the general health needs some stimulant, it is a much safer one, judiciously taken, than any of the ordinary wines of the present day, and all other forms of alcohol are unnecessary luxuries, far from being conducive to health or comfort.   Well-preserved fresh fruits are generally as beneficial as all kinds of dried ones are injurious and prejudicial.   During the long Arctic night, when from want of proper exercise and a monotonous mode of life the appetite fails, the nights disposed to sleeplessness, a glass of quinine wine twice a day is often beneficial ; in the proportion of about ten grains of quinine dissolved in twenty grains of citric acid, and added to a bottle of Marsala.

I should consider twelve men a sufficient crew for the sledge and boat, with the aid of about a dozen good Esquimaux dogs ; two of the mates as officers for sledge and boat, and the third mate, with the remaining eight men, to be left in charge of the vessel and house at Spitzbergen, to shoot reindeer and eider duck, and other wild fowl, as a supply for the ensuing winter, during the absence of the polar party.   A whale-boat of sufficient size to hold the party, or perhaps two smaller boats and two sledges might be found more manageable by a small crew. The size of the fore-and-aft vessel I have proposed may by some be considered small for such an undertaking ; but old Baffin in the *Discovery*, of only fifty tons, circumnavigated the bay which bears his name, and accomplished more in one season than more ostentatious expeditions have since done in many. In short, I would myself prefer a vessel of 100 tons to any other, as there would be fewer to feed.

PLAN FOR AN ATTEMPT TO REACH THE SOUTH POLE.

An enterprise far more difficult to encounter in all its bearings than the one to the opposite hemisphere, a climate vastly more severe, a navigation infinitely more intricate and dangerous in character, and removed so far from any base of operations ; with no known harbour for the security of the ships as winter quarters.

As the seasons are reversed in the Southern Hemisphere, the months for exploration will be December, January, February, and March; therefore the ships should take their departure from England about midsummer, and for this service there must be two ships of much larger dimensions, and in every way on a more extensive scale of equipment, than is needed for the north. They must be provisioned for three years, and well equipped with whale-boats, each ship having powerful engines.

The provisions and stores of a similar kind as already described for the North Polar outfit, only on a larger scale, and both the ships and equipment needing all the resources that can only be supplied from a Government dockyard. Therefore an expedition fitted out by the Admiralty would afford the best guarantee for success. Two frigate-built ships of about 1000 tons each, barkrigged, and well strengthened by stout doubling, with crews numbering 100 in each, picked seamen, and officers selected for their scientific attainments and zeal in this branch of the service, form desiderata of great importance. Astronomy, Geology, Zoology, and Botany, should each have its representative, and an artist skilled in the delineation of scenery, birds, and animal life generally, and the portraiture of savages. Large guns for firing signals, to prevent the separation of the ships amid the dense fogs of the Antarctic Seas.

I would recommend the meridian of Tasmania, but eastward of the track the *Erebus* and *Terror* followed, so as to attain that point of the Barrier where our own progress was finally arrested, and to follow it up, or whatever coast-line might turn up, and to explore every opening in the pack, barrier, or coast-line, holding out any chance of approaching the pole. On the passage out to Tasmania, if run short of coals for the engines, they may be replenished at Kerguelen's Land, as I have elsewhere stated in my account of that island, and only at the expense of a trifling *détour* to the southward for that purpose.

The ships, after refitting in the Derwent, and completing supplies at Hobart Town, might commence their voyage south in the month of November, and, if unsuccessful the first season, might bear up for Cape Horn and winter, and refit at the Falkland Islands, and in the ensuing summer try the track of Weddell; and lastly, failing on that meridian, try the one from the Cape of Good Hope, wintering and refitting there.

We know so little at present of the navigation of the Antarctic Ocean, that it would be presumption to point out any one particular course as preferable to another. Much must depend upon the movements of those vast packs of ice drifting around the polar area, at the mercy of the prevailing winds and currents, leaving temporary openings in the direction of the continent or the pole; and it is only by patiently circumnavigating that area during the season, and trying every promising opening, whilst tracing the circumference of the pack, that any successful advance in the direction of the pole can be hoped for. Carefully avoiding, by close attention to the direction of the winds and currents, that the ships are not caught and beset in the centre of one of these enormous packs, and the season be thus sacrificed, as was our experience in the *Erebus* and *Terror* for some six weeks.

As no one has yet wintered in the Antarctic regions, it would in the interests of science be most desirable to do so, could a harbour be discovered ; but therein lies the difficulty. All the breaks and openings along the whole line of coast are sealed up with the eternal ice and glaciers, literally a wall of ice extending along the whole seaboard—as far as our own examination of this vast continent went.

Yet, had I the command of an expedition in those seas, I would strain every effort to secure some well-protected nook, beyond the inroad and pressure of the packs in the coast-line of the Gulf of Erebus and Terror or Macmurdo Bay, for winter quarters.

From this spot what a sublime object of interest and awe must the burning volcanic Mount Erebus present throughout the long, dark winter's night, with its surroundings of everlasting ice and glaciers, emitting its dense smoke and flame, and possibly at intervals grand eruptions. What a glorious spectacle of its great Creator's work would it present; a field for thought and observation ! What with magnetism, and the opportunity of studying the Antarctic winter's sky, it would amply repay the outlay of such a voyage of discovery as I have endeavoured to call attention to ; and even if the main object failed of planting our flag on the pole, a nearer approach than hitherto, I think, at the least would be made.

Whales, seals, penguins large and small, and a variety of sea fowl abound on the ocean; but there is no vegetation, not even seaweed on the shores. The wandering albatross does not extend its flight so far. But I met with the stormy petrel at the Great Barrier, as I did in the north in the Wellington Channel.

END OF VOL I.

GILBERT AND RIVINGTON, LIMITED, ST. JOHN'S SQUARE, LONDON.

Printed in Great Britain
by Amazon

57981724R00292